# CURRENT PERSPECTIVES IN
# GEOLOGY

## 2000 EDITION

**Michael L. McKinney**

**Kathleen M. McHugh**

**Susan P. Meadows**

*University of Tennessee*

**Brooks/Cole**
Thomson Learning™

Australia • Canada • Mexico • Singapore • Spain • United Kingdom • United States

Acquisitions Editor: Nina Horne
Project Development Editor: Marie Carigma-Sambilay
Marketing Representative: Kelly Fielding
Editorial Assistant: John-Paul Ramin
Production Coordinator: Stephanie Andersen

Permissions Editor: Mary Kay Hancharick
Cover Design: Vernon Boes
Cover Photo: James W. Kay
Print Buyer: Vena Dyer
Printing and Binding: Globus Printing Company

*For more information, contact:*
BROOKS/COLE
511 Forest Lodge Road
Pacific Grove, CA 93950 USA
www.brookscole.com

*For permission to use material from this work, contact us by*
Web:    www.thomsonrights.com
fax:    1-800-730-2215
phone:  1-800-730-2214

Printed in the United States of America

10   9   8   7   6   5   4   3   2   1

ISBN: 0-534-37213-9

# Contents

# EXTENSION OF COPYRIGHT

# Preface

The Earth's geologic processes are nearly timeless, but the impact geology has on humans is very much a current issue. To represent the wide variety of topics upon which the study of geology touches human activity and inquiry, the editors of this anthology have collected articles from a number of general interest and science magazines.

The editors have carefully chosen articles to supplement material a student might encounter when taking a course in physical geology, historical geology, environmental geology, dinosaurs, and earth science. Often, there is an overlap of subject areas taught among these courses. For example, a considerable portion of an environmental geology course usually includes material on physical geology. To help readers identify these overlaps, the editors have divided the book into six parts. Parts one through three cover physical geology, parts four and five cover environmentally related topics, and the final part covers historical geology.

Each article opens with a brief overview and discussion of the issues or concerns generated by the topic. Following the articles, the editors then ask a few questions to help the readers focus on the issues and apply what they have learned from their geology classes.

## Acknowledgments

The editors and the publisher wish to express their sincere thanks to the many magazines, journals, and freelance writers for permission to reprint their articles.

# PART 1
# The Origin of Earth and Its Internal Processes

Within the last decade and longer there have been significant strides in our understanding of the earth's interior. The controversy over mantle convection seems to be approaching a resolution. Estimates of the outer core temperature are better constrained, and our understanding of the role of the core in the earth's magnetic field has made many progressive strides. Hopefully, the near future will make just as many advances in our understanding of the earth's interior.

# Dynamics of Earth's Interior

By Thorne Lay and Quentin Williams

The last 50 years have seen remarkable advances in our understanding of the thermal, chemical, and dynamical state of the Earth's deep interior. A host of observations and techniques, including analysis of many types of seismic waves, experimental determinations of equation of state and phase equilibria of Earth materials over almost the entire range of pressure and temperature conditions inside the planet inversions for the temporal history of the magnetic field, and numerical models of the geodynamic and mantle flow under realistic conditions, have enlightened us greatly about the deep reaches of our planet. While many mysteries remain, some of the most fundamental issues are on the verge of resolution.

In many ways, the late 1940s and early 1950s can be flagged as the start of the modern era in our understanding of the chemical and mineralogical properties of the mantle and core. While the gross layered structure of Earth was determined by seismology in the first few decades of this century, it was about 50 years ago that several fundamental tenets that have underpinned most subsequent studies of the properties of the mantle and core were definitively stated or demonstrated. These include the famous inference by Birch that high-pressure crystalline phase transitions likely generate the anomalous gradients in seismic velocity documented by Jeffreys between 400- and nearly 1000-kilometers depth; Bullen's assessment that the seismic velocity distribution in the lower mantle indicated that this largest region of the planet is homogeneous in chemistry and phase (except for the lowermost 200 kilometers); and a compelling case made from a combination of elastic data on iron, the seismic properties of the outer core, and compositional information on meteorites, that the core of the planet is iron-dominated, confirming a suggestion made by Dana and Wiechert in the late 19th century. The possibility of mantle convection was being promoted by Verhoogen (among others), but general acceptance of this idea had not yet occurred. The idea that permanent magnetization of the interior of the planet is required to generate Earth's dipolar field was in the process of being replaced by the magnetic field-generating magnetohydrodynamic models of fluid flow in the outer core, produced, for example, by Elsasser. Finally, the inner core of the planet (first recognized to exist in 1936) had been proposed to be solid and produced by a pressure-induced intersection of the geotherm with the melting curve of iron, although even the notion that this region is solid was unproven.

In spite of these first-order insights, profound uncertainties existed about the nature of deep Earth at the time that the American Geological Institute was founded. Many views of

From "Dynamics of Earth's Interior, by T. Lay and Q. Williams, Geo Times, November 1998, pp. 26-30. Reprinted with permission.

the composition of the upper mantle and the nature of the crust-mantle (Moho) discontinuity differed profoundly from those of today; the manner in which basalt was generated was unclear; the type of crystallographic transitions that might be present within the transition zone were a matter of complete speculation; and the available seismic observations were adequately explained by a laterally homogeneous, onion-like layered planet. Today, a richness of geophysical phenomena have been characterized that could not have been envisioned a half-century ago—from the plumbing systems delivering magma to mid-ocean ridges; to images of apparent mantle upwellings, and downwellings, including global maps of deflections of seismic discontinuities from which lateral temperature variations can be inferred, to continent-sized partially molten features at the base of the mantle; through to anisotropic structure and super-rotation of the inner core.

## Upper Mantle

With respect to the upper mantle, the idea that a glassy or magmatic basaltic substratum of ten to hundreds of kilometers of thickness was present at sub-crustal depth had largely run its course by the early 1940s. However, eclogite was frequently invoked as the dominant upper mantle rock, due to its chemical similarity with basalt and general compatibility with the seismic signature of the upper mantle. Given an Earth of meteoric composition, such a calcium and aluminum-enriched upper layer was anticipated to shift to a region dominated by magnesium silicates at depth. Birch, in his classic 1952 paper, actually excluded peridotite from being the dominant upper-mantle rock type based on its elasticity: This likely resulted from a lack of appreciation of the shifting phase assemblages from plagioclase-to garnet-peridotite at pressures below 2.5 GPa (depths shallower than 80 kilometers).

At present, a combination of detailed studies of basalt petrogenesis (spearheaded by Ringwood and co-workers), improved elastic data on minerals, and seismic velocity profiles (particularly of the oceanic crust and mantle) have resulted in a general consensus that the upper mantle is predominantly of peridotitic chemistry. This first-order revision of upper-mantle composition avoids the large degree of mantle melting necessary to generate basalt from eclogite, and limits the fraction of melting needed to generate basalt to ~20 percent and the usual depth of melt initiation to ~100 kilometers. Furthermore, the melting behavior of peridotite provides a natural mechanism for the genesis of harzburgite as the residue of basaltic melt extraction. Also, the appreciation of the importance of peridotite in the upper mantle has removed the gabbro-eclogite transition as a viable explanation for most observations of the Moho discontinuity—a popular interpretation of the late 1950s and '60s. Nevertheless, the gross similarity of the elastic properties of eclogite to the upper mantle continues to produce some ambiguity in the present-day interpretation of mantle structure: It remains difficult to completely preclude eclogite-regions from being at depth within the mantle of the planet.

## Deep Earth

Concepts about deeper regions of the planet have undergone even more profound revisions in the last half-century. In the 1940s, the transition zone of the planet was defined as a region of anomalous seismic velocity gradient between 400- and 1000-kilometers depth—a zone where the change in velocity with depth could not be explained by the effects of self-compression alone. The reasons for this shift were unclear, with the possibilities existing that a change in chemistry might occur or that phase transitions could shift the elasticity and density in this region. In 1936, the chemist J.D. Bernal had made the insightful proposal that olivine might convert to a denser spinel structure at high pressures; Birch followed in 1952 with the suggestions (acknowledged as from J.B. Thompson) that $MgSiO_3$-pyroxene might adopt the corundum ($Al_2O_3$) structure and $SiO_2$ might adopt the rutile ($TiO_2$) structure at high pressures. Remarkably, each of these ideas would ultimately prove to be correct (although the cation-ordered version of the corundum structure, $MgSiO_3$-ilmenite, has a rather small stability field relative to that of the perovskite-structured polymorph). Determining the phase equilibria of magnesium silicates at transition-zone conditions, however, required extensive developments in high-pressure technology.

Experimental capabilities in 1948 encompassed only crustal-level conditions, but today the full pressure range and most of the temperature range of deep Earth conditions can be stably achieved in laboratory experiments, with detailed structure and properties of the sample being characterized by X-ray diffraction and spectroscopic methods. In doing so, a sound physical and thermochemical basis has been provided for both the detailed seismic structure of the transition zone and the lack of structure in the bulk of the lower mantle: Essentially no phase transitions have been observed in mantle constituents at depths greater than ~900 kilometers. The synthesis of the high-pressure phases relevant to the transition zone was accomplished in the 1960s and 1970s, with high-pressure phase equilibria results continuing to the present day.

The high-pressure experimental results have advanced in tandem with improved seismic characterization of the transition zone. The refinement of the seismic structure of the transition zone into discrete seismic discontinuities at 410- and 670-kilometers depth (beginning with work in the 1960s) was readily placed into a mineralogical context—the former is associated with the transition from olivine to ß-phase (a phase with structural similarities to spinel), and the latter with a transition from $(Mg, Fe)_2SiO_4$-spinel to $(Mg, Fe)O$-magnesiowüstite and $(Mg, Fe)SiO_3$-perovskite. The depth of the former of these transitions provides, in conjunction with the pressure-temperature slope of the olivine to ß-phase transition, a key reference point for the temperature within the transition zone. Accordingly, the view of the transition zone of the planet has progressed from a featureless, high gradient zone of seismic velocities with poorly constrained mineralogy, temperature, and chem-istry to a zone whose complex mineralogy is well-understood, with the absolute temperature having fixed points, and for which there are only modest uncertainties in the bulk chemistry.

In turn, the recognition of plate tectonics coupled with improved observational and experimental constraints on the rheologic properties of mantle materials has moved mantle convection from the realm of speculation to one of the truly fundamental concepts in our understanding of deep Earth. Today, ideas about possible episodic material exchange between the upper and lower mantle of the planet, the genesis of plume-like upwellings, and the ultimate fate of subducted slabs each hinge on the critical recognition that Earth's mantle is an actively convecting dynamic system. The advent of seismic tomography in the 1980s profoundly impacted our understanding of mantle convection, as this procedure for imaging three-dimensional variations in seismic velocities revealed for the first time the deep structure underlying surficial plates. Today, the global resolution of lateral variations of upper-mantle velocity (controlled primarily by thermal variations but also by chemical variations) is approaching 500-kilometer-scale lengths, with 50-kilometer resolution in regions of dense seismic instrumentation. Patterns of high and low seismic velocity have been revealed throughout the mantle, with the strongest variations found in the upper 300 kilometers. Unexpected features, such as deep roots of high-velocity material extending 300-400 kilometers below cratons, have radically modified initial notions about plate tectonics and continental formation. Deep-seated upwellings under ocean ridges and beneath major volcanic hot spots are indicated by low-velocity regions and by deflections of transition zone discontinuities. Subducting lithosphere, illuminated only by deep seismicity in 1948, is now imaged as high-velocity tabular downwellings extending throughout the upper mantle, with possible aseismic extensions to depths of at least 2,000 kilometers. Numerical models of mantle flow that utilize the seismic images to constrain the distribution of buoyancy sources have accounted for long-standing mysteries about the shape of Earth's geoid, and detailed models of the entire mantle viscosity structure are beginning to converge. The unexpected observation that mantle heterogeneity is dominated by very large-scale structures has ironically provided support for some of the earliest conceptual models of a mantle convection system dominated by large patterns. Geochemical evidence for distinct reservoirs in the mantle persists, emphasizing the importance of plume flows from internal boundary layers as distinct from large-scale flows associated with oceanic plates. While the debate over whether the mantle has a layered convective system or not continues, a consensus seems to have emerged of significant material transport across the upper mantle-lower mantle boundary.

## Core-Mantle Boundary

Since the 1940s, ideas about the structure above the core-mantle boundary (CMB) have perhaps evolved more than those about any region of the planet. The view of this zone has shifted from an area characterized by a minor flattening of velocity gradients in Jeffreys' models (in 1949, Bullen inferred the presence of chemical heterogeneity of the lowermost 200 kilometers of the mantle based on this feature) to a zone recognized today to contain two laterally variable seismic discontinuities, at 5–50 kilometers and 130–400 kilometers above the CMB, with regions of anisotropic properties that appear to indicate the presence of texturing of material through lateral flow. The continually improving seismic images of the structure near the CMB indicate that it is an area in which the complexity of physical and dynamic processes rival those present in the lithosphere and shallow asthenosphere. Among the most basic of paradigms about Earth's deep mantle is that it is solid—a conclusion that appears not to hold for the absolute basal layer of Earth's mantle, where mantle and core meet. The recent discovery of a 5- to 50-kilometer-thick layer of dramatically depressed seismic velocities just above the core-mantle boundary is most readily explained by the presence of abundant partial melt at this depth—a wholly unanticipated result with profound rheologic implications for the lowermost mantle. Many of the seismologically characterized structural features near the CMB are still not well understood from mineralogic/petrologic or planetary evolution perspectives, but the simple demonstration of their existence has inspired a wide range of ongoing theoretical and experimental efforts to discern the properties of this enigmatic region of the planet—avenues of research that could not have been even dimly foreseen in the 1940s.

## Core

The core has long been viewed as a compositionally simpler system than the mantle (as Birch put, it, "predominantly iron"), but there is now certainty that a lighter component or components are alloyed with iron in the outer core at the ~10-weight-percent level. Furthermore, our estimate of the temperature of the outer core is now better constrained than ever before, based on high-pressure experiments, with the present minimum estimate of the temperature at the top of the outer core to be 4000°K, and the likelihood that it is significantly hotter—a result far better constrained than Verhoogen's 1953 range of 1500–6000°C for the base of the mantle or Gutenberg's 1951 estimate that the center of the planet was "probably closer to 2000°C than 5000°C." The elastic properties of high-pressure iron have been utilized to understand the observation that the inner core is anisotropic (seismic waves travel faster along the axis of rotation of the inner core than they do through its equator). As with the mantle, it appears that this lowermost region of the planet must be convecting and this results in sheared orientation of crystals—a realization not hinted at until the last decade.

Finally, the understanding of Earth's magnetic field has mushroomed over the last 50 years, with present-day numerical models of the magneto-hydrodynamics of the core system showing it is actually able to generate predominantly dipolar magnetic fields that undergo spontaneous reversals. The models bear many similarities to actual observations. Both the pivotal role of the solid inner core in modulating the fluid flow pattern of the outer core (and thus the character of the magnetic field) and the large-scale, complex turbulence of outer-core fluid flow are now routinely recognized features of deep Earth dynamics. The magnetic field has thus progressed from a feature whose genesis was only dimly understood to a first-order signature of the pattern of fluid flow within the core. It is fun to contemplate what comparable advances the next 50 years will bring.

## Questions:

1. What two things brought mantle convection from just speculation to an actual fundamental concept?

2. At what depth in the lower mantle are there no phase transitions?

3. What is the new understanding of earth's magnetic field?

Answers are at the back of the book.

## Activity:

Go to **http://pubs.usgs.gov/publications/text/unanswered.html**
Read only "What drives the plates?" Write a page summary.

## 2

Plate tectonics, discovered in the 1960s, changed the view of earth from a static entity to a dynamic planet. It also offered answers to many of geology's questions. In addition, the theory allowed for mantle plumes and hot spots which are responsible for the creation of the Hawaiian Island chain. These two discoveries, together, may explain the Laramide Orogeny, or the Rocky Mountains. The Laramide Orogeny has proven to be an enigma to many earth scientists due to its locality being far inland and its lack of volcanic activity.

# Mantle Plumes and Mountains

By J. Brendan Murphy, Gary L. Oppliger, George H. Brimhall, Jr., and Andrew Hynes

Traveling west on Sixth Avenue from downtown Denver, the Front Range of the Colorado Rockies looms 2,500 meters, above the mile-high city. Yet as magnificent as the panorama may be, it fails to do justice to the Laramide Orogeny, the process that started the Rocky Mountains' growth some 75 million years ago. Geologists estimate that the total uplift of the Front Range exceeded 7,000 meters. What drove the western half of Colorado to fracture and pile up to a height seven kilometers greater than that to the east? Earth scientists have long labored for a convincing explanation.

The theory of plate tectonics has offered clues. In a brief 30 years, it has revolutionized our understanding of mountain building. According to this theory, the outermost layer of the earth, the *lithosphere*, is composed of a mosaic of rigid plates that ride on a hot, pliable layer of the earth's mantle, the *asthenosphere*. As a consequence of circulation in the mantle, plates move with respect to each other at rates of a few centimeters per year. Over geologic time, this motion can account for the creation and destruction of oceans, the generation of mountain belts and sedimentary basins, the distribution of volcanic and earthquake activity, and the locations of ore, oil and gas deposits. Yet plate-tectonic theory has a tough time with the details of the Laramide Orogeny.

The conventional explanations of mountain building according to plate-tectonic theory all include horizontal plate motions and directly or indirectly depend on subduction zones, areas where oceanic crust descends back into the asthenosphere. Yet the Rocky Mountains lie 2,000 kilometers to the east of the current coastal margin, many times farther from an active subduction zone than mountain building generally takes place.

We propose that an additional mechanism of mountain building has been largely overlooked and may help explain not only the Laramide Orogeny but also other unusual geological features of the southwestern U.S. Our model involves the interplay of the horizontal motions of traditional subduction-related mountain building processes with vertical plumes of anomalously hot mantle ascending from thousands of kilometers below the earth's surface. Together these mechanisms may offer a convincing explanation for what long has been a geologically puzzling part of the world and may lead to better understanding of mountain building worldwide.

From "Mantle Plumes and Mountains," by J. B. Murphy, G. L. Oppliger, G. H. Brimhall, Jr., and A. Hynes, American Scientist, March/April 1999, pp. 146-153. Reprinted with permission.

## Horizontal Forces

In plate tectonics, as plates move apart, magma ascends from the asthenosphere, cools and solidifies to generate new lithosphere between the plates. This ongoing activity drives wedges of new lithosphere between the plates, separating them and generating a widening ocean. According to the theory, all oceans form in this way. The northern Atlantic Ocean, for example, has formed and progressively widened over the past 120 million years as the European and North American plates move apart. On a globe of constant radius, however, the divergence of plates and the construction of new lithosphere in some places must be compensated by convergence and destruction of lithosphere in others.

This is neatly accommodated by the recycling of oceanic crust in subduction zones. As it descends, the slab of lithosphere progressively heats up in the warmer ambient temperature of its surroundings. This eventually causes melting in the vicinity of the slab and in the overlying plate. These melts exploit weaknesses (such as fractures) in the overlying plate and ascend to the surface to produce volcanoes. In this way subduction leads directly to mountain building. The Andes are a modern example of mountains formed by such a process, and many of their highest peaks are either volcanically active now or have been in the recent past.

A second form of orogenic activity involves *microcontinents*, small islands or island chains located on oceanic crust. All modern oceans contain these islands; the Hawaiian chain is an example. Ultimately, when the tract of oceanic crust that separates these microcontinents from the continental margin becomes consumed by subduction, they will be swept to the margin and will collide with it. The impact results in deformation of rocks and igneous activity, which combine to form mountains. The coastal mountains of western North America formed in this way. Repeated collisions with many of these small landmasses over the past 400 million years has caused the North American Plate to grow westward by an average of 500 kilometers, extending from Baja California to Alaska.

In some instances subduction and convergence consume an entire ocean, and two continental land masses collide building mountains. Over the past 40 million years an ancient ocean called the Tethys was consumed by the collision of India with southern Asia and of northern Africa with southern Europe. The Himalayas and the Alps were pushed up in the Tethys's place.

None of these processes lends itself readily to an explanation of the Laramide Orogeny. Since the Laramide had no volcanic activity, conventional models of subduction do not apply. Likewise, it is clear that continents did not collide to form the Rockies. Furthermore, although collisions with microcontinents occurred during the time of the orogeny, these collisions were 1,200 kilometers away, at the very least, making this an unlikely explanation.

Geologists have been forced to invoke an unusual sort of subduction zone to explain the Laramide Orogeny, one that had an extensive subhorizontal zone, rather than the more typical angled one. This zone must have extended at least 1,200 kilometers into the continental interior, and the oceanic crust must have been anomalously shallow in order to avoid melting and the generation of magma. Although this mechanism is widely accepted, the reasons why such a subduction zone should exist have been elusive. We may be able to fill in some of those details.

## Plumes and Hot Spots

More than 30 years ago, Tuzo Wilson of the University of Toronto proposed mantle *plumes* to explain the formation of island chains. He suggested that Hawaii and several other volcanically active Pacific islands sit atop narrow columns, or plumes, of unusually hot rock and magma that ascend from deep within the earth. The interaction of plumes with the earth's rigid outer crust causes broad bulges or swells, and the melting induced by plumes provides the raw material for some of the world's most famous volcanoes. Volcanic centers above these plumes are known as *hot spots*.

In the example of the Hawaiian chain, the only active volcanoes are on Hawaii and the seamount Loihi, which is to the southeast of Hawaii and working its way toward the surface. Wilson noted that the islands are progressively older, less elevated and more eroded to the northwest along the length of the chain, and he interpreted this progression to be related to the westward motion of the Pacific Plate above " a jetstream of lava." Each volcano was born in the present position of Hawaii directly above the plume. But as the plate moved northwestward, each was cut off from its supply of magma below. As each volcano cooled and aged, it subsided and became progressively more eroded.

Implicit in this analysis is the fact that hot spots are relatively stationary and certainly move more slowly than the plates above them. Building on this idea, Jason Morgan of Princeton University proposed that three parallel island chains in the Pacific Ocean could have been formed by the motion of the Pacific Plate over three different hot spots.

Many investigators also think that the ascent of plumes is intimately associated with the breakup and dispersal of continents to form new oceans. Indeed, hotspot activity may have been an integral part of the breakup and dispersal of the supercontinent Pangea and the formation of the Atlantic Ocean. Don Anderson at the California Institute of Technology thinks that hot spots originate beneath large supercontinental land masses because continental crust conducts heat poorly compared with oceanic crust. By acting as an insulator, blocking the escape of heat from the mantle below, the supercontinent forces temperatures beneath it to rise, causing it to dome upward and eventually crack. Molten lava from the underlying asthenosphere rapidly ascends to fill the cracks, thereby driving the fragmented pieces of the former supercontinent farther and farther apart.

Nonetheless, hot spots are definitely not restricted to the locations of plate boundaries. The active Hawaiian volcanoes sit in the middle of the Pacific Plate at present, and mid-oceanic plate hot spots dot the globe. Thus their direct relationship to plate tectonics is unclear. Most earth scientists do accept that hot spots are the surface expression of hot columns of magma rising from a depth below the realm of plate tectonics, but just how deeply they originate is less certain. Recent evidence, however, suggests that these hot spots represent upwelling from near the core-mantle boundary, about 2,900 kilometers below the earth's surface. (See *American Scientist*, March-April 1995, pp. 134-147.) Thus plumes may be a phenomenon superimposed on plate motion rather than being a consequence of it.

The tell-tale signs of the origin and ascent of such features were recently revealed by seismic tomography, a procedure analogous to computer-aided tomographic (CAT) scanning of the human body by criss-crossing waves from an X-ray generator. This technique, which combines information from a number of seismic waves emanating from earthquake zones that penetrate deep into the earth, allows the construction of a three-dimensional image of much of the inner earth.

Seismic tomography, combined with laboratory and theoretical models, provides insights into the geometry of 9 plumes. An established plume has a relatively narrow central conduit in which hot mantle ascends, but it widens dramatically where it contacts the base of the lithosphere. Evidence from Hawaii and from the Yellowstone Caldera, which also resides above a plume, indicates that the plume's products are ponded at the base of the lithosphere and can "underplate" an area about 1,000 kilometers in diameter. This results in swelling and dynamic uplift of the lithosphere that is sustained as long as the plume remains active. Many of these regions are uplifted by as much as three kilometers above this hot, relatively buoyant material. Volcanic islands such as Hawaii rise more than another six kilometers above the dynamically domed sea floor, making them some of the highest mountains on earth.

## Hot Spots and Mountain Building

As a plate moves over a hot spot, the crustal swell is dragged in the direction of plate motion into an eccentric elliptical shape up to nearly 2,000 kilometers in length. Hot spots and their associated crustal swells must inevitably interact with continental margins. More than 40 hot spots lie beneath the modern oceanic crust, and no modern ocean could be consumed without at least some hot spots being overridden by continental crust. In accordance with a basic principle of geology known as *uniformitarianism*, which states that modern processes are typical of those that have occurred throughout much of geological history, no ancient ocean could have been consumed without overriding ancient hot spots. Furthermore, the overriding of hot spots by continental margins must be a common phenomenon.

We think that the overriding of a hot spot and its large, buoyant, elongate swell by a convergent plate margin would dramatically affect the geometry of the subduction process and therefore would profoundly influence the style of mountain-building activity at continental margins.

## A Modern Example

The geological evolution of the southwestern U.S. may offer such an example. From about 400 million to 75 million years ago this region experienced relatively normal mountain-building periods, including the Sonoma, Nevada and Sevier orogenies. Then, about 75 million years ago, a cycle of unusual tectonic processes began that continues to the present; we attribute these processes to the overriding of the Yellowstone hot spot by the North American Plate. Further, we conclude that the early manifestations of this event resulted in the Laramide episode of mountain building.

Before going into details, we must briefly review the geological history of western North America over the past 400 million years. During most of this time, mountain belts formed as the result of subduction and the episodic collision of microcontinents. Igneous rocks from that time have similar compositions to those in modern subduction-zone regions, and the style and distribution of rock deformation is also typical of such settings. As oceanic crust was consumed by the subduction zone, microcontinents collided with the margin. Specific collisions produced important and discrete episodes of mountain building known as the Antler (about 400-330 million years ago), the Sonoma (260-210 million years ago) and the Sevier (120-70 million years ago) orogenies. In addition, from about 100 million to 75 million years ago, ongoing subduction resulted in the emplacement of large granitic bodies that make up the backbone of the Sierra Nevada.

The breakup of Pangea, which began 180 million to 150 million years ago, probably accelerated these processes. Since that time, the North American Plate has drifted westward as the Atlantic Ocean formed and has progressively widened. Thus the western margin of the North American continent now lies at a location formerly occupied by oceanic crust of a wider Pacific Ocean.

About 75 million years ago, an unusual succession of tectonic events began. Between about 75 million and 40 million years ago, there was widespread deformation as vast portions of the continental crust were tectonically sliced and heaved on top of one another in the Laramide Orogeny, forming the early Rocky Mountains. The extent of this deformation (nearly 2,000 kilometers from the continental margin) is many times greater than the normal distribution of this type of deformation. During the same time, the region saw an almost complete lack of volcanic activity, a highly unusual situation for a typical subduction zone.

Both of these unusual features have been attributed to the presence of a subhorizontal subduction zone, rather than the more typical steeply angled zone that would have extended at least 1,200 kilometers into the continental interior. The deformation is associated with the interaction of this slab of oceanic crust with the overlying continental lithosphere. Because of the horizontal motion, the oceanic slab remained anomalously shallow and did not warm enough to generate magma. Although this scenario fits the available geological data, determining what would produce such a subduction geometry has proved difficult.

Equally enigmatic is the fact that after about 32 million years of quiescence, voluminous magma generation and associated volcanic activity began about 43 million years ago in northern Nevada and surrounding areas. This event may have had an important impact on modern economic development of the region. Many geologists think that this activity was directly responsible for gold mineralization in the area. Known as Carlin-type deposits (for their location near Carlin, Nevada), they qualify as one of the world's most productive districts, having yielded 50 million ounces of gold ($18 billion at $365 per ounce) since their discovery in 1962. Another 150 million ounces may be accessible over the next 20 years.

Magmatic activity continued in the region until about 6 million years ago. Yet despite the fact that the western margin of North America was still converging with the Pacific Plate, geologists working the vicinity of this magmatism have found convincing evidence that the area was extending dramatically at the time. In the early 1970s, John Proffett, then a graduate student at the University of California at Berkeley, estimated that the crust in the area may have been extended by more than 100 percent, to at least twice its original width. This estimate has been supported by the more recent work of Brian Wernicke at Caltech and his colleagues. One manifestation of this extension is the dramatic block faulting of the Basin and Range

Province. No consensus exists, however, on the underlying causes of the extension.

The chemical compositions of the igneous rocks generated at the initial stages of the magmatism 43 million years ago are typical of those produced by deep melting in the continental crust. They are relatively rich in elements, such as silicon and form products such as granite and rhyolite. About 18 million years ago, however, major eruptions of basalt, a volcanic rock relatively poor in silicon and rich in iron and magnesium, began to accompany the more silicic eruptions. The chemical composition of these basalts bears a strong resemblance to that of the basalts of Hawaii, suggesting that they may be related to a plume beneath the region.

Over the past 16 million years, volcanism produced a pronounced linear trough known as the Snake River Plain, which stretches from the northern Nevada border in a northeasterly direction across Idaho toward Yellowstone National Park in northwestern Wyoming. The ages of volcanoes along the Snake River Plain become progressively younger moving from west to east, just as they do in the Hawaiian Islands. This age progression appears to be related to the southwesterly motion of the North American Plate above a stationary plume. Thus 16 million years ago northern Nevada was situated where Yellowstone is today, and it has moved progressively toward its present location since then.

## The Model

Although evidence from the Snake River Plain indicates that the Yellowstone hot spot has existed for the past 18 million years or so, our model demands that it has existed at least for the past 75 million years. This is well within the typical life span of plumes such as the Hawaiian example, which shows no loss of strength after 76 million years. What might be the earlier history of the Yellowstone hot spot?

According to well-established plate reconstructions for the past 200 million years, the Atlantic Ocean lay much farther to the east 75 million years ago, and the western margin of the North American continent would have been 1,000 kilometers east of present-day Yellowstone National Park. Thus if the Yellowstone hot spot existed at that time and has indeed remained stationary since, it would have resided beneath an oceanic plate, as Hawaii does today. If so, it

would have generated similar ocean-island or seamount chains, the presence of which would be strong evidence for its antiquity.

The westward motion of the North American Plate and subduction along its continental margin would have destroyed direct evidence of the original Hawaiian-style island-chain geometry long ago. The subsequent motions of the islands, however, can be deduced. As previously discussed, such islands would have collided with the North American continental margin as microcontinents, a well known feature of orogenic activity along the western North American margin. Plate motions suggest that these collisions would not have been head-on; rather, they would have been highly oblique, resulting in deflection of the terranes northward along the margin. In fact, evidence of these former seamounts is found in the coastal regions of Washington and Oregon and as far north as the Yukon Territory of Canada.

Rocks in these regions bear all the hallmarks of their former oceanic Hawaiian style existence—including the nature of the volcanic eruptions, the chemical composition of the volcanic rocks and the presence of marine sediments. In addition, Stephen Johnson at the University of Victoria and colleagues working on the Yukon rocks were able to determine the latitude at which the volcanic rocks were formed. As iron-rich minerals crystallize out of magma and cool below 500 degrees Celsius, they behave as tiny magnets aligning themselves with the earth's magnetic field. Because this direction depends on the latitude of the rocks at the time of cooling, the iron-rich minerals preserve information as they freeze that can betray the latitude at which they formed. The ancient latitude recorded in the Yukon rocks matches the latitude of the Yellowstone hot spot, providing strong evidence for its existence at least 70 million years ago.

Given its oceanic position 70 million years ago and its modern continental position beneath Yellowstone, the hot spot and its crustal swell must have been overridden by the continental margin in the interim. The actual position of the continental margin at each stage during the intervening period can be determined from published plate reconstructions, and a hot-spot track can be developed. This record indicates that the hot spot would have been beneath the continental margin 55 million years ago and under northern Nevada between 40 and 30 million years ago. It then tracked northeastward across Idaho to its present position beneath Yellowstone

Although the commencement of the Laramide Orogeny 75 million years ago preceded the arrival of the hot spot beneath the continental margin, the crustal swell associated with the hot spot would have arrived sooner. Assuming a swell similar in size to that of the Hawaiian hot spot (elongated by 1,800 kilometers in the direction of plate motion), its arrival at the subduction zone may have preceded that of the hot spot by at least 15 million years, which is about the time the Laramide Orogeny began. The collision of this hot, buoyant, elevated oceanic lithosphere with the subduction zone provides an explanation for the onset of the Laramide Orogeny.

Recall that typical subduction zones are relatively steeply sloped because they subduct cold, dense oceanic crust, and the angle of descent into the asthenosphere is profoundly influenced by the density contrast between the crust and underlying mantle. In the vicinity of the hot-spot swell, the oceanic slab would have been considerably warmer and therefore more buoyant than normal, thereby producing an unusually shallow subduction. The buoyancy associated with progressive overriding of this swell may have aided in the gradual west-to-east shallowing of the subduction zone. This is consistent with geological data indicating that the age of deformation in the continental rocks is not uniform; rather, it gets successively younger to the east.

The lack of magmatism can also be explained by the hot spot. Assuming that the hot spot was similar in dimensions to modern hot spots in the Pacific Ocean, we determined the approximate location 50 million years ago of an associated crustal swell similar in dimensions to that of Hawaii. The hypothetical location of the swell agrees remarkably well with the position of the "magmatic gap." This suggests that the swell extended beneath the region of no magmatism and that its buoyancy contributed to the lack of magmatism.

As the North American Plate continued to migrate southwestward, the overridden plume would have been located beneath a progressively thicker cover consisting of the continental plate and the subducted oceanic slab. This additional cover would initially have reduced the ability of plume generated melts to rise to the surface. In addition, the shallowness of the subduction zone would have further inhibited melting of the mantle. Therefore, magmatism associated with both the hot spot and with subduction would have ceased.

Calculations of the stability of modern oceanic plumes show that they have difficulty penetrating even thin oceanic crust, let alone thick continental crust. When the hot spot was first overridden by the North American Plate, it would have entered an "incubation" phase. Although hot mantle continued to be pumped under the lithosphere by the plume, this material could not ascend to the surface. Nevertheless, the modern record of volcanic activity near Yellowstone clearly demonstrates that the hot spot did eventually break through.

During the incubation period, the ponded plume progressively assimilated the overlying slab of subducted oceanic crust. About 40 million years ago, when the hot spot would have resided beneath the Battle Mountain region of northern Nevada, the plume made its first contact with the overlying continental crust. Since continental crust has a lower melting temperature than does oceanic crust, voluminous magmatism would be expected, as is indeed observed there. The rich Carlin-type gold deposits of northern Nevada may also be related to plume activity, since the plumes may have ascended from the core-mantle boundary, a region thought to be unusually rich in gold.

At the same time, the well-documented establishment of a more typical subduction zone at the continental margin is also consistent with the model. The assimilation and severing of the subhorizontal subducted slab by the hot spot would necessitate re-establishment of a new subduction zone at the continental margin to accommodate continued convergence between the North American and adjacent oceanic plate.

About 18 million years ago the style of magmatism changed from one dominated by silicic rocks generated by crustal melting to a geologically world renowned example of contemporaneous magmatism of two contrasting compositions. Although the silicic crustal melts persisted, they were accompanied by the eruptions that were poor in silicon and rich in iron and magnesium and that bear a strong chemical resemblance to those of another plume—the Hawaiian Islands. We argue that the voluminous basaltic volcanic rocks across the

Columbia Plateau and the Snake River Plain were products of the Yellowstone Plume.

Finally, the rapid crustal expansion and magmatism in the Great Basin of Nevada that began about 30 million years ago has generally been attributed to the action of gravitational stresses associated with thick, weak continental crust. Our model, which places the Yellowstone hot spot beneath the Great Basin at the time of crustal expansion, provides a mechanism for generating this weakened crust.

## The Fourth Mechanism

The presence of hot spots beneath oceanic lithosphere implies that they must play an important role in orogenic processes. The Late Mesozoic-Cenozoic evolution of the southwestern U.S. may provide one example of the tectonothermal expression of such a process—with the Front Range being just one bit of evidence for it. Assuming that this model holds up to further scrutiny, we expect that many more examples will be found in the geologic record of the effects of horizontal plate motions combined with the vertical motions of mantle plumes. A few more pieces may thereby be added to the fascinating puzzle that is our planet.

## Bibliography

Bird, P. 1988. Formation of the Rocky Mountains, western United States: a continuum computer model. *Science* 239:1501-1507.

Hill, R. I., I. H. Campbell, G. F. Davies, and R. W. Griffiths. 1992. Mantle plumes and continental tectonics. *Science* 256:186-193.

Murphy, J. B., G. Oppliger, G. H. Brimhall, Jr., and A. J. Hynes. 1998. Plume-modified orogeny: an example from the southwestern United States. *Geology* 26:731-734.

Oppliger, G. L., J. B. Murphy and G. Brimhall, Jr. 1997. Is the ancestral Yellowstone hotspot responsible for the Tertiary "Carlin" mineral-ization in the Great Basin of Nevada? *Geology* 25:627-630.

---

**Questions:**

1. What is the evidence pointing towards Hawaii's creation by a mantle plume?
2. How long does the author's model call for Yellowstone's hot spot to have existed?
3. What are the two unusual features attributed to a subhorizontal zone under the Laramide Orogeny?

Answers are at the back of the book.

**Activity:**

Go to
http://volcano.und.nodak.edu/vwdocs/volc_images/north_america/yellowstone.html
Read the information about the Yellowstone hot spot. What is the evidence supporting the theory that the Snake River Plain is the result of a plate migration over a hot spot currently located under Yellowstone?

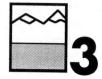

**3**

Scientists don't have to look to distant planets to find alien worlds. In fact, the earth's inner core is beginning to be seen as much different from what it was once thought to be. The inner core is usually described as an iron and nickel solid. Recently, discrepancies in the time it takes seismic waves to travel through the earth have been noted, leading some to believe that the inner core rotates slightly faster than the earth.

# The Globe Inside Our Planet
## Earth's Inner Core Is Turning Out to be an Alien World

By Richard Monastersky

In Dante's Inferno, the outcast Florentine poet took literary revenge on his political enemies by sentencing them to an eternity of torture without parole. They served their time locked inside Earth, writhing away in a circular, split-level hell, where the suffering grew more intense with each successively deeper and smaller layer.

Although modern geologists might scoff at Dante's portrayal of the world underground, the 14th-century writer was actually correct about the basic architecture of the planet. As geophysicists know today, the globe's interior consists of several spherical shells—an outer crust, a rocky mantle, and a metallic core—with the heat and pressure growing progressively more excruciating as the depth increases.

At the very center lies a solid sphere of iron as hot as the surface of the sun. Roughly the size of the moon, this so-called inner core remained for decades a terra incognita, hidden beneath 5,000 kilometers of solid rock and molten metal. Scientists knew so little about this distant orb that they tended to ignore it, preferring to spend their time exploring the regions closer to the surface.

"Up until just quite recently, the inner core has been thought to be an uninteresting, featureless place," says Bruce Buffett, a geophysicist at the University of British Columbia in Vancouver.

It turns out that researchers could have used a bit of Dante's imagination. Recent studies depict the inner core as a bizarre world locked inside our planet, where scientist's expectations often prove wrong. "It's become clear that the inner core is a lot more complex than we've given it credit for," says seismologist Kenneth C. Creager of the University of Washington in Seattle.

Two years ago, a pair of seismologists discovered evidence that the inner core is dancing to its own beat, spinning measurably faster than the rest of the planet. Now, researchers have found that the inner core appears to have a split personality, with one hemisphere manifestly different from the other. Both findings have flummoxed theorists who are trying to explain how such a strange body came to reside inside Earth.

"It's very puzzling," says Xiaodong Song, one of the researchers who gathered to discuss the inner core at a Boston meeting of the American Geophysical Union (AGU) in May. "As we learn more about the inner core, it shows more mystery. It's very difficult to explain." Song is a seismologist at Columbia University's Lamont-Doherty Earth Observatory in Palisades, N.Y.

Song and his colleague Paul G. Richards exposed some of the core's curiosities in 1996 with their study of seismic waves pulsing through the deepest section of the planet (SN: 7/20/96, p. 36). The waves came from earthquakes in the South Atlantic and passed through the inner core on their way to recording stations in central Alaska, taking about 20 minutes to make the journey.

The two researchers noticed something peculiar in recordings made over a 30-year period. The seismic waves took progressively less time to cross through Earth, even though they were traversing exactly the same distance. Vibrations leaving a particular spot in the South Atlantic arrived at the recording station about three-tenths of a second sooner in the 1990s than they had in the 1960s.

The difference, they surmised, comes from an odd characteristic of the way seismic waves pass through the center of the planet. Researchers in the late 1980s found that the inner core transmits seismic waves faster in some directions than others. Song and Richards realized that the orientation of these paths would change if the inner core were rotating slightly faster than the rest of the planet. They hypothesized that the inner core's motion has slowly shifted the fastest direction of seismic wave travel so that it now lines up more closely with seismic waves going from the South Atlantic to Alaska. As a result, these waves get to Alaska quicker.

Since then, two other studies have bolstered the concept of an independently rotating inner core, although researchers disagree on its speed. Song and Richards originally estimated that the inner core is rotating faster than the mantle and crust by 1.1° per year, so the central sphere would take a little more than 3 centuries to lap the rest of the planet. In 1996, researchers from Harvard and the University of California, Berkeley calculated that the rate could be as high as 3° per year, whereas last year Creager figured a value of only 0.2° to 0.3° per year.

The different estimates of rates and other unresolved issues cause some seismologists to question whether the inner core is truly setting its own pace. In the July 3 SCIENCE, Annie Souriau of the Observatoire Midi-Pyrenees CNRS in Toulouse, France, wrote a commentary titled "Is the Rotation Real?" Souriau argues that the evidence to date does not build a convincing case.

If the core's rotation is real, it places stress on geophysicists who wrestle with the subject of Earth's internal forces. They find the concept of core rotation hard to explain because the solid iron sphere should be locked in step with the rest of the planet by the almost indomitable force of gravity, says Buffett.

The gravitational link comes from slabs of cold rock that have sunk from Earth's surface deep into the mantle. Because these slabs are extra dense, they tug on the solid inner core and raise welts a hundred meters or so that stick up into the fluid outer core. The attraction between these bumps and the cold mantle slabs exerts a powerful enough force to prevent the inner core from turning with respect to the mantle, says Buffett.

For Song and other seismologists to be correct, the inner core must somehow slip from its gravitational shackles and rotate freely. Buffett suggests that the inner core could do so if it were far softer than previously presumed. "If the inner core was very deformable and very mushy, then you can explain the observations," he says.

In this case, the inner core's surface could reshape itself quickly. The bumps could migrate across the top of the inner core while staying positioned directly beneath the cold, dense patches in the mantle, Buffett told researchers at the recent meeting in Boston.

From the lowest estimates of the inner core's speed, Buffett calculates that the solid iron sphere would have to be far softer than rock but harder than the glacial ice that flows readily down mountain valleys. Faster spinning rates would require an even more malleable inner core that could keep remolding its surface at a geologically frenzied pace.

If Earth's heart is softer than previously thought, less solid still are theories about the history of Earth's solid iron center. Geophysicists have only vague guesses about when the core first started solidifying and how quickly it is growing today.

The core itself probably formed quite early, even before Earth reached its full size 4.5 billion years ago, says geophysicist Raymond Jeanloz of the University of California, Berkeley. Under a barrage of giant asteroids, the fetal planet would have heated up enough for most of its iron to melt and sink toward the center, forming a molten metallic sea. As the steel soup cooled, crystals of iron started dropping toward the center. The solid core was born.

Just when that happened remains "mostly speculation," says Jeanloz. The best guess comes from studies of Earth's magnetic field, which was born in the roiling currents of iron in the outer core (SN: 10/19/96, p. 250). Because most scientists think that a magnetic field with the current strength and configuration requires a solid iron core, the field offers a clue to the inner core's origin, says Jeanloz.

Geologists know that a strong geomagnetic field existed at least as far back as 2 billion to 2.5 billion years ago because it left a magnetic imprint in ancient rocks. This indicates the presence of an inner core by that time. For periods even more ancient, the rock record is sparse and harder to read.

One of the biggest questions nagging researchers today concerns the orientation of the iron crystals in the inner core. This issue surfaced a decade ago when seismologists found that earthquake waves traveling north-south cross the inner core faster than those going east-west—a property called "anisotropy." The explanation they offered was that iron crystals line up to form a distinct pattern or fabric, which quickens waves going with the weave.

Seismologists first assumed that the entire inner core had a uniform crystalline design. Japanese and U.S. scientists are now discovering, however, that the strong pattern appears in only half the inner core.

Satoru Tanaka and Hiroyuki Hamaguchi of Tohoku University in Sendai, Japan, first uncovered this rent in the core's fabric by studying two classes of seismic waves from earthquakes. One group dives through the mantle and outer core, ultimately passing through the inner core before heading back to the surface. The other group travels almost exactly the same path in the mantle but bends around in the outer core and never reaches the inner core. By comparing the travel times of the two types of waves, Tanaka and Hamaguchi isolated the effect of the inner core.

Tanaka and Hamaguchi noticed that seismic waves going under Asia and Australia took the same time to crisscross the globe, no matter if they were traveling east-west or north-south. But for waves crossing under the Americas, Europe, Africa, and the Eastern Pacific—a region that corresponds roughly with the Western Hemisphere—the north-south path took less

time than the east-west path. The difference was roughly 3 seconds out of about 18 minutes. The scientists published their results in the February 10, 1997 Journal of Geophysical Research.

Creager found similar differences in his study of core-crossing waves. The anisotropy in the Western Hemisphere is about 3 to 4 times the strength of the anisotropy in the Eastern Hemisphere, he reported at the AGU meeting this spring.

The two-faced nature of the inner core has so far stumped researchers. "It's very surprising," says Creag "it's hard to think of a physical mechanism that would cause these variations."

One answer might be that the temperature or chemistry of the eastern core differs from that of the western half. But when Creager averaged waves going in all directions across each hemisphere, he found no marked differences between the two sides of the core, indicating that they have the same general properties.

Instead, it appears that the pattern of the iron crystals in the Western Hemisphere lines up strongly in the direction of Earth's axis of rotation, whereas those in the Eastern Hemisphere show a more helter-skelter fabric, pointing in all different directions, with perhaps a slight bias toward the rotation axis, he says.

For now, geophysicists are still trying to answer the most basic question of why the iron crystals should align in the first place. Jeanloz proposed a decade ago that the solid iron of the inner core could actually be flowing slowly, carrying heat from its center toward its surface. The resulting currents would align the iron crystals as they flow—a process similar to what happens in glacial ice as it moves. If the convection is not uniform throughout the inner core, it could account for differences in the eastern and western halves, he says.

Others have questioned the idea of a churning inner core. "It's not clear why the inner core should be convecting. It's a metal and so is a good conductor of heat," says Michael I. Bergman, a physicist at Simon's Rock College in Great Barrington, Mass. If enough heat can escape from the inner core via conduction, then the solid iron would not need to flow.

In laboratory experiments, Bergman is testing an alternative hypothesis: The iron anisotropy develops because the crystals align themselves as they solidify from the molten alloy of the outer core. As liquid metal hardens, he says, it forms treelike crystals that grow outward into the remaining liquid. Theory suggests that the crystals should orient themselves parallel to Earth's equator because heat escapes fastest in that direction. Bergman is now testing that idea using a rotating sphere filled with salt water, cooled down to the point where ice crystals start to form, he reported at the AGU meeting.

By using an icy stand-in for the core, the physicist is unknowingly following The Inferno. Dante pictured the circles of hell leading down ultimately to a vast lake. In the middle of the basin sat Satan, beating his three wings so fiercely that they froze the water into a block of solid ice at the very center of the planet. Today's scientists don't expect to find anything so fanciful inside Earth, but they would agree that the inner core is turning out to be devilishly hard to fathom.

## Questions:

1. Why do some scientists question an independently rotating inner core?
2. Why does an independently rotating inner core present a problem to geophysicists?
3. What is anisotropy?

Answers are at the back of the book.

## Activity:

Go to **http://www.nsf.gov/od/lpa/news/publicat/frontier/3-97/3mysts.htm**
Read the article. If the earth's inner core is rotating, what will this discovery help explain?

According to plate-tectonic theory, the continents have continually changed their positions relative to each other throughout the history of the Earth. Sometimes the continents have moved together to form a giant supercontinent and other times they have moved apart to form many separate landmasses. Geologists know that supercontinents have formed at least twice in the Earth's history. Today the continents are fairly well separated, but they have been broken apart into more separate landmasses in the past.

Geologists have a very good idea of the path of the continents' movement back through the formation of the last supercontinent, Pangaea. Geologists are not sure of the exact path that the continents took before that. Some geologists think that the continents have separated and come back together into similar configurations. Others think that when the continents came back together before Pangaea they did so in completely different configurations. The following article offers some of the accumulating evidence for the latter view.

# Travels of America

By Tim Appenzeller

*Half a billion years ago, a large chunk of North America went missing. That chunk has now turned up in the Andes of Argentina.*

For Bill Thomas, the hills and hollows of the upcountry South are home ground. "I grew up speaking genuine Appalachian," says the tall, quiet geologist from the University of Kentucky. For nearly 30 years Thomas has been tramping around the Appalachians, trying to understand why they appear to stop in Alabama-only to resume hundreds of miles to the west, in Arkansas, as the Ouachita Mountains. That gap makes room for the broad Mississippi Valley and ultimately for the Gulf of Mexico, and Thomas has been asking how it formed. He never expected the answer would take him 4,000 miles south to the arid foothills of the Andes.

Then again, Ricardo Astini never thought he'd have to leave those foothills, called the Precordillera, for the thickets and piney woods of northern Alabama. Astini, a geologist at the University of Cordoba in Argentina, speaks genuine Spanish, a fast and animated version. He has spent a decade trying to understand how that tract of limestone hills—so weirdly distinct from its surroundings in fossils and rock types—ended up snuggled against the Andes in western Argentina. But by last year Astini was picking over rocks in Alabama, and Thomas was planning a field trip to South America.

What brought these two unlikely collaborators together wasn't an academic exchange program or yearnings for a drastic change of scene; it was the realization that they were working on the same problem. More than half a billion years ago, Thomas and Astini have now shown, a block of crust 500 miles square broke away from North America, drifted across an ocean, and welded itself to South America. The result was a gap in the North American coast that eventually became the Gulf of Mexico, and a large tract of foreign crust in South America that became the Precordillera.

The result, also, is support for a new map of the world of 500 million years ago. Geologists have tended to picture the continents as tango dancers, sometimes glued together, sometimes stepping apart, but always paired off with the same partner—North America with Africa, for example. But just about the time Thomas and Astini were tracing the history of the Precordillera and the Gulf of Mexico, another pair of geologists were conceiving a new view of ancient geography. In it, continents that would later be strangers faced each other across oceans that no longer exist. And in it, a land swap between the east side of North America and the west side of South America—once barely conceivable—was no more than routine.

Thomas, to start, was interested in only one sliver of that earlier world: the ancient East Coast. The southern Appalachians form a series of gentle curves, each one set a little farther to the west than its neighbor to the north. Those offsets, Thomas realized in the late 1970s, are a blurry image of sharp zigzags in the old edge of the continent—of the jagged rift along which Laurentia, as ancient North America is called, tore away from some other continent. Later an ocean called the Iapetus rolled through the gap; and later still, other blocks of crust slammed into Laurentia, shoving the ancient coast inland and raising the modern Appalachians.

All along the eastern flank of the mountains, from Virginia to Alabama, Thomas could see the stratigraphic signature of the rift that became the Iapetus. There were conglomerate layers that had formed when sand and gravel washed into the developing rift; there were limestones—consisting of fossilized sea creatures—that had been deposited after the sand and gravel, once the rift had widened into ocean. From the age of the limestones Thomas could tell that the process was already well under way by 540 million years ago. At every zigzag in the ancient coastline, the story was the same—except at the biggest one.

That one is at the southern tip of the Appalachians in Alabama, where a fault called the Alabama-Oklahoma transform joins them with the Ouachitas, 360 miles northwest near the Arkansas-Oklahoma line. Rocks retrieved from boreholes along the fault, mostly by oil prospectors, made it clear to Thomas that the Ouachitas are a continuation of the Appalachians, and that before they were mountains they too lay on the Iapetus coast. But there was one problem: there was no evidence that a rift had formed in the Ouachitas at the

same time—no 540-million-year-old limestones, for instance.

Apparently, Thomas realized, the great continental tear that created the edge of Laurentia didn't zigzag west through Oklahoma at first; it continued straight south from Alabama. Only later did it jump inland along the Alabama-Oklahoma transform, which may have been a line of weakness formed during the original continental breakup. Thomas thinks that jump occurred around 545 million years ago. That's the age other geologists have assigned to extensive layers of basalt and other volcanic rocks—now mostly buried under younger sediments—at the western end of the Alabama-Oklahoma fault. And that's when, according to Thomas, the land in Oklahoma started rifting, allowing magma to well up through the fractured crust.

The land to the east of this Ouachita rift was not a continent; that continent, North America's eastern partner, had already departed. What it had left behind was a 500-mile-square chunk of crust, stretching from southern Alabama west to Oklahoma and south beyond Houston, and bounded on the south and east by the Iapetus. As the Ouachita rift widened, this square of land broke away and began sliding east along its northern boundary, the Alabama-Oklahoma transform. Limestones buried in two ancient valleys just north of the fault, Thomas found, record how the ocean flooded into the valleys as the crustal block slid out of the way. By 515 million years ago, it had sailed clear of North America and was drifting free in the Iapetus.

The departed block left a gap that Thomas calls the Ouachita embayment—and that the rest of us know as the Gulf of Mexico. Although other landmasses later filled the gap—in the collisions that raised the Ouachitas—later still the added crust broke away along the original faults, creating the Gulf. Explaining the origin of that absence of land was Thomas's only goal at first. "This block was something I just needed to dispense with," he recalls. "I was asked several times, 'Where did it go? My attitude, somewhat cynically, was 'I don't care.'

"But now we think we know where it went."

Finding a bit of one continent in another isn't so startling these days. Since the late 1960s, geologists have known that Earth's surface is in constant motion, as new ocean floor congeals from magma at volcanic ridges, then trundles away from them at an inch or two a year. As it moves, the ocean floor can sweep along continents, as well as smaller landmasses and islands. The smaller bits eventually get plastered onto the leading edges of continents, giving them a fringe of "exotic terranes." California, for instance; is nothing but. Finding yet another terrane along the Andes would cause no particular stir.

But finding one that started off as a piece of North America is another matter. "To have said five years ago that the Precordillera was a part of North America was something most people just weren't willing to contemplate," says Eldridge Moores of the University of California at Davis. Few were willing to consider a map on which North and South America were close enough half a billion years ago to trade a piece of crust. Few people thought seriously at all about geography that ancient.

Until about five years ago, geologists rarely pushed their maps of Earth beyond Pangaea, the supercontinent that broke up some 200 million years ago. Because the oceans that opened then are the oceans of today, it's easy to trace how the continents were arranged in Pangaea, just by working backward from their tracks in the ocean floor. Those tracks say that the East Coast of North America once nuzzled up against North Africa, while South America's eastern shoulder fitted into the hollow of West Africa. "Before Pangaea, though, we don't have any ocean floor to guide us," says Ian Dalziel (pronounced dee-ELL) of the University of Texas at Austin. "So Pangaea is the oldest paleogeography that any of us will totally agree with."

When geologists have tried to look back before Pangaea, they have tended to re-create it again and again—to put eastern North America somewhere opposite North Africa, separated by an ocean that closed when Pangaea formed. In that picture, the Atlantic is only the latest in a series of oceans that have come and gone between North America and Africa. Dalziel calls this scheme "yo-yo tectonics."

And a few years ago he and Moores put the yoyo on the shelf. It started with Moores, who studies the Great Basin, the Sierra Nevada, and other parts of the American West. Like every other geologist who works in the region, he had seen signs that some matching landmass farther west had rifted away 650 or 700 million years ago. "You always worried, 'Where was that piece?'" says Moores. "People were always trying to find pieces on the Northern Hemisphere that might fit." None of them quite did.

Then in 1989, Moores ventured into the Southern Hemisphere—to Antarctica, on a field trip organized by Dalziel. "I knew almost nothing about Antarctica at the time," he says. "But there was a lot of talk on the ship about Antarctica, and they gave out these *National Geographic* maps. I like looking at maps, and I spent a lot of time looking at this one."

Illuminating as that map and the field trip itself were, though, there was no flash of insight on the ice sheet. That came several months later, back home in the Davis library. There Moores happened on a paper by two Canadian geologists proposing that 700 million years ago, Canada and Australia were joined together. The Canadians had stopped their analysis at the U.S. border, but Moores didn't. At the same time Canada was supposedly linked to Australia, he knew, Australia was linked to Antarctica. And Canada, as always, was closely tied to the United States. As Moores recalls it: "I thought, 'Aha.',

Moores guessed he had found North America's lost western partner. The part the Canadians cared about had once been connected to Australia, all right, but Moores's own professional territory had been connected to East Antarctica. Some quick research in the library convinced him that rocks reminiscent of a 1.8-billion-year-old belt in the American Southwest do indeed peep through the ice in the Transantarctic Mountains, the original edge of Antarctica. Moores called his new hypothesis SWEAT (Southwest U.S.-East Antarctica), and one of the first people he tried it out on was Dalziel.

"Eldridge made this leap of faith and put pen to paper and faxed me this map and asked me, 'Is this crazy?'" recalls Dalziel. Dalziel thought not; he had been musing along the same lines. And now he started thinking about North America's long-lost *eastern* partner—the landmass that had rifted away 540 million years ago to create Bill Thomas's zigzag coast.

Geologists had generally assumed that partner was Africa, which later slammed back into North America to raise the Appalachians—the yo-yo model.

But in Peru, of all places, Dalziel had found rocks that contradicted that assumption. They seemed to have been deformed in a mountain-building episode more than a billion years ago—at the same time as rocks in Labrador. If

Labrador and Peru were thousands of miles apart at the time, as the standard scenario required, that would have been a remarkable coincidence. But it made perfect sense, Dalziel realized, if the northeastern corner of North America had once nestled into the sharp bend in Peru's coast—if South America rather than Africa had been North America's eastern partner before Pangaea. Using software that allowed him to play continental matchmaker on a computer, Dalziel compared the outlines of the potential partners and also checked the relict magnetism in their rocks, which is a rough indicator of their latitude at the time the rocks formed. "I asked myself, 'Was it paleo-magnetically reasonable for Laurentia to be down next to South America?'" Yes, indeed, it was.

Thanks to Moores and Dalziel, then, there was a whole new map of Earth before Pangaea. Until 750 million years ago, in this view, North America was near the South Pole, wedged between Antarctica and Australia on one side and South America on the other, in a supercontinent called Rodinia. After that, according to Dalziel's simulations, North America went through Houdini-like contortions to escape from its partners, culminating in a 250-million-year-long end run up the west side of South America. Clearing the northern end of that continent, it faced off for the first time against North Africa, which was to become its neighbor in Pangaea.

The end run worked on a computer screen, but was there evidence for it in the real world? Dalziel proposed that geologists look for some. He said they might find North America's "calling cards"—pieces of crust it had deposited on other continents during its wanderings. As it happened, Argentine scientists had already found one years before.

Even now the Precordillera is a realm apart. For nearly 500 miles, north to south, it rises from the vineyards of western Argentina, in pale cliffs of limestone and shale, to peaks that pierce 14,000 feet. Here and there the crumpled terrain is cut by a braided river carrying snowmelt from the Andes, which nourishes a fringe of grass and willows. Acacia bushes and cactus claim the rest of the level ground, however, and there's little enough of that: the Precordillera is a desert tipped on its side. Most of it is a geologist's dream, with exposed layer cakes of rock that rise

a thousand feet and more and reveal hundreds of millions of years of history.

More than half a billion years ago, according to those rocks, the land here was not a fractured, buckled desert. Long before the geologic violence that uplifted the Andes, the Precordillera lay flat, forming shallows and tidal flats in a warm inland sea. And the fossils that now pepper its limestone cliffs—crustacean-like trilobites and the winglike shells of brachiopods—indicate that sea was nowhere near South America. In such features as the shape of a trilobite's head or the flare of a brachiopod's shell, the fossils differ subtly but unmistakably from typical South American ones. As early as 1965, according to geologist Victor Ramos of the University of Buenos Aires, Argentine paleontologists were saying, "Jesus Christ, those trilobites are North American."

The fossil experts had little idea how the intruders might have gotten there. But Ramos came up with one. In 1981, at a symposium in the United States, he heard speakers explain how to recognize exotic terranes from the traces of seafloor along their boundaries. "I opened my eyes, because the kind of evidence they were talking about was the kind we had," Ramos recalls. "The thing that had amazed me was the pillow lavas on both sides of the Precordillera." Pillow lavas form at seafloor volcanoes when magma is rapidly quenched in cold water. Their presence on both sides of the Precordillera suggested it had once been surrounded by ocean. To Ramos, it had to be a terrane.

The North American fossils indicated where that terrane had come from. So did its thick limestone layers, which resembled ones in the northern Appalachians—or so Ramos thought. In 1984 he proposed that the Precordillera might have broken free of North America's East Coast half a billion years ago and collided with South America. Later it became landlocked when another chunk of crust swept in from an unknown source to the west.

At a time when Moores and Dalziel had not yet redrawn the pre-Pangaea map, Ramos's scenario was mind-boggling: it required the Precordillera to have sidestepped thousands of miles across the globe. In Argentina, Ramos's proposal "caused a big debate," recalls Ricardo Astini. "Lots of geologists were mad at him because it was a revolutionary idea." In North America the response was mainly silence.

But Astini and his paleontologist colleagues at the University of Cordoba, Luis Benedetto and Emilio Vaccari, were actually working in the Precordillera. They had seen its strangeness for themselves and understood that it seemed to call for strange explanations. In the mid-1980s, they set out to test Ramos's idea. They soon found rocks and fossils that traced the details of the Precordillera's surprising voyage.

The fossils showed the voyage was swift. Until the early Ordovician Period, 480 million years ago, they are quintessentially Laurentian. Then their aspect starts to change. The trilobites and brachiopods that filled the shallow seas of the Precordillera for the next 15 million years or so are as distinctive, to paleontologists' eyes, as the plants and animals of certain islands are today. The Precordillera was apparently a kind of Madagascar—a large island in a now-vanished ocean (albeit an island that was mostly water-covered). By 465 million years ago, though, the island had docked in South America: the fossils from then on look identical to those found on the continent.

From the rocks, Astini and his colleagues extracted a blow-by-blow account of the collision. On the east side of the Precordillera, its leading edge dipped under the South American coast, forming a submarine trough that collected thick beds of sediment that can still be seen today, now crumpled and uplifted. On the west side, earthquakes produced by the collision shook loose bus-size blocks of limestone, sending them hurling into the sea; today those blocks can still be seen, trapped in other sedimentary rocks. In the central Precordillera, gaps in the rock sequence show how the collision thrust the limestone well above sea level, where it could be worn down by weather. Finally, not long after the Precordillera snuggled up against South America, gravel deposits sifted down onto its eastern edge from mountain glaciers on the continent; South America was in the grip of an ice age 440 million years ago. North America was still tropical. But the Precordillera's career as a tropical reef was over.

Working in its parched landscape, Astini and his colleagues could piece together its whole journey—except for its starting point. Ramos had suggested the northern Appalachians, but the rocks and fossils there didn't seem a perfect match. Astini's group started looking at rocks from elsewhere in the Appalachians. Then in 1992 they got a lucky break. The North-South connection was made for them by Christopher Schmidt, a geologist from Western Michigan University.

"Schmidt came to Cordoba to work on another topic," Astini recalls, "and after a while we met each other. I talked to him about our project and the probable foreign origin of the Precordillera."

Bill Thomas takes up the thread: "Chris and I had been working together, and he had become pretty familiar with my story" about the origin of the Gulf of Mexico. "Ricardo was describing the Precordillera to Chris, and Chris said, 'Hey, I know where it came from.'"

"Chris encouraged me to be in contact with Thomas," adds Astini. "I read his 1991 paper on a terrane that went out of Laurentia that has no name. It should be somewhere in the world, I thought."

So Astini began comparing strata from the southern Appalachians with those of the Precordillera; he measured the size of the missing piece of North America against the Precordillera; he considered the mirror-image histories of departure and arrival. He knew right away he was seeing the beginning of a beautiful intercontinental friendship.

"When you look at some of the sections in Alabama where the southern Appalachians outcrop, or you look at samples from boreholes, they really match one-to-one what we have in the northern Precordillera," Astini says. "It's incredible—the colors, the thicknesses of the rock all match. Where you have green shales, we have green shales. Where you have red shales, we have red shales. It's really incredible to travel so far and have the same strata." It becomes credible, though, if you accept that the Appalachians and the Precordillera were once connected.

Last fall that idea received about as much ratification as a scientific hypothesis can hope for. Dalziel and three other geologists organized a meeting to discuss how the Precordillera might fit into the geography of the ancient world. Researchers came from all over the world to the town of San Juan, on the eastern edge of the Precordillera. They took field trips out into the limestone mountains to see the fossils and strata for themselves; they listened to Ramos, Astini, Thomas, and their colleagues present their cases. And in the end they reached a unanimous verdict: the Precordillera was the long-lost piece of the Gulf Coast. As one geologist, George

Viele of the University of Missouri, put it: "If this isn't Bill Thomas land, we have a lost continent floating out there like the Flying Dutchman."

The pre-Pangaean Earth, meanwhile, was beginning to look like Moores-Dalziel world. It was hard to see how North America's southeast coast could have handed land off to South America's west coast if the two hadn't passed by each other. And it was hard to see how North America could have made an end run around South America without having previously been part of Rodinia, the southern supercontinent, in the way Moores and Dalziel had envisioned. Just how close the two continents passed after separating is in dispute; Dalziel is convinced they actually collided again, exchanging the Precordillera in the process. But most researchers at the conference were convinced by Astini's fossil evidence that the Precordillera had crossed the Iapetus on its own, as an island. From the time it took for the fossil changeover, they even estimated how wide the ocean must have been: 1,200 to 1,800 miles.

The enthusiasm for North-South land swaps ran so high in San Juan that the Precordillera came to seem but the clearest example of the process. Ramos now thinks the crust straight west of the Precordillera, in the high Andes, was also once a southern extension of North America, which followed 100 million years behind and locked the Precordillera into South America. Some terranes may even have traveled the other way. A tract of alien rock trapped trapped in the Piedmont of North Carolina, for instance, looks suspiciously South American to some researchers, as does the Oaxaca region of southern Mexico. "It sounds as though we had to have some traffic police," Moores joked at the meeting. "We had all these pieces that would have collided if they hadn't kept to the right."

"What I think things may have looked like," he adds, "is what you see in the western Pacific today. And what you see are chunks of continents, and volcanic island arcs, and you see arcs colliding and continents running into arcs and this incredible complexity that's changing very fast." Fast-changing complexity is not what the world of half a billion years ago used to look like; it used to be a blank. But that was before the new era of North-South cooperation.

## Questions:

1. What two mountain chains in North America were once connected by a now-missing chunk of land?
2. What is an "exotic terrane"?
3. What were the implications of the fossils in the Precordillera of Argentina?

Answers are at the back of the book.

## Activity:

Using a world map or atlas, find where the Andes Mountains, Ouachita Mountains, and the Appalachian Mountains are located. What is now found between the Ouachita and Appalachian Mountains?

# PART 2
## Earthquakes and Volcanos

5

Volcanoes make lots of sounds before erupting; could these sounds be predictable? Could these sounds help scientists monitor and predict eruptions of active volcanoes? There is now a computer model which could answer these questions.

Geophysicists currently use seismometers to pick up low-frequency vibrations, but these waves radiate through many layers, fissures, fractures and cracks distorting and scattering the waves. These measurements should be easier and more accurate using a microphone attached to a computer.

When magma flows into the volcano's internal chamber and rises, its pressure drops and gas bubbles are produced. When the bubbles reach the surface of the magma they explode, sending sound waves throughout the chamber. The microphone, positioned at the mouth of the volcano, records the sound. The sound is then fed into a computer model, which provides data on the volcano's internal pressure, gas content, and so on. This model may soon make it possible for scientists to forecast eruptions.

# What's This Volcano Trying to Tell Us?

By Daniel Pendick

On the steamy, tropical slopes of Arenal volcano in Costa Rica, clouds of moisture hang above the trees like a blanket. There's danger in the air too: the volcano's dark cone looms over nearby towns and villages, and has been erupting almost continuously for three decades. Not on the face of it a place that anyone in their right mind would want to climb.

But in 1992, geophysicist Milton Garces did just that, scrambling up Arenal's rainforested slopes to cock an ear at the volcanic event at the summit. "I sat there in the fog and the rain and listened for about a week," he says. "I heard explosions, I heard hissing, I heard this remarkable chugging sound, like a train. Through all that time I was thinking that I could use these sounds to find out what is happening inside the volcano."

Not many people would have lingered so long on an erupting volcano. For Garces, however, Arenal's angry growls sounded like a whole new way of studying volcanoes. Do the sounds volcanoes make before they blow their top change in some predictable way? Could those sounds help scientists monitor active volcanoes and forecast their deadly eruptions?

Inspired, Garces teamed up with Michael Buckingham, an expert in underwater acoustics at the Scripps Institution of Oceanography in San Diego, California, and set to work developing a computer model that could help answer those questions. The result is the equivalent of a stethoscope for volcanoes. Like a doctor listening for the irregularities in a heartbeat that would betray a faulty valve, they are homing in on the subtle variations in the boiling hearts of volcanoes that show, for example, whether a river of fresh magma charged with explosive gases is surging up from the depths.

Garces, now based at the University of Hawaii at Manoa, has listened to volcanoes from Antarctica to Italy. And his latest results confirm that a volcano's deafening clamour has a

From "What's This Volcano Trying to Tell Us?" D. Pendick, New Scientist, February 20, 1999, pp. 26-30. Reprinted with permission.

promising future in the troubled area of predicting eruptions.

Volcanoes first caught his attention while he was studying underwater acoustics. In 1989, Garces went to the Scripps Institution to work with Buckingham. Here, he became involved in research aimed at detecting volcanic eruptions on the seabed. Underwater volcano hunters, he learnt, wanted to distinguish these eruptions from the confusion of background noise in the ocean. But what is it that makes a volcano sound like a volcano, and what could these noises reveal?

Garces became intrigued by the question, but realised that studying the acoustics of underwater volcanoes wasn't simple: most eruptions occur kilometres down on the seabed, and learning their secrets would be difficult and expensive. There had to be another way.

Volcanoes on dry land produce a cacophony of noise, ranging from high-pitched whooshing and whining to low rumbles of infrasound with frequencies of less than 20 hertz, outside the range of human hearing. "These are the waves that physically shake you," says Garces—much like the roar that you feel in your chest when a jet passes close overhead.

Infrasound, Garces learnt, can reveal a volcano's innermost secrets. Some infrasound is generated near the surface by exploding bubbles of gas and rock—volcanic "belches" that throw magma and other debris into the air. But infrasound also comes from the very heart of a volcano, produces by the turbulent fluid dynamics of magma flows deep beneath the surface. Here, geophysicists believe that oscillating streams of magma or gas bubbles expanding and contracting with changes in pressure may give out deep growls and bellows. Since the magma has a different density to its surroundings, the molten rock traps these sounds and acts like a waveguide, channeling them up through the magma-filled passage called the conduit to the volcano's vent.

These waves of infrasound force the surface of the magma to vibrate, says Steve McNutt, a volcano seismologist from the Geophysical Institute at the University of Alaska in Fairbanks. "It behaves like the head of a drum, radiating sound into the air," he says.

This infrasound is a complex mixture of low frequencies. And because some of it comes from deep inside a volcano, Garces believes that it contains vital information about the processes that trigger eruptions—maybe even telling you how likely a volcano is to blow its top: "There's just an ocean of information in these low frequencies."

Geophysicists already eavesdrop on low-frequency vibrations using seismometers. Many of the processes that generate infrasound inside a volcano also send out waves known as volcanic tremor that race through the ground. Measuring these waves can tell geophysicists much about a volcano, but so far, attempts to use changes in the amount of tremor as a warning of eruption have given mixed results. "It's been useful, but there are big variations in how much tremor you see before an eruption," explains Bruce Julian, a seismologist at the US Geological Survey at Menlo Park in California. "Sometimes you see none at all."

## Cracks and fissures

No one knows why this is. But Garces believes that part of the problem lies with features in the ground beneath a volcano that scatter and distort the seismic waves. "The waves are radiating through a very complicated path that has layers, fissures, and cracks," says Garces. On the other hand, he argues, if you "listen" to a volcano's rumblings by replacing seismometers with microphones, the task becomes considerably easier. "Do your measurements in the atmosphere and you get a much more accurate measure of the physics of the volcanic system," says Garces. "The air is a much simpler material."

The first step in deciphering the complex language of a volcano was to model its physical structure mathematically, using parameters such as the size of the conduit, the rigidity of the bedrock and the gas content of the magma. The model generated an infrasound spectrum that they could compare with the real thing: tweak the parameters until the two spectra matched, and they should have a good idea of what goes on inside the volcano.

Model number one was based on simple physics. To start with, they decided to recreate the short, sharp explosive blasts that occur at the summit of an erupting volcano. The first moment after a blast, a strong shock wave radiates in all directions, like the sonic boom of a jet. This is followed by a low, infrasonic rumble that tapers off gradually.

Garces and Buckingham represented the conduit of a volcano as a cylinder of rock filled with magma. The sharp, percussive boom of the initial blast comes from a sudden burp of gas that explodes near the surface. This creates acoustic waves that ripple out into the magma, bounce around inside the conduit and then die away—producing the long, low rumble that subsides slowly.

Mathematically, this model is almost identical to an organ pipe. Blowing air into an organ pipe creates a broad spectrum of vibrations that bounce back and forth inside. Some vibrations have wavelengths that match the length of the pipe exactly, forcing it to vibrate in resonance. These become louder, while other vibrations with wavelengths that don't match the length of the pipe are lost. Decode this spectrum of sounds and you can learn all about the organ pipe that created them.

By 1994, the two were ready to test their model against real explosions. They chose one of the world's most studied volcanoes, on Stromboli, an island among the Italy's Lipari group. The volcano on Stromboli has been active more or less continuously for about 2000 years. Because it is so well understood, there were real numbers to feed into the model. For example, to represent Stromboli's upper conduit where the explosions occur, Garces used a cylinder 10 metres wide and 100 metres deep. He then added other basics, such as the physical properties of the magma and the strength of the rock surrounding it.

This accounted for the organ pipe, but the model also needed acoustic waves for resonance—and this required explosions. The source of explosive eruptions at Stromboli is believed to be a constriction in the conduit, about 60 metres below the volcanic vent. The constriction acts like the nozzle on a spray can. As magma flows upwards through it, the pressure drops suddenly and volcanic gases come out of the solution. This creates gas bubbles that expand quickly. As they grow, changes in pressure force them to oscillate rapidly, blasting acoustic waves into the conduit. "It's like high explosives detonating underwater," says Buckingham. They based this aspect of the model on research on underwater explosions.

Later that year, Garces traveled to Stromboli to record some real explosions to compare with their simulation. With help from the Vesuvian Observatory in Naples, he set up a listening station about 150 metres from an active vent on the west side of the volcano. For eight days, Garces recorded thousands of booming explosions and the long rumble of tremor with an infrasound microphone.

Back in the lab, Garces and Buckingham tweaked the parameters of the model slightly—adjusting the strength of the explosions, for instance, and the amount of gas dissolved in the magma. They were able to re-create the infrasound fingerprint of Stromboli's explosions in striking detail. Buckingham admits that he was surprised that they got such a close match, especially as they were using such a simple model. "Actually, I was amazed," he says.

Garces and Buckingham feel pretty confident that their organ pipe model is on the right track. The four strongest peaks at low frequencies matched, as did the peaks at higher frequencies. It also matched a conspicuous valley in the spectrum that related to how far down in the magma column the explosion source lies. "It's not just one feature—it's a combination of them," Buckingham says. "Sure, you can get one of them to work, but to get more than one is more than coincidence."

But even then, something was missing. While Garces was listening to Stromboli with his microphones, a network of seismometers on the island were picking up the steady background hum of volcanic tremor. The whole island was shaking.

When Garces compared the spectral fingerprint of these tremors to the explosions he recorded, he discovered that the patterns were completely different—as if one throat was singing with two voices. How could this be? "This puzzled me for a very long time," Garces admits.

Eventually, he came up with an explanation. Perhaps the magma in the conduit had separated into two layers, like a thick film of oil floating on top of water. If the densities were sufficiently different, acoustic waves would be reflected at the interface. This would turn the volcanic conduit into two separate organ pipes, one of top of the other. Reverberations from explosions at the surface would be trapped in the upper pipe, while infrasound from tremors deep in the earth would be locked in the lower pipe.

Small gas bubbles, released far below the conduit, are the key to the layering effect, says Garces. The most abundant gases in magma are water vapour and carbon dioxide. As the magma rises, its pressure drops and, just like a freshly-opened bottle of champagne, the gas comes out as bubbles. Carbon dioxide leaves the magma first since it doesn't combine readily with molten rock. But the magma only relinquishes its grip on the water in the last few hundred metres of the conduit. The result is a light, frothy layer of magma that is filled with bubbles of steam, sitting on a layer of denser molten rock.

## Loud and clear

But even this complex picture is an oversimplification. In 1997, Garces went back to Arenal with two seismologists from the University of California in Santa Cruz, Michael Hagerty and Susan Schwartz, to record more tremors and explosions.

What they found was curious. Arenal's frequent explosions of gas and ash always came through loud and clear, but the low growl of the tremor was fickle. Sometimes it seemed muffled or undetectable, as if a thick blanket had been thrown over the volcano, and at other times the low rumblings were clearly detectable.

Garces believes Arenal may operate in two modes. When fresh magma flows slowly up into the conduit, the frothy, spongy layer has time to form on top as the gases fizz out. As they'd already discovered, this traps infrasound beneath it in the denser layer. But if the flow of magma is more vigorous, there's no time for the frothy layer to form. That means that the infrasound from tremor can pass freely through the conduit to the vent, and the microphones can pick it up clearly. It seems that a volcano can behave like one or two organ pipes, depending on how fast magma flows through the volcano.

Complex behaviour like this seems to present Garces's model with a problem. There are thousands of volcanoes on the planet, and they're all as individual as people. They erupt different kinds of magma with different acoustic properties and their plumbing systems are as varied as the sounds they make. "I think volcanoes do all kinds of things to make noise," says Julian. If no two volcanoes are alike, what

chance is there of using infrasound to forecast eruptions?

Fortunately for the researchers, it seems that there's a very good one. The infrasound signature of a volcano is especially sensitive to the amount of gas locked up inside its magma. And this gas is critical to the way a volcano behaves. Gases dissolved in the magma provide the driving force behind an eruption ("When volcanoes get violent", *New Scientist*, 26 October 1996, p 28). The danger is that this gas will come out of solution very fast, burst through the vent in one gigantic blast, spewing lava, rocks and other debris onto nearby towns or villages. "It's the most important parameter. The more gas trapped inside, the more explosive an eruption might be," says Garces.

This is why Garces's acoustic model is so promising. If infrasound tells you that the volume of gas in the magma is increasing rapidly, it could warn you of an impending explosive eruption. "It would be extremely important if you could tell that sort of thing," says Julian. New data seem to confirm that forecasting eruptions using infrasound may soon be possible. In May last year, Garces spent some time recording the boom of infrasound on Sakurajima volcano on Kyushu, one of Japan's southern islands. He found that he could pick out clear changes in the character of the infrasound spectrum before an eruption. "You could definitely see changes in the acoustic signal," he says. "If you can correctly interpret these changes, you could issue a warning."

In the next few years, Garces predicts, volcano models will emerge that can translate volcano sounds into information almost instantaneously. Such a volcanic stethoscope could convert the spectral fingerprint of tremor and explosions into the kind of measurement that would mean a lot for those living close to a volcano—so if the concentration of gas in the magma was rising, for instance, they would be put on alert. "It'll be a couple of years before we can do that," he says. "But we'll get there eventually."

After years of struggling to understand what Arenal was telling him that day in 1992 when he first climbed its slopes, Garces is busier than ever. In the past year he's churned out a flurry of new papers, and thanks to a tour of active volcanoes in Japan, he has a pile of high-quality acoustic recordings to test against the model. And even though no one knows yet whether his model reproduces exactly what happens inside volcanoes, Garces feels he's at least taken the most important first step—listening to volcanoes. "Basically it's been a labour of love for the last five years, trying to get people to understand that this can help us. It was quite frustrating," he says. "But now they're asking me to work with them."

---

## Questions:

1. Name the two most abundant gases in magma.
2. What are volcanic 'belches?'
3. What is the first step in deciphering the complex language of a volcano?

Answers are at the back of the book.

## Activity:

1. Go to the Website: **http:/volcano.und.nodak.edu/volc_of_world.html**
   Here you can explore Earth's volcanoes, see which volcanoes are currently active, watch volcano movie clips, etc. Spend some time browsing through this Website, then list a few of the currently active volcanoes and where they are located. Then write a short paragraph describing a volcano on another plant or on a moon.

2. Simulate a volcanic eruption via the Internet. Go to the following Website: **http:/volcano.und.nodak.edu/vwdocs/kids/fun/volcano/volcano.html**
   Click on: Simulate a volcanic eruption. Be patient, because it takes a few minutes to load. Play around with the parameters to change how the volcano erupts.

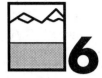6

How do you predict an earthquake? That is the important question that scientists have been trying to answer at least since the beginning of this century. Many scientists from all over the world have worked on this problem. They have scoured the records of past earthquakes, set up monitoring networks throughout the world, sampled gases and water and animal behavior and who knows what else. How many successful predictions have been made? One. The author of the following article suggests that accurately predicting earthquakes may be impossible. If this is the state of earthquake prediction at the present time, where should we go from here? Should we continue to search for ways to predict earthquakes? Should we put our efforts into earthquake-proof construction and building codes? These are some of the questions that you should consider as you read the following article.

# Faulty Premise

By Midori Ashida

*After decades of effort and hundreds of millions of dollars spent, short-term earthquake prediction could be a disaster in the making.*

At 5:45 A.M. on January 17, 1995, a geological fault that had slept for hundreds of years under the sea near Kobe, in central Japan, suddenly began to rupture. The break opened southward onto Awaji Island, then rushed northward to the middle of Kobe. The entire event took ten seconds.

Houses in Kobe and neighboring cities, squeezed onto a 1.7-mile-wide plain between the Rokko mountain range to the north and Osaka Bay to the south, collapsed instantly, killing thousands of people. Those who survived pulled themselves out of the rubble and searched in the darkness for their families or helped dig out their neighbors. No police cars or ambulances came to help them. There was still an hour before dawn.

As day broke, people saw that the city had collapsed. A 600-meter segment on an elevated highway had fallen, tossing cars off. Railroad tracks were twisted like wire. Department stores, banks, hospitals and office buildings were smashed. Fires had started in several places and spread gradually and steadily, burning for three days and killing hundreds more people trapped under their houses.

Because telephone and data lines were smashed, it was noon before people outside the area realized the extent of the disaster. Japan's Self-Defense Forces—the equivalent of the American National Guard—arrived only in late afternoon. By then, Kobe, a beautiful port city famous for its futuristic city planning, had become an endless pit of rubble and a burning inferno. That night television cameras taped bizarre scenes of victims patiently waiting in lines to buy a bottle of water and a small rice ball or a Danish pastry, which they would share with their families. Coffins covered with white cloth lined classrooms and temples. Men and women, children and the elderly flooded into school gymnasiums and government offices for shelter, or stayed in the park.

Days and weeks passed, but the condition of the victims scarcely improved. Nearly 800 people died of illness because of the poor living conditions in the shelters. In the end, some 6,300 people were killed, 35,000 were injured and about 415,000 families lost their homes. Estimates of the amount of damage range as high as $200 billion. The Kobe earthquake was the most expensive one in history.

It should not have been so.

Situated on the border of major tectonic plates, Japan is earthquake-prone. But the government has expended nearly all its earth quake-disaster efforts on a system based on a flawed premise: that earthquakes can be predicted. The government started its national earthquake-prediction program in 1965 and since then has spent more than $1 billion on research for the program. Under the 1978 Large-scale Earthquake Countermeasures Act, a warning system as well as emergency protocols have been established. But those operational earthquake-prediction counter-meas-ures apply only to cities in the Tokai district, some seventy-five miles west of Tokyo, where the next "big one" was predicted to take place.

Kobe in retrospect, was a disaster waiting to happen. Although there had been no major earthquakes for decades, seismologists and geologists knew that the area had many active faults. For more than ten years the densely populated city had eagerly expanded its land area—mostly loosely compacted sand and river sediment—by adding landfill and reclaiming marshland. The sandy sediments amplified the ground tremor, and artificial landfill caused soil liquefaction.

Lulled into complacency by the government's prediction system—though that system did not cover Kobe—the city spent little on practical protective measures. Old wooden houses with heavy roofs had no reinforcement. The city had no fire-prevention measures, and the government had no crisis-management system. The governor did not know how to ask the Self-Defense Forces for help. The police could not control the traffic jams, which prevented rescue crews from reaching the city.

The tragedy, when it came, was only the latest installment in a long history of false hopes and failed predictions. For centuries people had tried to divine impending earthquakes in freakish weather, in oddly behaving animals and in the arrangements of the planets. Those hopes

have only heightened as earthquake studies have grown scientific. Since the mid 1960s, hundreds of millions of dollars have been poured into prediction projects around the world. The results, however, have been more than discouraging: earthquakes, most seismologists now believe, are simply too complex for short-term predictions. Japan's early-warning system is further proof of that failure, and the source of its crowning irony. In the seventeen years since the system was established in the Tokai district, earthquakes have killed about 400 people in Japan, not including the Kobe earthquake. All of them took place outside of the district.

Until the aftermath of the 1906 San Francisco earthquake, the origin of earthquakes had been a mystery. Then Harry F. Reid of Johns Hopkins University in Baltimore, Maryland, studied the region in the vicinity of the San Andreas fault in California and discovered that the western side of the fault had shifted toward the north-northeast before the earthquake took place. He also noted significant horizontal shearing in the fault. Earthquakes, he concluded, are cased by sudden slippage in faults in the upper part of the earth.

That slippage, seismologists now know, is a consequence of the movement of huge segments of the earth's outer layer, called plates, which are slowly driven around the earth's surface by convection currents in the planet's interior. As the two sides of a fault move in opposite directions, stress accumulates slowly in the ground. When the stress finally reaches a certain critical point, the weakest part of the stressed rock suddenly ruptures, releasing the stored elastic energy as the two sides of the fault rebound to a less strained position. The violent tremors of an earthquake are the seismic waves radiating from the slipping fault. Sometimes, the energy released exceeds that of a nuclear explosion.

According to Reid's view, known as the elastic rebound theory, after an earthquake takes place, stress again starts to accumulate in the rock. Thus, in its simplest form, the theory seems to imply that earthquakes should take place at regular intervals. If that were the case, reliable long-term earthquake forecasts could be made merely through extrapolations from past seismic activity.

Unfortunately, things are not so simple. Astronomers can predict planetary orbits accurately because they can rely on the theory of gravitation, but seismologists have no such well-established theory for predicting earthquakes. That leaves them with two choices: either give up on prediction or adopt a strictly empirical approach. Workers opting for the latter sift through vast amounts of data in the hope of finding some phenomenon that always takes place before a large earthquake but never takes place otherwise. First, investigators study past precursor phenomena associated with large earthquakes. Then they try to detect the phenomena and determine whether earthquakes follow.

Reid was optimistic about the prospects for prediction. The strains that precede an earthquake, he believed, could be measured by a line of piers built at right angles to the fault: if "the surface becomes strained through an angle of about 1/2000," he wrote, "we should expect a strong shock." The U.S. Coast and Geodetic Survey undertook such measurements in the early 1920s, but noted no sure sign of strain change. Other intensive surveys were done in southern and central California, but again, no regional strain change was observed.

At around the same time as Reid's work, Akitsune Imamura, a seismologist at Tokyo Imperial University, studied the catalogue of earthquakes that had struck Japan since the fifth century. Those earthquakes, Imamura noted in a 1928 paper, had recurred at periods of between one hundred and 150 years in the Tokai district of central Japan. Seventy-four years had passed since the last great event, he went on, so Tokai would probably be hit by a major earthquake in the near future.

Imamura maintained that the great earthquakes of Tokai's past were "accompanied by conspicuous topographical changes over an extensive area." By observing those changes, he wrote, he would be able to say whether an earthquake was coming. In the late 1940s he set up seven private observatories in Tokai, and professed to have detected an uplift of the land just before the magnitude 7.9 Tonankai earthquake in 1944. But his findings were not generally accepted by seismologists, and earthquake prediction research had almost petered out by the time it was revived in the mid-1960s.

Three factors led to great advances in seismology in the 1960s. First, the U.S. government deployed a global network of 120 seismic stations in sixty countries to detect underground nuclear tests. As a byproduct of its primary mission, the network generated abundant seismic data on earthquakes throughout the world. Second, the emerging computer technology enabled seismologists to analyze the massive quantities of new data. Third, the newly accepted theory of plate tectonics offered an explanation for the basic dynamics of earthquakes.

Those advances engendered strong hopes that science would at last be able to mitigate the consequences of earthquakes. In 1965, a year after a huge earthquake hit Alaska on Good Friday, an ad hoc panel on earthquake prediction chaired by Frank Press, then a professor of geology at the California Institute of Technology, issued a report that called for a ten-year earthquake-prediction research program. Similar programs were started in China, Japan and the U.S.S.R.

In 1971, at an international meeting in Moscow, Russian seismologists said they had detected changes in the velocities of seismic waves before earthquakes. One of the seismologists impressed by the Russian data was Christopher H. Scholz of the Lamont Doherty Earth Observatory of Columbia University in Palisades, New York. Scholz proposed a physical model that supposedly linked observable changes in the earth's crust with the occurrence of an earthquake.

Scholz's hypothesis, known as the dilatancy theory, is based on a well-known phenomenon observed in rocks in the laboratory: As stress builds, microfractures in the rock close, decreasing the rock's volume. After all the fractures have closed, the stress-strain relation becomes linear, which is characteristic of a solid. Then, as stress continues increasing to about half the fracture point, the rock begins to crack and expand its volume, the phenomenon known as dilatancy. Initially the cracks fill with air, but gradually groundwater seeps in, increasing the pore pressure and weakening the rock. According to the dilatancy theory, that action could lead to several precursory phenomena in the field: an increase in the volume of the rock, causing the ground to uplift and tilt near active faults; the release of water-soluble gases, typically radon; a change in the velocity of seismic P and S waves; an increase in the electrical resistance of the rocks; and a change in the frequency of small, local earthquakes.

Excited by the variety and specificity of the potential precursors, many seismologists tried to detect them, and many claims of success made headlines. The most famous precursor took place in China. On February 4, 1975, officials of the Manchurian province of Liaoning warned people of a strong impending earthquake. That evening, a magnitude 7.3 earthquake struck the town of Haicheng. Initially the Chinese government announced that only a few people had died, but in 1988 the official figure was corrected to 1,328.

There are good reasons for skepticism. Given the fragility of Chinese buildings, the number of deaths is remarkably low. Moreover, the Haicheng earthquake took place during the Cultural Revolution, and Chinese seismologists were under orders from Mao Zedong to learn how to predict earthquakes—or else. According to Robert J. Geller of Tokyo University, such political pressure might well have caused them to exaggerate, or even to tell outright falsehoods.

Nevertheless, many Western observers continue to credit the Chinese with a successful prediction. "If you ask whether they really evacuated, the answer is clearly yes," says Lucile M. Jones, a seismologist at the U.S. Geological Survey in Pasadena, California, who studied the Haicheng prediction for five years. "But if you ask, Did they do any better than random success? Could we do this by just guessing? The answer is, We don't know." The circumstances surrounding the prediction were unusual, adds Jones. "There were 400 earthquakes in four days in an area that hardly ever had them. So they guessed that these were foreshocks and evacuated on that basis. Well, that's not a bad guess. But that doesn't mean it's a successful, repeatable earthquake prediction."

Luck has not smiled again on the Chinese seismologists. They failed to predict the Tanshang earthquake of July 27, 1976, which killed more than 250,000 people. The next month, they predicted an earthquake in Guandong Province and evacuated people, but nothing happened.

In the U.S., President Jimmy Carter signed the Earthquake Hazard Reduction Act of 1977, which emphasized short-term prediction. Not a single earthquake has been successfully predicted since. How could so many people so badly misplace their confidence?

"It was a sociological phenomenon," says the seismologist Robert L. Wesson of the USGS in Reston, Virginia. In the mid-1970s, as a young investigator at the USGS in Menlo Park, California, Wesson thought he observed anomalous velocity changes in seismic waves before an earthquake in Bear Valley. When he and a colleague published their observation in a scientific journal, however, other seismologists challenged their work. Wesson now admits that the anomalies he saw were artifacts, not true precursors. As he notes today, many other anomalies have been reported, but none have been observed to arise systematically before other earthquakes.

Last May yet another strong precursor candidate was eliminated. Retrospective studies have shown that foreshocks, or small earthquakes, are detectable between five and ten days before 70 percent of the earthquakes of magnitude 7.0 or greater. Rachel E. Abercrombie of the University of Southern California and Jim Mori of the USGS in Pasadena have studied foreshock activity recorded in California and Nevada for more than a decade. But they have noted no systematic correlations between the foreshocks and the magnitude of the subsequent main shock. Furthermore, says Mori, "as far as we can tell, there is almost no way of distinguishing what is a foreshock and what is not a foreshock."

Another short-term prediction technique has recently been the subject of heated debate. Since the early 1980s three Greek physicists, Panayotis Varotsos and Kessar Alexopoulos of the University of Athens and Kostas Nomicos of the Technical University of Crete, have asserted that they can predict the location, time and magnitude of a number of earthquakes. Their method, called VAN after the initials of their last names, measures anomalous electric currents in the ground. The VAN group says that changes in such currents indicate changes in the stress on rock just before an earthquake.

For the past seventy years, no earthquake with a magnitude greater than 6.0 has taken place in the Cascadia subduction zone, which runs from northern California through Oregon and Washington into British Columbia. Seismologists attributed the low seismic activity to the youth of the oceanic plate and the large amounts of sediments that work as lubricant. Then, in the early 1980s, the geologist Brian F. Atwater of the USGS moved to Seattle. Although the land is uplifting by three millimeters a year, he discovered, the coast of Oregon and Washington has risen only a few meters. Atwater suspected the subsidence was due to large earthquakes in the past, and so he dug in several places along the coast. What he discovered was sobering: at least six giant earthquakes have struck the region in the past 7,000 years. Moreover, huge numbers of trees died near the sea level about 300 years ago. Alerted to Atwater's discovery, Kenji Satake of the Geological Survey of Japan and his colleagues studied the tsunami record of that time in Japan. From computer simulations of the tsunami, they concluded last January that the most recent Cascadia earthquake took place at 9:00 p.m. on January 26, 1700, and had a magnitude of about 9.0.

The public and the governments of Oregon and Washington are now on notice that they are not free of earthquake hazard. Since the late 1980s local governments have revised building codes and started reinforcing bridges and other important infrastructures. When, exactly, an earthquake will hit is anyone's guess. But when one does, people in the Pacific Northwest—unlike the citizens of tragic Kobe—might not be caught entirely off guard.

## Questions:

1. Why are there so many earthquakes in Japan?
2. When and where was the only successfully predicted earthquake?
3. When was the last large earthquake in Cascadia, and how was the date determined?

Answers are at the back of the book.

## Activity:

Research an earthquake that occurred in your region or a neighboring region. Write at least a one page summary describing the earthquake, the damage done, and the citizen's reactions. Did there seem to be any forewarning to the quake?

**7**

Preserved under seven meters of volcanic ooze and ash, the sixth-century village of Ceren, El Salvador, is a scientific gem for geologists and archaeologists. Studying this Salvadoran volcanic eruption helps scientists to understand geological processes and to predict and prepare for future events.

# Clues from a Village: Dating a Volcanic Eruption

By Lawrence Conyers

*Anthropology*

During the early evening hours of a late summer's day sometime in the late sixth century, the lives of villagers in a small farming community in El Salvador were violently altered. Only 500 meters away from their houses, a relatively small but powerful volcanic eruption began along an active fissure zone. The eruption spewed out ash and cinders, which rained down on the countryside, burning everything within a radius of a few kilometers.

Although inhabitants of this volcanically active part of Central America were no strangers to such events, their proximity to the vent and the speed and violence of the eruption must have been terrifying. The populace quickly abandoned the village, never to return. They left behind everything important in their lives, including a wide range of utilitarian and ceremonial objects. Within a few days, their houses were covered by as much as seven meters of tephra. Whole fields of growing corn, orchards, gardens, and storerooms filed with agricultural products were buried and preserved under volcanic ash.

The cataclysmic event, which obliterated all visible signs of the village, produced one of the best preserved archaeological sites in the Western Hemisphere. Named after the nearby town of Ceren, the site's archaeological remains and associated volcanic stratigraphy record a "snapshot in time" that is helping us reconstruct both the lives of these ancient people and their short, but violent encounter with the volcanic forces that shaped this part of Central America. Using a multidisciplinary approach, archaeologists and geologists have determined the rapidity and magnitude of the eruption as well as the time of day and season in which it occurred.

## Archaeology at Ceren

The Ceren site was discovered in 1976 when the area was being bulldozed for the construction of grain storage silos. In a bulldozer cut, workers saw in profile the remains of what appeared to be an adobe house platform with standing columns at its corners. Singed but still well-preserved roof thatch along with pottery shards and other artifacts lay on the floor of the structure.

Payson Sheets, an archaeologist with the University of Colorado in Boulder, happened to be surveying the region at the time and was informed of this unusually well-preserved feature. Sheets identified the pottery as coming from the "Classic Period." Radiocarbon assays later confirmed this finding, dating the pieces to A.D. 590, plus or minus 90 years.

Under Sheets' direction, a large contingent of archaeologists, geologists, geophysicists, paleo-botanists, and other scientists have been excavating and studying the Ceren site since

1979. To date, 15 structures have been identified by excavation; 12 of these have been completely excavated. Using ground-penetrating radar, I have identified an additional 22 buildings and many other buried features nearby.

Excavations and geophysical surveys show that the village consisted of many individual households of three or more clustered structures. Each household typically included individual buildings for sleeping and eating, cooking, and food storage. Household clusters were separated by gardens which produced a variety of food, medicinal, and spice crops. Flowers and small orchards were also present. Wellworn footpaths connected individual households to each other, to the village center, and to outlying fields.

Nonresidential structures included a large domed sweat-bath, which could hold as many as 20 people, and a well-built communal building where corn beer may have been dispensed. One structure may have been used for ritual activities by a shaman, while a nearby building provided space for communal food preparation and distribution. Excavated structures for storing food (called bodegas) still contain ceramic vessels full of seeds, cribs with dried corn, and hanging bunches of chile peppers.

Large fields of corn surrounded the village; most of the usable ground was under some sort of cultivation. (Plants encountered during excavation occur as hollow cavities in the volcanic overburden. Once unearthed, they are filled with plaster and preserved as casts.) Corn was planted in bunches of three to four plants, in rows aligned along ridges. Near one household stood an orchard containing guayaba trees (which produce a small green fruit the size of an apricot) as well as avocado, nance, and cacao trees. The majority of the plant remains discovered were domesticated varieties, indicating that the prehistoric inhabitants of Ceren had considerable knowledge of horticulture and a varied diet. Their staple food, however, was probably corn.

## The Volcanic Eruption

The volcanic stratigraphy at Ceren has been studied extensively by C. Dan Miller of the U.S. Geological Survey's Cascades Volcano Observatory in Vancouver, Wash. He concluded that the tephra units that buried the village, especially during the initial stages of the eruption, were deposited as wet, relatively low-temperature surge deposits. These deposits flowed into the village at high speeds, collapsing many of the less substantial building walls while encasing and preserving delicate plants and other organic material in wet ash. Because the initial layers of ash were emplaced at temperatures near 100°C, many of the plants growing in the community were not ignited.

Ash surge deposits at Ceren are interbedded with block and lapilli layers deposited at temperatures approaching 575°C. These air-fall units consisted of many ballistic bombs, which rained down on the village, collapsing roofs, crushing pottery, and starting fires.

The volcanic vent, called Loma Caldera, is located along an active north-trending fault that projects south toward San Salvador Volcano, one of the larger composite cones in the country. Today, Loma Caldera is an eroded tuff ring, which may have partially collapsed during the later stages of its eruption.

Analysis of the ancient land surface in outcrops and archaeological excavations indicates that seismic shocks, slumping, and faulting occurred just before the eruption—events which may have warned the people that they needed to flee. No human bodies have been found at Ceren, suggesting that all the inhabitants may have escaped. But it is also possible that the villagers fled only a short distance before being overcome by the first quickly moving ash flow.

Numerous bodies of birds, which were probably blown out of their nests, have been found on the buried "living surface" of the village. Mice are preserved in roof thatch, and a domesticated duck tied up in one of the bodegas was also encased in ash.

## Dating the Loma Caldera Eruption

Geologists and archaeologists are usually content to date prehistoric events with a precision of a few centuries, or decades at best. Standard radiocarbon dating techniques, for example, tell us that Loma Caldera erupted sometime between 500 and 680 A.D. That 180-year interval is about as precise as we can be with respect to the actual year of the event. But the excellent preservation of botanical remains and artifacts at Ceren allows us to identify other temporal aspects of the Loma Caldera eruption far more precisely. We've been able to determine the season of the year, and even the time of day, when the first ash flow was deposited.

One of the best indicators of the season at Ceren is the maturity of the cultivated corn which was preserved in growing position. Four fields have been exposed, each containing plants at different stages of growth. In the tropics, the maturity of growing corn is almost wholly a function of the timing of the rainy season. In El Salvador, the summer rainy season usually begins in May and ends in October; 95 percent of yearly precipitation falls during this period. All of the yearly corn crop must be grown during these months.

In one of the preserved fields, large mature ears of corn had developed, while in an adjoining field only juvenile plants with ears 15 to 20 centimeters long were found. In another field, the corn had been recently harvested, and in a fourth, mature corn stalks had been purposely bent over with the ears still attached—a traditional method of field drying that is still used by some farmers during the middle to late rainy season.

The different maturities and harvesting schedules of corn indicate that the rainy season was well advanced at the time of the eruption. Corn was nearly ripe in one field. Farmers had just finished harvesting a field and had replanted another, probably in the hope of growing a second crop during the same growing season. A mature crop was drying in a fourth field. Assuming that corn takes 120 to 140 days to reach maturity and that a typical rainy season started in May, a September eruption is most likely.

Other botanical indicators also point to an eruption in the rainy season. For example, chile peppers were found drying inside bodegas rather than outside, where they would have been hung during the dry season. Guayaba fruit, which was nearly ripe, was blown out of trees during the initial stages of the eruption. In El Salvador, guayaba usually begins to ripen in late August or early September.

One of the most delicate botanical remains discovered at Ceren was a small cacao tree, which had blossoms growing from its trunk. Cacao blossoms form during the rainy season and usually open soon after sundown. They stay open all night to allow pollination by ants, and then dry up during the heat of the day. The preserved cacao tree with its open blossoms thus suggests that the eruption occurred not only during the rainy season, but probably after dark.

The placement of certain artifacts in or near houses indicates that the eruption began during the early evening. Field workers had returned from their fields, storing agricultural tools under the eaves of houses. Fires in the cooking structures had died out, and cooking pots had been stored away, some with the remains of the evening meal still inside. Sleeping mats were found preserved in the rafters of the domicile buildings: they had not yet been placed on the raised adobe sleeping platforms for the night. From these artifacts, we can conclude that workers had returned from their fields and that their evening meal had been cooked and consumed. But they had not yet retired for the night when the eruption of Loma Caldera disrupted their lives forever.

A reconstruction of this eruption using both geological and archaeological evidence shows that during a warm tropical evening in late summer, soon after the inhabitants of Ceren had eaten their evening meal, a violent earthquake occurred, accompanied by faulting and slumping of the ground. As the frightened inhabitants fled into the dark, leaving their possessions behind, a hot glowing cloud of ash surged into their village, burying everything in its path.

The volcano continued to erupt for a number of days, covering the village with alternating beds of ash and coarser volcanic bombs and cinders. The village was covered by as much as seven meters of tephra, leaving a wonderfully preserved time capsule of rural sixth-century Central America for archaeologists and geologists to study.

**Questions:**

1. When did the eruption at Ceren take place, and how was this date determined?
2. When during the year did the eruption occur, and how was this determined?
3. When during the day did the eruption occur, and how was this determined?

Answers are at the back of the book.


**Activity:**

Name the four most common radioactive-stable element pairs used in radiometric dating. What are their half-lives? What types of rocks are they used for, and, if applicable, what occurrence does the resulting date indicate? Also, state if there are any restrictions when using any of the element pairs.

8

Most natural disasters can be forecast, but earthquakes, so far, cannot. Predicting earthquakes may not be able to save structures from destruction, but it certainly could save lives. Proponents of earthquake prediction have abandoned the black box approach and are beginning to focus their attention on hazard assessment which can offer alarm systems as earthquakes arise, earthquake probability, and damage assessment.

# Like a Bolt from the Blue

By Robert Coontz

For an article in Science magazine, its title was blunt: "Earthquakes cannot be predicted." The text of the piece, which appeared in March last year (vol 275, p161), was just as uncompromising. The authors, four well-known scientists specialising in the physics of the Earth's interior, were calling for nothing less than the end of a line of scientific endeavour. Earthquake prediction, they proclaimed, is doomed to failure. Anyone who funds it should look for other ways to spend the money. Game over.

To some geoscientists, this was familiar stuff. For most of the past ten years the lead author, Robert Geller, chief seismologist at Tokyo University, had been ripping into earthquake prediction research at conferences and in journal articles, and his standard targets were ready with standard rebuttals to this latest attack. Still, it was clear that Geller and his coauthors had touched a sore spot. For scientists who study natural disasters, earthquakes have always been the ones that got away. Hurricanes, tornadoes, floods, even volcanic eruptions can be forecast, monitored and tracked, their basic mechanisms observed and modelled with some degree of confidence. But earthquakes have remained cryptic and slippery—earthborn Godzillas that spring from the depths, strike at will and then rumble out of existence, taking their secrets with them.

Geller's fury was mainly roused by the gargantuan Japanese prediction programme, which he says has swallowed up billions of yen over the past few decades and produced nothing in return. In spite of this, the Japanese research continues, along with work in Greece, Russia and China. But back in Geller's native country, the US, earthquake prediction is the research that dare not speak its name. It peaked sometime between the death of Elvis Presley and the advent of Elvis Costello and has been declining ever since, a victim of its early excesses. Yet the prediction scientists still have some tricks up their sleeves—if they're allowed to keep playing. Prediction might not be the only game in town, they argue, but the horrific cost of earthquakes makes it a gamble worth taking.

## Long-range Forecasts

The critics of prediction are not calling for an end to earthquake research per se. After all, it's their line of work too. Nor do they object to using science to forecast the future. They are quite comfortable with long-term estimates of earthquake probabilities, such as the 30-year seismic hazard maps that the US Geological Survey prepares for 20 different segments of the San Andreas Fault. Such efforts (which most experts now prefer to call earthquake forecasts) are based on measurements of ground movement along faults, historical records of

From "Like a Bolt From the Blue," by R. Coontz, New Scientist, 1998, pp. 36-40. Reprinted with permission.

earthquakes and evidence of ancient landslides, toppled trees and the like—scientifically sound methods, as even the fiercest opponents of quake prediction agree.

What draws their fire are efforts to make short-term earthquake predictions, hours to months in advance. The idea is that earthquakes may be heralded by warning signs—precursors powerful enough to be detected, sensitive enough to distinguish a big earthquake from a small one and dependable enough to catch most dangerous quakes without too many false alarms. The issue is cut and dried: if useful precursors exist, earthquake prediction is possible; if they don't, it's not.

Twenty-five years ago, scientists in the US thought they were hot on the trail of such precursors. They looked for them in patterns of small earthquakes, movements of the Earth's crust, electromagnetic signals emitted by rock being squashed or split by tectonic forces, changes in level of water in wells and in the amount of radon gas seeping from the ground. The research went into overdrive when scientists theorised that several promising candidates might stem from a common cause. Lab results showed that when rock is stressed to near its breaking point, it becomes shot through with a network of microscopic fractures that fluff it up like a pillow. That expansion—called dilatancy—along with the subsequent draining and refilling of groundwater in the cracks, looked as if it could explain a host of changes in electrical conductivity, the speed of seismic pressure waves, even the elevation of the ground in earthquake-prone areas. If a grand unified theory of earthquake precursors were close at hand, could prediction itself be far behind?

Overseas, it looked as though prediction might already be becoming a reality. In 1975, after an unusual swarm of small quakes rattled Haicheng in Manchuria, China, the city authorities alerted residents in time to save lives. Some 90 per cent of the city's buildings were destroyed or damaged, but casualties were low. Even without a grand unified theory, it looked as if prediction was working.

Then the bubble burst. When scientists took measurements in the field, they found that dilatancy-diffusion applied only to small portions of the crust—too small to explain the precursors. Worse, when they double-checked the precursors themselves, many of them proved to be measurement errors or wishful thinking. In China, the success at Haicheng turned out to be a fluke. The following year, an unpredicted earthquake levelled the city of Tangshan in Shandong just 450 kilometres away and killed at least 250,000 of its million inhabitants. To this day, scientists disagree about whether the Haicheng prediction was more than a lucky guess.

Searching for still subtler signals, scientists in the US decided to concentrate their efforts on one big experiment: a patch of ranchland bristling with instruments near the hamlet of Parkfield in central California, where earthquakes measuring 6 on the Richter scale seemed to strike every 22 years with near-clockwork regularity. By watching carefully for precursors, they hoped to predict the next quake, which experts estimated had even odds of striking by 1988 and a 95 per cent chance of striking by 1993.

Fourteen years after the Parkfield Earthquake Prediction Experiment began, the expected earthquake still has not come. Elsewhere in the US, prediction research has ground to a virtual standstill, a casualty, its proponents say, of the field's unfulfilled early promise. "We were very naive," says Max Wyss, chairman of the earthquake prediction subcommittee of the International Association of Seismology and Physics of the Earth's Interior. "Much of earthquake prediction research was done very poorly."

The main problem, says David Jackson, science director of the Southern California Earthquake Center in Los Angeles and a coauthor of Geller's article in Science, was the philosophy that you didn't need to understand very much about earthquakes to predict them. That black-box approach to science has been largely abandoned in the US, but it still flourishes in Robert Geller's adopted country, Japan. And, say Western critics, all Japanese prediction researchers have to show for their ever-increasing budget (worth $185 million last year) is a record of unrelenting failure.

Most quake prediction scientists in the US read Geller's broadsides as coded attacks on Japanese research priorities—attacks with which, by and large, they agree. But in the US, they say, the target of the criticism is hardly worth hitting. At the US Geological Survey, shrunk by downsizing and straitened by policies favouring research with immediate practical applications, prediction is distinctly démodé. "It's a bad word," says John Langbein, chief scientist for the Parkfield experiment (whose operating cost of $1 million per year makes up only about 2 per cent of the USGS's budget for earthquake research). It is much the same all over, says long-time prediction enthusiast Paul Silver, a seismologist at the Carnegie Institution of Washington. "Nobody likes to be unsuccessful, so there's been a shift in emphasis toward things we know how to do."

These things include long-term forecasts of quake probabilities, studies of how the ground will shake during future earthquakes, and warning systems to detect major quakes as they arise. One such system, a network of sensors known as TriNet, is being installed in Southern California. The five-year, $20 million project will help emergency managers assess damage just minutes after an earthquake strikes. Future versions may sound an alarm in the seconds while seismic waves are still on the way. The new watchword is hazard assessment: finding where the danger lies and building cities to survive quakes, whenever they choose to strike.

Some researchers in the US, however, refuse to give up on prediction. The blackbox approach has not worked, they say. But that doesn't mean that you should abandon the whole idea. "The emphasis from the beginning should have been to understand the processes that accompany the occurrence of earthquakes,'" says Allan Lindh, a USGS geophysicist who led the lobbying campaign to set up the Parkfield experiment. "When you understand things well enough, then you start to get better at predicting them."

The problem is that quakes are fiendishly hard to study. For a start, Lindh says, "there are 10 to 15 kilometres, of dirt, rock and fluids between us and where the action is." Scientists can study that action indirectly by measuring seismic waves, electromagnetism and strain—the movement and deformation of rock under tectonic stresses—but their remote readings, distorted by intervening layers, are a poor substitute for being there. Furthermore, big earthquakes, which are the ones of most interest to forecasters, are rare and strike capriciously in space and time. And everything happens fast. The rupture begins, or nucleates, in a small patch of fault and spreads at speeds of kilometres per second—too fast for current instruments to pick up the details. The longer the rupture, the bigger the earthquake. But are nucleation zones too

small to be detected? Are they all the same size, or do they vary? Is the size related to the length of the rupture? No one knows.

## Impossible Reality

Faced with these and other difficulties, some critics argue that prediction is out of the question. "I actually think that it's impossible in principle," says Lucile Jones, a USGS seismologist working on the TriNet project. The problem, she says, is that rather than predicting every quake, you need to predict which of the many that occur every day will turn into large, damaging ones. The magnitude of an earthquake, she points out, depends on the length of the rupture. That, in turn, depends not on where the rupture starts, but where it stops. And where it stops, some earthquake scientists believe, can be just about anywhere. A rupture is not a single event, but a series of tiny ones—a cascade that gathers momentum as it goes, like a landslide. According to some models, a rupturing fault resembles a pile of sand so delicately poised that there is no way to tell whether the next grain you add will drop quietly into place or trigger a landslide.

In chaos theory, that kind of delicate balance is called "self-organised criticality". What it implies for earthquake prediction, opponents say, is dire: there is no way to tell in advance whether a rupture will be short or long, and thus there is no way of predicting whether the earthquake will be imperceptible or colossal. "If that's true," says Jones, "earthquake prediction as people want it is impossible."

## Open Question

That's a big "if." For one thing, chaos alone does not necessarily imply unpredictability. "Planetary orbits are chaotic, but that doesn't mean I can't predict the orbit of Mars," says Lindh. According to Charles Sammis, a geophysicist at the University of California, Los Angeles, who is working on chaos-based methods for analysing earthquake statistics, prediction is impossible only if all of the Earth's crust is in a state of self-organised criticality all the time. But that is as unrealistic as assuming that the Earth's surface has a uniform temperature. Sammis says the criticality conditions vary widely from place to place and fluctuate over time. A big earthquake, for example, can render a whole region temporarily noncritical—just as a big landslide can flatten out the lower slope of a pile of sand. The challenge is to identify where earthquakes are predictable at any one moment. That, he says, is an open question.

It's one thing to show that quake prediction is "not impossible," and quite another to come up with a reliable precursor. But tantalising evidence of at least one such predictor has already turned up in the laboratory. If you simulate a fault by sawing a chunk of granite in half and then squeezing the halves back together with a hydraulic press, slightly before the miniature fault ruptures, nearby rock begins to shift almost imperceptibly. James Dieterich, now director of earthquake hazards at USGS's Western Region headquarters in Menlo Park, California, described this "premonitory creep" mathematically in the early 1980s. Along real faults, he says, such slip might conceivably warn of impending earthquakes if it is large enough to be detected. Unfortunately, Dieterich's experiments leave this question open.

In 1997, seismologists at the USGS and at Stanford University in California found subtle evidence of such creep in seismic data from earthquakes around the world. But their observations might also have resulted from a landslide-like cascade of tiny ruptures. Another candidate appeared in August, when creepmeters near San Juan Bautista, California detected an unusual creep of a fraction of a millimetre directly over the spot where, a week later, a quake nucleated that reached 5.1 on the Richter scale. Dieterich won't swear that it was premonitory creep, but says it's the best documented case yet seen.

Whether or not premonitory creep is the key to earthquake prediction, many scientists believe that any precursors that do turn up will have at least two things in common with it: they will be furtive—researchers have long since refuted all the blatant ones—and they will appear as anomalies in measurements of strain.

That's only natural, Silver says: most would-be precursors investigated so far, including changes in well levels and in groundwater geochemistry have been by-products of compressed rock.

So instead of investigating a ragbag of indicators, it makes sense to measure strain directly. The trouble is that existing receiver networks for the Global Positioning System of satellites are not sensitive enough. The GPS is perfect for tracking the slow, steady accumulation of strain that raises mountains and provides a basis for 30-year earthquake forecasts. Individual earthquakes, however, are more likely to be heralded by much shorter changes in strain, lasting hours or weeks. Such short-term changes, or transients, slip right past the GPS but can be detected by strainmeters—underground barometers capable of sensing a change in rock pressure of one part per billion. Strainmeters are expensive to install so they tend to be scarce. According to Paul Silver, Southern California will soon boast hundreds of GPS receivers, but it has only five strainmeters.

Silver would like to remedy that. In the March/April 1998 issue of *Seismological Research Letters* he proposed a network of up to 2000 new monitoring stations to record seismic waves, strain, strain transients and ground shaking along the Pacific coast. The "plate boundary deformation network" would run for about 20 years at a cost of $25 million a year. By the end, he says, scientists would have racked up data from enough quakes to know for sure whether they are predictable.

More and better measurements are all very well, says Jackson, but even if the precursors exist, it would take a century or more to log enough big earthquakes to find out how they work. That doesn't worry prediction advocates. They have long realised that effective tools for prediction will not arrive overnight, but they still insist it should be done.

Think of cancer research, says geophysicist Mark Zoback of Stanford Unversity. Finding a cure is clearly a very hard problem, but that's not to say all effort should be stopped. "How would the public feel if we abandoned the quest for a cure and just issued pronouncements on health like try not to smoke and stay out of the sun?"

If we give up on prediction research, he says, we're in danger of violating the public's trust. "We shouldn't promise something that we can't deliver, and if in the end quakes are not predictable, so be it. But it's too easy to work on problems that can be solved rather than ones that should be solved. This is stuff that's worth doing, and worth doing well."

## Questions:

1. What is dilatancy, and why is it not a precursor to earthquakes?
2. What is the problem with the old black box approach to predicting earthquakes?
3. What is premonitory creep?

Answers are at the back of the book.

## Activity:

Presently, prediction of earthquakes is very much like rolling the dice. Roll a set of dice 30 times. Each time guess what number you will roll beforehand. How many times did you roll the number you guessed? Figure out your probability of rolling the number you guessed on the dice. How does this compare with the rest of the class?

# PART 3
# External Earth Processes and Extraterrestrial Geology

**9**

The bedrock beneath much of the Earth's surface is composed of sedimentary rocks deposited long ago, when sea level was higher than it is today, and the sea covered much of the continental area of the Earth. At other times, such as the Ice Age, when much of the ocean's water was tied up in glaciers and ice sheets, sea level was lower, exposing land that is now covered by water. Sea level has risen and fallen many times throughout the history of the Earth. By looking at seismic records of continental margins, scientists have been able to trace the changes in sea level over the last 250 million years. Since the 1975 creation of curves describing the rise and fall in sea level, many people have expressed skepticism about their validity. New research helps confirm the curves. But what causes the changes in sea level over time? That question remains unanswered.

# Ancient Sea-Level Swings Confirmed

By Richard Kerr

*Geologic benchmarks long touted by Exxon scientists apparently do record changes in global sea levels, but the driving force behind the oldest sealevel shifts remains mysterious.*

Back in the 1970s, the oil giant Exxon offered the world's geologists what the company saw as a precious gift. By analyzing the jumble of sediments laid down on the edges of the continents as the seas advanced and retreated, Exxon researchers had charted the rise and fall of sea level over the past 250 million years. If authentic, such information would indeed be valuable, for the ups and downs of the ocean hold a key not only to finding the world's oil and gas deposits, but perhaps also to tracking the waxings and wanings of the ice sheets—and the climate changes that drove them. But outsiders were dubious about the curves, in part because the supporting data were proprietary. So skeptical academics have struggled for the past 20 years to determine whether Exxon's gift was geological treasure or merely fool's gold.

Then, last month, oceanographers returned from drilling nearly 3 kilometers of core from the Straits of Florida and reported preliminary data that match Exxon's curves. Together with other, recently published results, the cores provide strong support for the contention that at least for about the past 40 million years, the records of changing sea level bestowed by Exxon are indeed a prize worth having. Even some early doubters are now won over. "I'm saying, a little sheepishly, 'By golly, those Exxon guys seem to have gotten it pretty close to being right,'" says oceanographer Gregory Mountain of Columbia University's Lamont-Doherty Earth Observatory, who has been critical of the Exxon curves. Mountain, Kenneth Miller of Rutgers University, and colleagues recently reported evidence in support of the curves from seven core holes drilled off New Jersey. Bilal Haq, a former Exxon researcher who is now director of the Marine Geology and Geophysics Program at the National Science Foundation, is delighted with the endorsement.

"Ken Miller and his colleagues were some of the biggest critics of the curve when it first came out," says Haq. "Now they are the biggest supporters."

But doubters remain. And even Haq readily concedes that much of the promise of the Exxon sealevel curves—particularly that of the most ancient records—has yet to be fulfilled. The problem is that researchers can see no mechanism to drive the oldest of the global sea-level changes. All they can think of are ice sheets—which are hard to envision in the warm climate that prevailed before about 50 million years ago.

## The Ocean's Dipstick

The Exxon curves were born back in 1975, when Peter Vail, now at Rice University, and colleagues at Exxon Production Research Company in Houston claimed they had found the geologic equivalent of an oceanic "dipstick" preserved on the continental margins. Each time the sea retreated, the shoreline moved toward the edge of the continental shelf. The researchers argued that erosion of the exposed shelf created a distinctive gap in the geologic record, and that such gaps could be recognized in the radarlike seismic images of the sediments beneath the sea floor today. The team used these erosional gaps or unconformities as a sort of low-water mark on the dipstick of the continental margin's sediment pile. Once dated at a single site, these marks could be recognized elsewhere.

Exxon scientists scanned continental margins around the world, found many unconformities having the same ages, and concluded that only global falls of sea level could be responsible. Furthermore, some of the ups and downs of sea level were very rapid— taking only a million years to rise or fall tens or even hundreds of meters—and they concluded that only fluctuations in the size of major ice sheets could add or withdraw water from the ocean so quickly.

Those inferences add up to an impressive package of knowledge—assuming that the curves really contain all the goodies that the Exxon workers claimed. But academic researchers noted that other, more local mechanisms could also move shorelines back and forth across the continental margins. In particular, tectonic forces could have pushed the margins themselves up and down—in effect moving the dipstick itself. "We have problems [even] today figuring out what sea level [change] is because we can't work out whether the land is moving or the sea is moving," notes Christopher Kendall of the University of South Carolina. "We have nowhere to stand." Such local tectonic forces could have moved shorelines at different times at different places, without a global change in ocean volume. If so, the Exxon curves might be counterfeit rather than real.

The problem was compounded by the fact that Exxon researchers couldn't release the proprietary seismic and well data behind their curves. So academic researchers went in search of their own records from continental margins, hoping to independently confirm—or rebut— the Exxon curve.

The latest such study to be fully analyzed drew on the Ocean Drilling Project's (ODP's) 1993 cores from offshore New Jersey as well as two drill holes on the New Jersey coast. As they reported in *Science* (February 23, 1996, p. 1092), Miller, Mountain, and colleagues combined several dating methods to determine the age of 10 unconformities occurring between 10 million and 36 million years ago. Their results generally match Exxon's for that time. "I think the [Exxon] curve has done a very good in getting the timing of global sea-level changes," says Miller. "They have about the right number [of unconformities], and [they're] about the right age."

But this single site in New Jersey does not make an airtight case, especially because the Exxon curves themselves relied heavily on data from this area. So although the curves match, the shoreline change could have been driven by local tectonic motions. The latest results from ODP Leg 166, however, sample a different area—off the Bahama Bank in the Straits of Florida. In addition to being far from New Jersey, this site had the added attraction of continuous deposition, as the deep straits accumulate sediment even during sea-level low stands. That and more abundant microfossils allow researchers to date low stands to within 200,000 years rather than the 0.5 million to 1 million years typical of offshore New Jersey, says marine geologist Gregor Eberli of the University of Miami, a co-chief scientist on Leg 166.

Drilling of Leg 166 wrapped up only last month, so complete results won't be out for years, but preliminary analysis supports the Exxon curve. "In some places we were spot on," says Eberli. "In other places, especially when you go back beyond 10 million years ago, we have different times than [Exxon] has." But he notes that the global nature of the sea-level changes in earlier times gets additional support from recent data from offshore Brazil. There, Vitor Abreu and Geoffrey Haddad of Rice University, using well data provided by the Brazilian oil company Petrobras, tracked sea-level changes that correlate very well with the Florida data, Eberli says. The mismatches between his own results and Exxon's are understandable, he adds, given that the most up-to-date Exxon curve is now almost 10 years old: "We will refine their curve."

This double-barreled documentation of the curve hasn't yet swayed all doubters, though. Andrew Miall of the University of Toronto, for example, remains a staunch opponent. "I don't think this is good science at all. There are so many events in the Exxon curve and the margin of error in dating is so large that you could correlate anything with it," he says. Indeed, Miall has shown good correlations between the Exxon curve and randomly generated sets of events.

"Andrew's point is well taken," says Miller. Matching a sea-level change from one site to the Exxon curve is inevitably subjective, he notes, so there has been a tendency to make matches where none exist. But, he says, "we're nailing the timing … At some point, it's reasonable to say these changes are correlated and [therefore] they are casually related." Kendall agrees: "Whereas Miall is scientifically correct—it is difficult if not impossible to date all of these things perfectly—what we find is that it seems to be working."

## A Mysterious Mechanism

Even if the Exxon curve is a faithful record of global undulations of sea level, it's likely to spark another controversy, over what's driving sea-level change. Researchers have presumed that the answer is the melting and growing of ice sheets. But the Exxon curve pushes the glacial explanation to the breaking point, for the curve rises and falls in a rapid rhythm throughout the past 250 million years—and Earth was thought to be too warm for ice sheets for much of that time.

And while researchers have been able to link the Exxon curve and ice volume during the recent past, the links peter out at earlier times. To measure past ice volume, researchers analyze the oxygen-isotope composition of carbonate sediments. As glacial ice grows at the expense of seawater or melts into the ocean, it changes the isotopic composition of seawater and the carbonate skeletons of marine plankton.

Now the Leg 155 group has correlated these changes in oxygen isotopes with their New Jersey sea-level changes and with the Exxon curve, back to 36 million years ago. And in a paper in *Geology*, Miller and James Browning of Rutgers extend the link between isotopic changes and the Exxon curve to at least 43 million years ago. Abreu's analysis of isotope data also shows signs of ice-driven sea-level change, up to 49 million years ago. But before that, while the world was experiencing the warmest heat wave of the past 65 million years, both groups find that the correlation falls apart, leaving no mechanism to drive sea-level changes.

Yet the evidence for rapid, global change in sea level continues to accumulate. Heather Stoll and Daniel Schrag of Princeton University have used strontium preserved in carbonates to track the exposure of continental margin sediments during the period of relative warmth 90 million to 130 million years ago, when oxygen isotope records are unreliable. When falling sea level exposes sediment to leaching by fresh water, the amount of strontium in the world ocean increases. In work presented at last fall's meeting of the American Geophysical Union, the researchers found that seawater strontium doubled in a few hundred thousand years, suggesting rapid sea-level drops of 30 to 50 meters, and the drops coincide with major falls in the Exxon curve. Stoll and Schrag also turn to a glacial explanation, suggesting that ice sheets may have temporarily grown large enough to lower sea-level—a provocative idea, given signs in the fossil record of balmy, high latitude climes.

If glaciers didn't drive sea level up and down, what did? The jostling of tectonic plates has been suggested; Kendall has even speculated that meteorite impacts might have done the job in torrid times, by changing tectonic stresses. But there's little evidence for such theories. "People start having problems" with the Exxon curve in earlier times, concedes Haq, "because the mechanism is still unknown." Geologists may now be willing to accept Exxon's gift, but they haven't yet unwrapped all its meanings.

---

## Questions:

1. What is one of the key geological features used to make the sea-level curves?
2. What does Andrew Miall see as one of the problems with the sea-level curves?
3. What mechanism may have driven sea-level change for the last 40 to 50 million years, and why does it not work for earlier times?

Answers are at the back of the book.

## Activity:

View the three maps of the Americas with different sea levels. Would your city be under the sea if sea level rose sixty-five meters?

What impacts on coastal ecosystems will a rising sea level have?

# 10

Satellites, in addition to their role in modern global communication, have become an important tool for earth science. One area where satellite data can be very useful is in the study of the ocean, called by many "the last frontier. " Oceans cover 71% of the Earth's surface area, and reach depths of over 11 kilometers. The great pressures in the deep ocean make study by humans or remote submersibles impractical or impossible. With so much of the Earth unavailable to direct observation, new, high-resolution remote sensing can have an important impact on our understanding of oceanic processes. The recent release of U.S. Geosat data, and the launching of the European Space Agency's ERS-1 satellite, have provided high-resolution data enabling scientists to see the ocean floor in unprecedented detail. The time-consuming analysis of these data has already begun.

# The Seafloor Laid Bare

By Tom Yulsman

*Top-secret data recently declassified by the Navy have enabled scientists to view the seafloor almost as if the oceans had been drained completely of water.*

Deep within the Pentagon, in a suite of rooms built like a safe, scientists and U.S. Navy officers gathered in top secrecy to discuss something of enormous strategic and scientific value: a satellite that could, in effect, peer through thousands of feet of seawater and lay the seafloor bare.

The year was 1977, and the seafloor was Earth's final frontier, a place shrouded in greater mystery than the surface of Mars. According to Bill Kaula, a geodesist at the University of California, Los Angeles, and a participant in the Pentagon meeting, scientists knew at the time that much of the seafloor's topography is mirrored on the surface by slight variations in sea level. This is due to the subtle gravitational pull exerted by seamounts and other hidden features of the deep.

Since 1969, Kaula had championed the idea of charting the seabed's gravitational highs and lows, and in a rough sense, its topography, using measurements of sea level taken from space by satellite. By 1977, the Navy had realized that knowledge of those highs and lows might give the United States an advantage in the Cold War against the Soviet Union by helping American submarines find their way through the murky depths.

Eight years and one false start later, a satellite called Geosat roared off the launch pad. For the next 18 months, it carried out its mission for the Navy, spinning a dense web of tracks over the oceans while conducting a detailed global survey of the height of the sea surface. To the dismay of scientists who knew just what a treasure trove this sea-surface altimetry data could be, the Navy kept it all secret. So while space probes were beaming back incredibly detailed views of other planets, public knowledge of the seafloor remained fuzzy.

All that changed dramatically this past November. Nearly 20 years after Bill Kaula's meeting in the Pentagon safe, David Sandwell of the Scripps Institution for Oceanography and Walter Smith of the National Oceanic and Atmospheric Administration convened a press

Reprinted with permission from Tom Yulsman, Associate Professor, School of Journalism and Mass Communication, University of Colorado at Boulder.

conference in Washington to unveil a ceiling-to-floor map of the global seafloor.

Based in part on data the Navy had finally decided to declassify, the map revealed an oceanic landscape painted in false but dramatic fluorescent color.

"It's like we pulled the plug out of the bathtub," Sandwell says. "We've drained the oceans."

Technically speaking, Sandwell and Smith have not charted the physical topography, or bathymetry, of the seafloor. But stunning topographic details do emerge from the map's gravitational highs and lows. The map, Sandwell says, "is a beautiful confirmation of plate tectonics." Jagged ridges snake sinuously through the middle of ocean basins, marking the seams in Earth's crust where magma wells up to create new seafloor. Fractures sweep across vast swaths of territory, recording the direction tectonic plates have taken as they've crept steadily outward from the mid-ocean ridges. At the far end of tectonic plates, away from the ridges, scimitar-shaped trenches cut deeply into the crust along subduction zones. Here seafloor plunges directly into Earth's interior. And stippling the seafloor like the barbs on some sort of exotic tropical fruit are literally thousands of seamounts.

To be sure, these features have long been standard in seafloor maps. "But what's new is that we finally have a clear and uniform picture," Sandwell says. The gravity data reveals all discrete seafloor features larger than about 3,000 feet high and six miles across—a first. While this resolution may seem coarse compared to satellite images of the land, it's actually a vast improvement over prior global maps.

"Now anything the size of a modest mountain or larger cannot hide from us, whereas before areas the size of Oklahoma were unknown," Smith says.

Because Geosat was able to survey regions that no boat has ever charted, Sandwell and Smith's map also contains many surprises, including thousands of volcanic seamounts never seen before. About half of the seamounts visible in the new map, Sandwell says, were previously unknown.

Given all that, it's not surprising that the map received widespread coverage from a national press corps not usually enamored of geophysics. But a fascinating footnote to the larger story of the Cold War was absent from much of the coverage. Why did it take so many years for the Navy to release this kind of data, and what was the Navy doing with it for all those years? That story, along with the science behind the new map, is told here in detail for the first time.

• • •

The Navy's decision in 1995 to release the satellite data was the last chapter in a story that extends back to the meeting at the Pentagon. Shielded from spies by thick walls and a heavy door secured with a combination lock, Kaula and his colleagues discussed the upcoming mission of a predecessor to Geosat called Seasat. High on the agenda, Kaula says, was a discussion of what to do with Seasat's findings. Data from NASA projects like Seasat are usually made public. But the Navy worried that this would serve up a feast of sensitive military information to the Soviet Union. Making the case for science, however, Kaula argued that locking the data "behind the screen" would deprive researchers of extremely valuable information on the structure and evolution of ocean basins.

After a series of meetings, Kaula and his colleagues convinced the Navy that Seasat probably wouldn't tell the Soviets anything they didn't know already. The final recommendation: The data should be made public.

NASA launched Seasat in 1978. As it orbited, it measured how long it took for radar pulses beamed toward the sea to reflect back. With this information, Seasat calculated the height of sea level along its track.

Since massive objects on the seafloor, such as a volcano, exert greater gravitational pull than their surroundings, they tend to pull seawater toward them. With a greater volume of water around and above them, these massive objects are mirrored at the sea's surface by a slight mounding of water superimposed upon the choppiness of waves. Trenches, on the other hand, have slightly less gravity than their surroundings and therefore create a slight deficit in water at the surface.

The bumps and depressions on the sea surface vary by as much as 300 feet from the level the sea would take were there no gravitational effects. Because these features are many miles across, their slopes are so gentle they're imperceptible to any boat that cruises over them. But Seasat saw them clearly. And with accurate tracking of the satellite's position, it was possible to use Seasat's altimetry data to chart variations in the gravity of the seafloor.

According to Kaula, the satellite performed well for three months. But then an electrical glitch turned it into a useless hunk of space junk. Some scientists say the Navy sabotaged Seasat when it realized just how good the data were, good enough for the Soviets to benefit after all. Although a review panel blamed a poorly designed component, suspicions did not die easily.

Before it failed, Seasat did manage to return a small set of data that NASA made public. In 1983, William Haxby of the Lamont-Doherty Earth Observatory processed this data to make the world's first gravity map of seafloor features based on satellite altimetry. Because Seasat's tracks over the oceans were spaced no closer than 50 miles apart, the map was poor in detail. But in large scale, it did show trenches, mid-oceanic ridges, fracture zones and other tectonic features. And it gave geophysicists like David Sandwell a taste of what they could expect from Geosat, if only they could get their hands on the data. It was now the early '80s, and the Reagan military buildup was in full swing. There was no longer any debate about secrecy: The Navy would not release Geosat's data.

• • •

Launched in 1985, Geosat performed flawlessly from its polar orbit about 500 miles high. Beaming an 1,000 pulses of radar down to the sea surface each second, it measured sea level with an accuracy of an inch along tracks no more than five miles apart.

Yet despite this success (not to mention discussions in the safe about how seafloor gravity data could help submariners), the Navy claims today that it really wasn't all that interested in what Geosat could reveal about the seafloor. Theoretically, at least, that kind of information should have been invaluable, because the gravitational tug exerted by seamounts and other topographic features can throw submarines off course and thereby impair their ability to hit the bull's-eye with nuclear missiles. But Ed Whitman, technical director of the Office of the Oceanographer of the Navy, says Geosat's data simply wasn't detailed enough to be of much use in helping a submarine compensate for these gravitational effects.

"We need for some of our weapons very precise measurements of local gravity," he says. "Altimetry-derived gravity has never been able to provide that."

Whitman describes another way that Geosat's data was useful. Sonar, he explains, has a hard time penetrating boundaries between relatively warm and cool waters such as those at the edges of currents and eddies. So if U.S. Navy submarines can find such interfaces, they can hide from their adversaries behind them. Whitman says Geosat's data has helped subs do just that. How? Warm water expands, subtly elevating sea level.

One could imagine, however, that the Navy has reason to be less than forthcoming. After all, why tell your adversaries how you've used formerly top-secret information?

David Sandwell probably knows more about Geosat's data than any other civilian. Although he professes no personal knowledge of how the Navy used that data, he does say that Geosat's gravity readings were detailed enough to help improve submarine and missile navigation.

Because today's missiles typically carry multiple warheads, each warhead can't be too big. This means that each one has to be extremely accurate to insure that it can take out its target. To plunk a warhead down on top of an enemy missile silo from a firing point in a submarine thousands of miles away, the sub has to tell the missile's guidance system exactly where it is at launch time.

Submarine navigators can't see where they're going in their windowless craft. And there are no radio beacons for them to follow. Instead, submarines are equipped with inertial navigation systems that sense changes in direction and speed and then analyze this data to keep track of the vessel's path through the murky depths. But according to Sandwell, when an uncharted seamount tugs on the sub, the nav system is fooled into thinking the vessel has turned onto a new heading when in reality it has only been tugged slightly to the side. As the sub travels for hundreds of miles, this subtle error is magnified exponentially. As many such errors accumulate, a sizable error is introduced in the nav system's fix on the sub's location.

In other words, when it comes time to fire its missiles, the submarine doesn't know exactly where it is. With a slightly inaccurate navigational fix entered into the missile's onboard computers, the warheads may not be able to hit their targets precisely.

But Sandwell says that an accurate map of gravity anomalies plugged into the nav system's brain could prevent these errors altogether, at least in theory.

"Although most of what I say is speculation, it is based on physics and the accuracy of the data," he says.

No matter how the Navy used Geosat, the fact that it kept the full set of data secret for 10 years shows just how sensitive the information was. But then in 1991, something happened that would make secrecy moot: The European Space Agency launched ERS-1, a satellite much like Geosat. ERS-1 measured the topography of the sea's surface with the same accuracy as Geosat, and the Europeans felt no compulsion to keep the data secret. When the measurements from ERS-1 were made public in 1995, the Navy no longer had a compelling reason to keep Geosat's bounty under tight wraps.

And so in July of 1995, it released all of Geosat's data. (It had released parts of the data set earlier, but had withheld most of the data covering the area north of 30 degrees south latitude.)

This was the moment Sandwell had been waiting for since he was a student of geophysics in 1978, the year that Seasat failed. But he and his partner Walter Smith had to overcome some major hurdles before they could transform the altimetry measurements of Geosat and ERS-1 into a global gravity map. Not the least of these problems was the effect of tides. To subtract this from the sea level measurements, the scientists used mathematical models to estimate how much tides had caused the surface of the ocean to move with respect to the center of Earth as the satellite passed overhead. Smith and Sandwell were also helped by having two sets of data, which allowed them to reduce errors introduced by tides and waves.

• • •

Geosat and ERS-1 conducted, in essence, a gigantic scientific reconnaissance survey—one that would have been done without those satellites. Although ships carrying sensitive gravity meters can provide extremely detailed measurements of discrete areas of the seafloor, they are slow and expensive to operate. According to Marcia McNutt of the Massachusetts Institute of Technology, the data

obtained by the $80 million Geosat over 18 months would have cost $5 billion to collect using a ship operating continuously for 100 years.

Because of this expense, huge swaths of the seafloor, particularly in the remote southern oceans, were unknown to science before Sandwell and Smith released their global map. In the past, scientists frequently had to grope about in the dark to find things of scientific interest on the seafloor. Now the map is helping them focus their efforts.

"It can tell you where to go to maximize the use of your ship time," Sandwell says.

He speaks from personal experience. Using snippets of Geosat data declassified in 1987, he got a detailed view of faint stripelike variations in the gravity of the seafloor in the South Pacific. These faint lineations had first been discovered by William Haxby in Seasat's data. Scientists had proposed several explanations for the lineations. According to one, they were the result of a basic tectonic process: the movement of tectonic plates over small, individual jets of volcanic material, called mini-hot spots. As these jets erupt material onto the moving seafloor above them, they create stripes of high gravity volcanic rock.

Other scientists proposed that the lineations were caused by alternating areas of upwelling and downwelling of material in the mantle, a region of hot rock beneath the crust. And still others theorized that the seafloor was being pulled apart in this region and that the lineations were, in essence, stretch marks.

Intrigued, Sandwell surveyed the site from a research vessel in 1992 and 1993. He and his colleagues dredged rocks from the seafloor and mapped the area in detail using sophisticated sonar equipment. They discovered a series of volcanic ridges, later dubbed the Pukapuka Ridges, oriented in a narrow line running from the East Pacific Rise on a southeast-northwest axis. Sandwell says the evidence collected at the site supports the idea that the ridges formed when lava erupted out of cracks that opened as the seafloor was being stretched. In a paper in the *Journal of Geophysical Research*, he proposes that the stretching is occurring as the tectonic plate is being sucked down into trenches to the west.

If Sandwell is right, many features previously. attributed to hot spots may have to be reappraised.

The gravity map may also produce as many questions as it helps answer. One example involves the mid-oceanic ridges where tectonic plates spread apart. Scientists have identified three types: slowspreaders, which have a deep, broad valley running down their middles, fast-spreaders, which do not, and an intermediate type. Since all the midoceanic ridges in the world are connected, scientists have wanted to know whether the transition between fast and slow-spreading rates is abrupt or gradual.

In helping to answer that question, the new map has raised another. According to Sandwell, the Southeast Indian Ridge, revealed now in greater detail than ever before, appears to have fast-spreading and slow-spreading segments. Significantly, the transition between the different segments is very abrupt: slow and fast segments join without any transition. "This was a big surprise," Sandwell says, one that remains to be explained.

Scientists aren't the only ones who are excited by the new gravity map. Commercial fishermen are already using it to locate shallow banks where they might find good hunting. Petroleum companies are using it to locate promising areas to prospect for oil. And engineers from Honeywell have considered using the gravity data to reprogram inertial navigation systems for jetliners, which can be thrown off course by long trenches and seamount chains in the Pacific.

As much as Sandwell and Smith's map has already revealed about Earth's final frontier, much more still lies hidden away behind the screen of secrecy. Geosat was just one of many projects the Navy has undertaken to study the oceans. Reams of data are still top secret, including information on sediments, the chemistry and temperature of the oceans, and the magnetic properties of the seafloor. The Navy has also conducted highly detailed topographic surveys of the seafloor.

A government advisory panel called Medea recently concluded that much of this data can be released without compromising national security. Panel member John Orcutt, a geophysicist at Scripps, is confident that the release of Geosat will not be an anomaly. "I feel that most of the data outlined in the report should be released," he says. "But the process for doing so is formal and, as with any undertaking with a government bureaucracy, this takes time. I do feel that the Navy is approaching this issue in good faith for the most part and that real progress on releasing much of these data will be made in the next few years."

Get ready for more remarkable revelations.

## Questions:

1. How did satellites determine the elevation of features on the sea floor?
2. What is the resolution of the satellite data?
3. How does Sandwell think the newly discovered lineations are formed?

Answers are at the back of the book.

## Activity:

Using maps with seafloor topography, try to outline all of the tectonic plates on a sheet of paper with just outlines of the continents.

11

Much attention has been directed towards the study of Mars. Interest in the role of meteorite impacts on earth and other planets includes atmospheric effects. On July 4, 1997, the Mars Pathfinder landed on the surface of Mars to investigate the Martian atmosphere, geologic processes, and its interior. The landing confirmed a presence of running water in the history of the planet, and rocks believed to be deposited by ancient floods that could offer information about their origins were the focus of the Pathfinder mission.

# Rocks at the Mars Pathfinder Landing Site

By Harry McSween, Jr., and Scott Murchie

The metallic tetrahedron, with its protective cover of inflated airbags, hit the ground of Mars at 60 kilometers per hour, bounced 16 times across the red desert floor and rolled end over end, finally coming to a stop amid a field of boulders. There was, of course, no one there on July 4, 1997, to watch the ungainly (many thought improbable) landing of the Mars Pathfinder spacecraft, but more than a hundred million miles away earthlings paid a great deal of attention. Pathfinder, designed principally as an engineering demonstration, proved to be a public-relations bonanza for NASA. But the spacecraft also delivered scientific instruments that investigated the structure of the Martian atmosphere and interior, its weather, and the geologic processes that sculpted the landing site and created the pervasive red soil. And for the very first time, rocks on the Martian surface were analyzed using a robotic rover.

Rocks, christened with nicknames like *Barnacle Bill*, *Casper* and *Yogi*, were clearly the scientific centerpiece of the Pathfinder mission. Even the choice of the landing site, as described by project scientist Matt Golombek of the jet Propulsion Laboratory, was predicated on its having an abundance and variety of rocks. The confluence of *Ares Vallis* and *Tiu Vallis* (two ancient river valleys) was once the largest known floodplain in the solar system, a location specifically selected because of the variety of rocks thought to have been carried by catastrophic floods down the outflow channels and dumped near their mouths. Images of the Pathfinder landing site revealed a Marsscape of ridges and troughs strewn with rounded pebbles and boulders, resembling huge depositional fans on the earth. These characteristics, along with near-by teardrop-shaped hills streamlined by floods confirm the geologic setting inferred earlier from orbital photographs.

The rocks were ripped up by the raging Ares and Tiu floods, presumably eroded from the *ridged plains uni* through which the rivers coursed and perhaps from the bedrock underneath. In the stratigraphic nomenclature adopted for Mars, the ages of these two units, as mapped by Sue Rotto and Ken Tanaka of the U.S. Geological Survey are Hesperian and Noachian, respectively. Both are thought to be billions of years old, dating roughly from the time of formation of the oldest preserved continental rocks on the earth. Near the Pathfinder landing site are a number of younger impact craters that excavated the sedimentary deposits and lofted fragments, or ejecta, across the area.

Sorting out the geologic processes and events that created the local martian rocks was the petrological mission of the Mars Pathfinder.

From "Rocks at the Mars Pathfinder Landing Site," by H. McSween, Jr. and S. Murchie, American Scientist, January/February 1999, pp. 36-45. Reprinted with permission.

Through petrological investigations of terrestrial rocks, geologists routinely unravel complicated histories, determine the timing of various events and glimpse processes such as melting of our planet's deep interior or deposition of sediments on an ocean floor. Close examination of a rock's physical appearance, identification of the minerals it comprises and measurement of its chemical composition allow scientists to determine how the rock formed and whether it was subsequently modified. In designing the instrument package for the Mars Pathfinder, it was expected that a combination of observations and analyses would be sufficient to specify the origins of the martian rocks. The mobility required for this task was provided by the Sojourner rover, which might be considered a geologist of sorts (the Geological Society of America has named Sojourner as its first non-human Society Fellow). Although all the instruments aboard the lander and rover functioned as designed, understanding how rocks at the Pathfinder site formed has proved to be a challenge.

One difficulty encountered in interpreting the Pathfinder data is that the rocks, like nearly everything on Mars, are coated with fine red dust. Even the martian atmosphere owes its peculiar salmon color to suspended dust particles. Here we describe ways that multispectral images by the Imager for Mars Pathfinder (IMP) and chemical analyses by the rover-mounted Alpha Proton X-ray Spectrometer (APXS) can be used to peer through this dust coating, and the rocks that are revealed suggest a planet more petrologically complex than previously supposed.

## Textural Observations

The texture of a rock—its size, shape and the packing and orientation of its constituent particles—is usually an excellent indicator of its origin. For example, volcanic rocks are fine-grained or glassy, sedimentary rocks tend to be layered, and rocks formed by an impact are commonly *breccias* composed of angular fragments. Rock shape is sometimes related to texture, because the way a rock breaks into smaller pieces usually reflects its internal features. Rocks at the landing site exhibit a variety of shapes—rounded, sharp and angular, flat and tabular, and even irregular and lobate—which hint at different textures.

Textures are best observed directly, however, by close examination of freshly broken surfaces. This is one reason why geologists arm themselves with rock hammers and hand lenses. Unfortunately, all images of the Pathfinder site are confined to unbroken rock surfaces that have been molded by wind, partly obscured by dust and perhaps weathered, making textural interpretations ambiguous. Another complication is that the image resolution (1 to 5 millimeters per pixel) afforded by the IMP, which was mounted on the stationary lander and thus could not be moved closer to rocks in the field of view, is not sufficient to identify small features such as individual mineral grains. Miniature cameras (with resolution of about 1 millimeter per pixel) were mounted on both the front and rear of the rover, and these could be positioned close to rocks. However, even images from these cameras are unable to resolve individual grains in the rocks.

The most abundant textures on many rock surfaces at the Pathfinder site are pits. Some may be holes formed by gas bubbles, called *vesicles*, that commonly decorate lavas. As magma ascends toward the planet's surface and the pressure on it decreases, the solubility of dissolved gases is reduced so that the magma effervesces, much like the release of carbon dioxide bubbles when a can of soda pop is opened. The presence of vesicles would indicate that the rocks are volcanic. In this instance, the pits were probably eroded out of vesiculated rocks by windblown sand. Many of the rocks have elongated pits resembling features called *flutes* that form during sand abrasion. The flutes have a consistent orientation from rock to rock, suggesting a prevailing wind direction. Many rocks in Antarctica, the most Mars-like surface environment on earth, have pits that were initiated from corrosion by salt and enlarged by wind erosion.

A few rock surfaces show pits accompanied by rounded bumps of comparable size and shape. The bumps could be large crystals or fragments that are relatively resistant to erosion. The occurrence of rocks exhibiting both bumps and pits is suggestive of large particles having been plucked from a sedimentary rock. A conglomerate is a coarse sedimentary rock formed by rounded pebbles cemented together with finer materials; in a breccia, the large particles are angular. If the bumps on Pathfinder rocks are pebbles, the pits could be sockets formed by their removal. The occurrence of conglomerates would be important because they are formed during flooding like that which shaped the landing site.

Breccias can form by a number of processes, including the welding of fragments produced by impacts. Linear features, typically appearing as repeated light and dark bands spaced a few millimeters apart, are also seen on some rocks. These could be stress fractures, sedimentary layering, or aligned vesicles or mineral grains.

Overall, though, the observed surface textures are sufficiently equivocal to prevent definitive statements about the origin of these rocks. Based on these observations the rocks at the Pathfinder site might have formed as lavas that flowed onto the martian surface, as coarse sediments carried and deposited by water or as materials that were pulverized and melted by meteor impacts.

## Mineralogical Observations

Designed by Peter Smith of the University of Arizona, the IMP contains a rotatable wheel with 12 filters, through which the Pathfinder scene was imaged at different wavelengths of light. The filters range from 440 to 1,000 nanometers, covering parts of the visible and ultraviolet portions of the spectrum. To obtain color images of the martian surface, images at red, green and blue wavelengths were combined. However, the filters were also used to measure the spectra of sunlight reflected from the rock surfaces. When exposed to light, atoms in certain minerals absorb specific wavelengths, so the corresponding reflected light is partly missing those wavelengths, resulting in dark regions (*absorption bands*) in its spectrum. The wavelength positions of these absorptions are "fingerprints" that can be used to identify some of the minerals present. The distinctive coloration of the Red Planet results from minerals rich in oxidized (ferric) iron, and these commonly have strong absorption bands centered at wavelengths less than 440 nanometers, with an edge that extends beyond 750 nanometers. Some iron-bearing minerals also absorb at longer wavelengths. For instance, pyroxenes, (which are ferrous iron magnesium-calcium silicates that are abundant in many igneous rocks) generally have an absorption band near 900 to 1,050 nanometers, with longer wavelengths corresponding to higher calcium and iron contents.

Spectra of large regions of the martian surface, measured using telescopes on earth, show that the occurrence of ferric minerals in bright red regions covered with dust give the planet its distinctive coloration. Many of the dark gray regions thought to be dominated by rocks show the spectral fingerprint of pyroxenes with moderate calcium and iron contents. A notable exception is Acidalia, the vast dark region adjacent to the Pathfinder site.

It was hoped that the spectra of Pathfinder rocks would allow determination of their mineralogies. Athough rocks at the Pathfinder site have surprisingly varied color properties, they are nearly lacking in mineralogically diagnostic absorption bands. Most rocks are relatively gray, and (like Acidalia) lack a pyroxene-like absorption band within the IMP's wavelength range. There are three plausible options to explain the absence of this band. The rocks may be solidified glasses, of either impact or volcanic origin, and thus lack crystalline pyroxene. Glasses also have a diagnostic absorption band, but it lies at 1,100 nanometers, beyond the wavelength range of IMP. Alternatively, the rocks may contain an opaque mineral that does not transmit light and consequently hides pyroxene. In terrestrial rocks the iron oxide magnetite often disguises the presence of pyroxene. Or the pyroxenes may be so rich in iron and calcium that their characteristic absorption falls at longer wavelengths than can be detected by IMP. If IMP's wavelength range extended just a few tens of nanometers farther, one of these three possibilities might have been indicated.

Large rounded boulders at the Pathfinder site, thought to be the oldest materials present, are partly covered by a distinct maroon rind. This rind is absent from more abundant angular rocks, suggesting that it formed only during the site's earliest history. The spectra of the rind are different from other materials, with a distinct ferric absorption band near 900 nanometers. The spectral shape is most easily explained by one of two common ferric iron oxides, maghemite and ferrihydrite. To IMP, both minerals appear nearly the same, but the conditions under which they formed may not be alike. Ferrihydrite, a mineral which gives a rusty color to shallow lake bottoms on earth, forms in liquid water. Maghemite likewise can form in water, but it can also form by weathering of precursor minerals without significant water present. The maroon rinds constitute tantalizing but inconclusive evidence for the idea that long ago the Pathfinder site was submerged under water. If IMP had just one additional filter at a diagnostic red wavelength, the two minerals could have been distinguished, and this ambiguity in understanding the rocks would have been eliminated.

Erosion and transport of dust by prevailing winds have had an enormous effect on color properties of the rocks. Gray rocks have flat spectra, without any distinctive absorption bands. Surfaces of both gray and maroon rocks are partly coated by windblown dust. Rock faces oriented away from the prevailing northeast wind direction are most likely to be draped by dust and tend to be red. Upwind faces are scoured by the wind and usually are gray. On large rounded boulders, the maroon rind is preserved on downwind faces and mostly abraded off upwind faces. In fact, except for the maroon rinds, nearly all color variations in rocks can be explained simply as effects of a thin red coating of dust on gray rocks.

One group of rocks that excited the Pathfinder scientists soon after landing has a bright pink color and tabular shapes. These rocks raised hopes of identifying sedimentary rocks. When the APXS analyzed the pink rock *Scooby Doo*, its composition was found to be identical to the soil. The rover also drove onto the rock and turned a wheel in place to try to scratch its surface, but the rock was unaffected. Laboratory experiments show that compacting Marslike soil rich in ferric iron causes the soil to appear bright pink. Rocks like *Scooby Doo* may simply be compacted soil, with sufficient strength to resist abrasion by a spinning studded wheel.

Color variations unfortunately do not provide evidence of the Pathfinder rocks' origins. There is no clear indication of color banding due to sedimentary layering. Only a few small rocks exhibit mottled color variations that might indicate that they are compositionally heterogeneous. Even those rocks with "bump-and-socket" textures are not spectrally mottled, which may cast doubt on the identification of them as conglomerates or impact-produced breccias.

A rough chronology for rocks at the landing site has been proposed based on the observed spectral features. The oldest rocks at the site are the large, rounded boulders with maroon coatings, deposited by floods and likely stained by reactions with liquid water. Sometime later, smaller gray rocks and broken fragments of maroon boulders were added to the scene, probably as ejecta from nearby impact craters. Erosion by windblown particles then partially removed the maroon rinds from some boulders. At one time windblown dust apparently accumulated several inches deep, but more recently winds have sculpted and partly denuded the site.

## Chemical Analyses

The rover-mounted APXS contained tiny pellets of radioactive curium, which bombarded the analyzed rocks with alpha particles. Three kinds of interactions of the target materials with alpha particles (producing backscattered alpha particles, protons and x rays) allowed analysis of the elements that the rock comprises. The measurements only determined the chemistry of the outermost rock layer, no more than a few hundred micrometers in thickness. Because of difficulties encountered in analyzing samples in the presence of the martian carbon dioxide-rich atmosphere, only data from the x-ray mode are currently available. Rudi Rieder of the Max Planck Institut für Chemie in Mainz, Germany, and his colleagues have so far reported preliminary chemical analyses of five Pathfinder rocks and six soils.

When element abundances measured for the rocks are plotted against each other, they tend to form a nearly linear array, with soil analyses clustering at one end of the array. The most straightfoward explanation of these linear chemical trends is that they are mixing lines, representing rocks contaminated with varying amounts of adhering dust. This idea is strongly supported by a relation between measured composition and color. The martain soils are rich in sulfur and are very red (which can be expressed as a high red/blue spectral ratio). Gray (spectrally bluer) rocks coated with increasing amounts of dust become progressively redder and richer in sulfur.

Mars is home to some of the solar system's largest volcanoes. It is logical to assume that a significant portion of its surface is covered with volcanic rocks, or perhaps sedimentary materials derived from them. Erupting lavas contain relatively little sulfur because the solubility of sulfurous gases in magma is low under normal conditions. If we assume that the Pathfinder rocks (or the fragments they comprise) are volcanic in origin, we can estimate their composition by plotting various chemical components against sulfur and extrapolating the straight lines they define to "zero" sulfur. The composition of the rock can be estimated by graphically removing the sulfur-rich dust coating in this way.

The interpretation of linear chemical trends as mixing lines has a consequence that was not anticipated in planning the mission: All the analyzed rocks must have nearly the same composition. The Pathfinder location was originally touted as a grab-bag site containing samples carried by floods from the ancient martian crustal highlands to the south. However, the distance to the highlands boundary is some 800 kilometers, much farther than rocks have been carried by even the largest terrestrial floods. It seems more likely that the rocks are locally derived, eroded and tumbled along perhaps for kilometers rather than hundreds of kilometers. The rocks in the neighborhood are still old, to be sure, but they might not be derived from the ancient highlands. If the floods did not carry these rocks for great distances, they might not have been mixed with rocks from other sources. However, it is worth noting that, owing to the difficulty of positioning the APXS sensor head against rock faces, only a few large rocks at the site were analyzed. Some smaller rocks may well have different compositions and origins.

## Andesites or Icelandites

The sulfur-free rock composition is surprisingly rich in silicon. (Geologists commonly describe elements in rocks as if they were oxides because they are usually combined with oxygen to form minerals; in this case silicon becomes silicon dioxide, or silica.) Employing a widely used chemical classification scheme for volcanic rocks, the high silica content of the sulfur-free rock corresponds to andesite. The existence of rocks having andesitic composition on Mars came as a surprise. By far, the most abundant terrestrial lavas are basalts, with less than 52 percent silica (by weight). Basalts are also thought to be very common on Mars, based on spectral similarities between its rocky regions and a few basaltic meteorites that are believed to be martian samples.

The chemical composition of a rock can be recast into mineralogy using a calculation scheme called a "norm." Assuming that a rock consists of some mixture of common igneous minerals, the norm is the proportion of those minerals that most accurately explains the abundances of the elements measured. The norm may not necessarily correspond to the actual minerals the rock comprises, but the equivalence is usually good for igneous rocks. The calculated normative mineralogy of the sulfur-free rock suggests that it is dominated by pyroxenes, feldspars (sodium-potassium-calcium aluminosilicates) and quartz (silica), with minor amounts of magnetite and ilmenite (iron-titanium oxides). The relative proportions of these minerals are consistent with those of igneous rocks. Of course, the rocks could still be sedimentary or impact produced, but in those cases perhaps formed from fragments of andesitic lava.

The term "andesite" is a loaded word for geologists, because on earth it has important implications for plate tectonics. Collisions of plates cause the edge of one slab of rock to be thrust downward, or subducted, beneath another. Andesitic magmas erupt above subduction zones, such as the one that underlies their namesake, the Andes Mountains of South America. Deeply subducted slabs of oceanic rock are heated, and as they are metamorphosed they release water which rises into the overlying mantle. The addition of water lowers the melting point of the solid mantle rock, producing basaltic magma. As this magma ascends towards the earth's surface it begins to solidify, and some crystals settle out of the liquid and are left behind. This process, called fractional crystallization, changes the composition of the remaining magma, driving it towards higher silica contents. These magmas may also incorporate some of the enclosing crust, which tends to contain silica in greater abundance, further increasing the magmas' silica content. By a combination of fractional crystallization and assimilation of crustal rocks, andesitic magmas are produced. Water is apparently central to the process, as it initiates melting in the first place and is concentrated in the resulting magma, which in turn affects its fractional crystallization

After the first analyzed Pathfinder rock (Barnacle Bill) was described at a press conference as having the composition of andesite, many geologists were asked by reporters to explain the implications of this finding. Given the geologists' terrestrial experience with this rock, it is understandable that there appeared numerous media reports suggesting that martian andesite may imply a planet with a wet interior, and perhaps with plate tectonics—in short a world more like our own than had previously been supposed. But is that really a necessary conclusion from these results?

There are several alternative, perhaps more likely, explanations for the existence of andesitic rocks on Mars. One, offered at the Barnacle Bill press briefing, is that these rocks were formed by repeated melting events. Partial melting produces a liquid with a higher abundance of silica than the rock that was partly melted. If this magma crystallizes and then is partially melted a second time, the silica content of the resulting magma is increased further. Thus, andesitic magmas might be produced in a series of melting steps, with each event further concentrating silica. A protracted process of this sort is what is envisioned for the formation of the earth's continents. The continental crust is much richer in silica than oceanic crust and, in fact, has roughly the composition of andesite. Perhaps parts of the ancient crust of Mars formed in a similar manner, as they were remelted and refined by protracted volcanism and by large impacts.

Another possibility is that the Pathfinder site is not petrologically representative of the martian crust at all, but instead contains some unusual rocks that, although interesting, constitute a minor part of the surface. Fractional crystallization of basaltic magmas on earth is not restricted to subduction zones. In fact, anywhere basaltic magma occurs, it may undergo partial crystallization with separation of the crystals and liquid. In other settings, however, fractional crystallization occurs under different conditions, usually nearer the planet's surface (that is, at lower pressures) and with less water available. These conditions promote changes in the order in which minerals appear, so fractional crystallization results in slight differences in the chemistry of the resulting silica-rich magmas. Andesitic liquids are still generated, but they have higher iron and lower aluminum contents at any given silica content. Such volcanic rocks were first recognized in Iceland and are now referred to as icelandites. The Pathfinder sulfur-free rock is also rich in iron and depleted in aluminum, and, in fact, icelandite provides its closest terrestrial analogue. Therefore, these martian rocks may have formed simply by fractional crystallization of basaltic magma, perhaps the same magma that crystallized to form most of the ridged plains unit through which Ares Vallis has been cut.

## Lessons Learned

Rocks at the Mars Pathfinder landing site are difficult to classify, principally because of the limited spatial resolution of the Pathfinder cameras and because of coatings of windblown dust that obscure the rocks' textures and spectral features and contaminate analyses of the rocks' chemical compositions. An obvious lesson for future Mars explorers is the need to carry devices that can expose fresh rock surfaces for observation and analysis. At the present time we do not know whether a whisk broom may be sufficient or if a chipping or coring device may be necessary. Another lesson is that future imagers should have sufficient resolution to reveal individual mineral grains, with enough filters to distinguish minerals like maghemite and ferrihydrite.

We believe that the physical, mineralogical and chemical characteristics of the analyzed Pathfinder rocks, as best they can be determined from available data, are consistent with volcanic rocks of andesitic composition. Some rocks might consist of volcanic rock fragments that have been eroded and deposited by wind, water or impacts, but without much chemical modification. The maroon rinds on the oldest rocks record an ancient environment, perhaps the wet one suggested by landforms at the site. After the catastrophic floods deposited the oldest boulders, the only major changes have been the addition of small angular rocks to the site, perhaps as ejecta from nearby craters, and the redistribution of dust by the wind. The Pathfinder site is actually stunning for how little it has changed over billions of years.

If these rocks are samples of the ancient martian crust derived from local basement outcrops, the sulfur-free rock composition may suggest the existence of crustal materials similar to those that make up the earth's continents. Such crustal rocks would presumably have formed in an analogous manner, by repeated partial melting and solidification that progressively increased the silica contents of magmas and the rocks that crystallized from them. The compositional similarity between the sulfur-free rock and terrestrial icelandite, however, suggests another origin: simple fractional crystallization of basaltic magma. Fractional crystallization is an inefficient process, because a large amount of basaltic magma must be processed to produce a relatively small fraction of andesitic liquid. If the Pathfinder rocks are icelandites, then it is likely that the bulk composition of the martian crust would be basaltic rather than andesitic. Because the existence of martian rocks having andesitic compositions can be readily explained by recurrent partial melting or fractional crystallization, speculations about possible roles for water and plate tectonics in martian petrogenesis are probably unwarranted.

## Bibliography

Golombek, M. P., R. A. Cook, H. J. Moore and T. J. Parker. 1997. Selection of the Mars Pathfinder landing site. *Journal of Geophysical Research* 102, E2:3967-3988.

Golombek, M. P., R. A. Cook, T. Economou, W. M. Folkner, A. F. C. Haldemann, P. H. Kallemeyn, J. M. Knudsen, R. M. Manning, H. J. Moore, T. J. Parker, R. Rieder, J. T. Schofield, P. H. Smith and R. M. Vaughan. 1997. Overview of the Mars Pathfinder mission and assessment of landing site predictions. *Science* 278:1743-1748.

McSween, H. Y. Jr., S. L. Murchie, J. A. Crisp, N. T. Bridges, R. C. Anderson, J. F. Bell III, D. T. Britt, J. Brückner, G. Dreibus, T. Economou, A. Ghosh, M. P. Golombek, J. P. Greenwood, J. R. Johnson, H. J. Moore, R. V. Morris, T. J. Parker, R. Rieder, R. Singer and H. Wänke. In press. Chemical, multispectral, and textural constraints on the composition and origin of rocks at the Mars Pathfinder landing site. *Journal of Geophysical Research*.

Rieder, R., H. Wänke, T. Economou and A. Turkevich. 1997. Determination of the chemical composition of Martian soil and rocks: The alpha proton x-ray spectrometer. *Journal of Geophysical Research* 102, E2:4027-4044.

Rieder, R., T. Economou, H. Wänke, A. Turkevich, J. Crisp, J. Brückner, G. Dreibus, and H. Y. McSween Jr. 1997. The chemical composition of martian soil and rocks returned by the mobile Alpha Proton x-ray Spectrometer: Preliminary results from the x-ray mode. *Science* 278:1771-1774.

Rotto, S., and K. L. Tanaka. 1995. Geologic/geomorphic map of the Chryse Planitia region of Mars. *U. S. Geological Survey Miscellaneous Investigations Series*, Map 1-2441.

Smith, P. H., M. G. Tomasko, D. Britt, D. G. Crowe, R. Reid, H. U. Keller, N. Thomas, F. Gliem, P. Rueffer, R. Sullivan, R. Greeley, J. M. Knudsen, M. B. Madsen, H. P. Gunnlaugsson, S. F. Hviid, W. Goetz, L. A. Soderblom, L. Gaddis and R. Kirk. 1997. The Imager for Mars Pathfinder experiment. *Journal of Geophysical Research* 102, E2:4003-4025.

Smith, P. H., J. F. Bell III, N. T. Bridges, D. T. Britt, L. Gaddis, R. Greeley, H. U. Keller, K. E. Herkenhoff, R. Jaumann, J. R. Johnson, R. L. Kirk, M. Lemmon, J. N. Maki, M. C. Malin, S. L. Murchie, J. Obserst, T. J. Parker, R. J. Reid, R. Sablotny, L. A. Soderblom, C. Stoker, R. Sullivan, N. Thomas, M. G. Tomasko, W. Ward and E. Wegryn. 1997. Results from the Mars Pathfinder camera. *Science* 278:1758-1765.

---

## Questions:

1. What do petrological investigations offer to geologists?
2. What are some of the possible origins of the Pathfinder rock, *Barnacle Bill*?
3. What are some of the problems encountered with classifying these Martian rocks?

## Activity:

Using viable internet resources, write a page or more comparing Earth's terrain and atmosphere with htose of Mars. The U.S. Geological Survey and NASA sites will be helpful.

# 12

Scientists now believe they know where to look for indicators of life on Mars, past or present. They have been studying a small piece of igneous rock that fell to Earth from Mars 13,000 years ago for approximately 14 years. Evidence from this rock sample indicates that liquid water may be or may have been present on Mars. Scientists also know that some organisms are capable of surviving under very extreme conditions and, in fact, thrive in those conditions. One such condition is the extreme heat of a hydrothermal ocean vent. With this information scientists have a better idea of where to take samples on Mars.

# A Martian Chronicle

By Andrew Knoll

On a clear night confronted with the vast starry blackness of the sky, it is hard not to wonder about the possibility of life beyond our planet. But despite that primeval human curiosity, facts have been elusive. The planets and moons of our solar system have proved devilishly hard to fathom. The investigations that have been launched—the Pathfinder mission to Mars last year is the most prominent recent example—have yielded images and data, but no evidence of life.

Small wonder, then, that scientists and the public alike were exhilarated by the stunning announcement of August 7, 1996: members of a research team led by the geochemist David S. McKay of the NASA Johnson Space Center in Houston, Texas, had discovered what they called convincing evidence for an ancient, microscopic form of life in a small piece of igneous rock that had fallen to earth from Mars.

The rock had been part of the Martian crust, which formed by cooling 4.5 billion years ago. About 16 million years ago, the impact of a meteorite uprooted the rock from the Martian surface and launched it into space. The rock wended its way through space until it fell into the gravitational influence of the earth: from that moment it was only a matter of time before it spiraled ever closer to our planet and finally fell to the surface—as it happened, onto the Uan Hills region of Antarctica, directly south of New Zealand, some 13,000 years ago. There it remained until an international team of geologists that searches each year for Antarctic meteorites picked it up on December 27, 1984. The rock languished for a decade, however, before investigators ascertained that it was from Mars, and more time passed while the analyses suggesting life were done.

It has been nearly two years since ALH84001, as it is known to planetary scientists, burst into the public consciousness. President Clinton announced the story from the White House that the meteorite might hold clues to life on Mars—and that event triggered a degree of interest in planetary exploration not seen since the glory days of NASA a quarter century ago. The interest, in my view, is entirely justified, and yet the four-pound-three-ounce potato-shaped rock that started all the excitement is probably not the "smoking gun" its proponents originally maintained it was.

The reported evidence for life within the rock is based primarily on cracks that developed

This article is reprinted by permission of The Sciences and is from the July/August 1998 issue. Individual subscriptions are $28 per year. Write to: The Sciences, 2 East 63rd Street, New York NY 10021.

early in its history, probably between 3.6 and four billion years ago. The cracks contain small patches, or blebs, of carbonate minerals deposited when warm water trickled through. The blebs are associated with chemical, mineralogical and microstructural features that on earth are often the products of biological activity. But continuing analyses, performed with an unprecedented degree of sophistication and precision, have failed to confirm (or definitively reject) the biological hypothesis. According to the most recent studies, particularly the ones conducted by a team led by the geochemist John P. Bradley of MVA Inc. in Norcross, Georgia, the Mars rock does not constitute compelling evidence of extraterrestrial life.

Those negative findings do not mean that the idea of life on Mars is a fantasy. On the contrary: reasonable inferences about the environment of Mars—liquid water likely once existed there, for instance, and water may persist below the surface as permafrost—combined with striking new discoveries of terrestrial organisms that thrive in habitats once thought inhospitable to life, lend credence to the idea that our celestial neighbor might now or in the past have supported a biosphere. Where, then, should the organizers of the next mission to Mars direct the spacecraft? And once there, how best to sample the planet for life.

The hunt for life on Mars has a long, and some would say credulous, history [see "Mission Impractical,'" by Norman H. Horowitz, March/April 1990]. At the end of the nineteenth century the American astronomer Percival Lowell built a special Mars observatory near Flagstaff, Arizona. Training his telescope on the red planet, Lowell discerned what he took to be a system of artificial canals for channeling water from the polar ice caps to more temperate regions of the planet. The canals, he theorized, had been built by intelligent beings as part of an ingenious strategy for surviving in the inhospitable Martian climate. Lowell's ideas inspired everything from advertising campaigns to serious scientific proposals for communicating with the canal builders. The public imagination was aroused—enough so that a slightly-too-realistic 1938 radio dramatization of H. G. Wells's *War of the Worlds* caused a mass panic in which millions of people became convinced that hostile Martians had invaded the earth.

In the 1950s a theory arose that vegetation existed on Mars, based on a wave of darkness that had been observed to envelop the planet on a seasonal basis. In time, however, both Lowell's canals and the vegetation theory were relegated to the realm of science fiction; sharper images of the planet made from satellite-mounted instruments portrayed a bleak landscape formed over the ages by volcanism, asteroid impacts and erosion.

This much is certain: conspicuous, intelligent beings have never evolved on Mars. But it was only recently, geologically speaking, that modern human beings evolved from what began 3.5 billion years ago as a stirring in the primeval oceans. The history of life on earth is principally the history of microorganisms, and so, as McKay and his colleagues rightly recognized, if one is to find signs of life on Mars, one should look not for little green men but for bacteria.

The creatures whose traces McKay and his colleagues asserted they had evidence for would have lived on Mars billions of years ago, when Mars is thought to have had a warm, wet climate quite different from the cold, desertlike one of today. McKay and his coworkers based their assertions on the presence of four mineral and organic substances detected, in close spatial association, in the Allan Hills meteorite:

• The precipitated carbonate minerals, which, as I noted earlier, are much like carbonate deposits associated with biological activity on earth;

• Distinctive grains of the mineral magnetite, which resemble the crystals formed within some terrestrial bacterial cells;

• Complex organic molecules, interpreted as evidence of the breakdown of biomolecules; and perhaps most striking, minuscule rodlike structures, interpreted as microfossils.

Essentially, McKay has advanced a series of arguments that, by his own acknowledgment, are not conclusive individually. Taken together, however, they portray an environment that might well have harbored life. McKay's approach is like that of a prosecutor who argues a criminal case on the basis of strong circumstantial evidence: no one saw Jones kill his wife, but based on the fact that he had threatened her, taken out a large insurance policy in her name and lied about his whereabouts on the fatal evening, the jury might find him guilty. Then again, it might not.

The carbonate precipitates are a good example of the logic in question. By themselves, they imply only that the cracks in the Allan Hills meteorite once served as a conduit for fluids that were supersaturated with carbonate minerals. Carbonate minerals—a group that includes common limestone—often form as a result of bacterial metabolism. For example, it is widely accepted that in marine sediments sulfate-reducing bacteria can change the pH of the water, causing calcium carbonate to settle out and accumulate as limestone. But carbonate minerals can also form under circumstances in which no biological influence is possible—deep in the interior of the earth, for instance, at temperatures of hundreds of degrees Fahrenheit. The temperature at which the Allan Hills carbonate minerals accumulated is much debated, but at least some investigators echo McKay in asserting that they formed within a temperature range conducive to life. Even if that is so, however, the only valid conclusion is that the carbonate precipitates are consistent with the presence of life—not that they require it.

The magnetite crystals, McKay's second piece of evidence, have equally controversial origins. Magnetite, a mineral made of oxygen and iron, is found within the bodies of a variety of organisms, from homing pigeons to whales, and is thought to act as a kind of internal compass, helping them navigate by sensing the magnetic field of the earth. Among those species are certain kinds of bacteria, which harbor a chain of crystallographically uniform magnetite grains within their cytoplasm. When such bacteria die, the magnetite crystals can be preserved in sediments. McKay and his colleagues argue that the Allan Hills magnetite crystals accumulated in just that way.

One problem with that scenario is that the magnetite crystals in the meteorite come in different sizes and shapes. Hence, a biological origin would imply that the traces were left by many different species of magnetite-containing microorganisms: an ancient bacterial community of surprising complexity and diversity. But Bradley and his colleagues have recently reported a finding that is even more damaging to the Martian life hypothesis. At least some of the magnetite grains in the meteorite have defects in their crystal structures—defects that are known to form only at unlivable temperatures of several hundred degrees.

McKay's third piece of evidence for ancient life on Mars is the presence of complex organic molecules in the meteorite. Known as polycyclic aromatic hydrocarbons, or PAHs, the molecules are found in small concentrations almost everywhere on earth. They often form through the breakdown of biological molecules: the burning or decomposition of plant matter, for instance. PAHs occur naturally in coal and oil (which accrue from the altered remains of plants, animals and microorganisms) and are released by smokestacks and automobile engines. PAHs also occur, however, in carbonaceous meteorites formed in the outer solar system, so their association with biology is not obligatory.

Critics have suggested that the PAHs in the Allan Hills meteorite may actually have originated on earth. Antarctic ice contains small amounts of them, and so in the thousands of years the meteorite was lodged in the ice sheet, meltwater from freeze and thaw cycles could have percolated through cracks in the rock, leaving PAHs behind. Indeed, the presence in the meteorite of carbon 14 and minute amounts of terrestrially derived amino acids—reported this past January—have confirmed that groundwater did infiltrate the rock after it landed on earth.

That line of criticism, however, now seems relatively moot. Recent investigations by the cosmochemist Simon J. Clemett (a member of McKay's team) and his colleagues at Stanford University have argued convincingly that, regardless of any terrestrial contamination, most of the PAHs in the Allan Hills meteorite did come from Mars. Clemett's team found that PAHs were more abundant deeper inside the meteorite, where the terrestrial contamination would not have reached.

A Martian origin for the PAHs, however, does not necessarily prove they have any connection with life. A team that included the geochemist Luann Becker, then at the Scripps Institution of Oceanography in La Jolla, California, compared ALH84001 with other Antarctic meteorites, including a second meteorite thought to have originated on Mars. The second meteorite is much younger than the Allan Hills rock. It appears to have formed relatively late in Martian natural history, long after liquid water ceased to be a prevalent feature and thus during a lifeless period on the planet's surface. Yet despite those differences, the second meteorite carried a suite of PAHs similar to that of the Allan Hills rock. Becker and her colleagues concluded that the PAHs in ALH84001 and its cousin came from some nonbiological source either during the rocks' relatively recent sojourn in Antarctica or millions of years ago, when carbonaceous meteorites and interplanetary dust particles were bombarding Mars. Ultimately, the lesson must be that because PAHs make up less than 1 percent of the organic carbon in the meteorite, and because they are not always associated with life, they are probably not terribly useful as biological markers.

What, then, about McKay's fourth piece of evidence: the elements interpreted as microfossils? The structures look like simple bacterial cells, except that they are exceedingly small: compared with them, Escherichia coli bacteria are as elephants to mice. In fact, the microbiologist Kenneth H. Nealson of the jet Propulsion Laboratory in Pasadena, California, calculated that the cell-like structures in the Allan Hills meteorite are so small that they could be made up of only a handful of molecules—whereas a one-celled organism on earth is made up of millions.

But McKay and his team have a clever rejoinder. The images they generated with a scanning electron microscope exploited the cutting edge of technology, showing unusually small objects at remarkable resolution. No one really knows, they contend, what could lie in store when comparable methods are applied to earthbound niches. Perhaps living among us are terrestrial bacteria so small they have managed to elude detection. Indeed, recent images made by the geologist Robert L. Folk of the University of Texas at Austin show minute round structures in many terrestrial limestones—though the origin of those spheres remains obscure.

To bolster its argument, the McKay team calls attention to recent reports that Finnish biologists have isolated unprecedentedly small bacteria from the human bloodstream. Such reports are intriguing, but not directly relevant. Like viruses, bacteria living parasitically in a host that fulfills most of their biochemical needs can afford to be remarkably small; free-living microorganisms cannot, because of the number of molecules needed for metabolism and reproduction.

It could be, of course, that the rodlike Allan Hills microstructures record life in some early stage of evolution—before the rise of molecular complexity—and thus are unlike anything known today on earth. Or perhaps they are the remnants of more conventional cells that shriveled during postmortem decay. But further work by Bradley and his colleagues discourages those ideas. The Bradley team has concluded that the structures trumpeted by McKay as microfossils are probably mineral grains instead, unrelated to any biological process. With transmission electron microscopy—which shows the interior structure of objects, not just their surfaces—Bradley and his group found magnetite crystals that were remarkably similar in size and shape to McKay's alleged microfossils.

McKay and colleagues counter that the history of the carbonate precipitates in the Allan Hills meteorite is complex and that Bradley and colleagues were not looking at the same structures. That possibility notwithstanding, Bradley and colleagues have still more recently noted that the alleged microfossils may actually be mineral ledges whose wormlike appearance has been accentuated by the treatment process that electron microscope samples undergo.

It is all too easy to become mired in an endless debate whose boundaries are constantly shifting. A Bradley says "Gotcha," and a McKay replies "You're not looking at what I was looking at." In this case, though, the burden of proof ultimately rests with McKay and others who maintain that ALH84001 carries traces of life. It is hard, of course, to disprove such claims definitively. But to hold up under cross-examination, the circumstantial evidence must be much more tightly constructed.

However adaptable, however variable life might be, its essential ingredients are well established: water, nutrients and a source of energy. Furthermore, those ingredients tend to be useful only in certain forms. Most of the familiar creatures on land, for instance, from ferns to giraffes, survive only on freshwater at temperatures from just below freezing to about 120 degrees F., and their energy comes, either directly or indirectly, from sunlight.

But life on earth is turning out to be far more diverse and tenacious than anyone suspected. In the past two decades workers have shown that even apparently barren habitats support remarkable communities. Many microorganisms, it turns out, can live at extraordinarily high temperatures up to 235 degrees F. (113 degrees Celsius). Furthermore, such microorganisms can extract the energy they need not from the sun but from chemical reactions in which hydrogen, sulfur or other simple substances are combined metabolically with carbon dioxide, oxygen or compounds of nitrogen and sulfur. Such compounds are readily available in hydrothermal environments such as the geysers and hot springs at Yellowstone National Park in Wyoming.

More recently, bacteria in abundance have been discovered in fluid-filled cracks in igneous rocks—sound familiar—more than a kilometer below the earth's surface. Some members of those unanticipated populations live by metabolizing the hydrogen gas given off in chemical reactions between minerals and water. Such populations are the first human contact with what may turn out to be a huge reservoir of biological activity within the outer crust of our planet. And diverse microorganisms also live at hydrothermal vents all along the oceanic ridge— a single system that sutures the planet along a continuous 46,000-mile-long path, like the stitching on an enormous baseball.

Described as hyperthermophilic because of their ability to survive and reproduce at high temperatures, many of those unfamiliar creatures cannot grow at the temperature used to pasteurize milk (162 degrees F.)—not because it is too hot but because it is too cold! The remarkable micromenagerie includes:

• *Sulfolobus acidocalderius*, a lobed ball between one and two microns in diameter that thrives in a hot acid environment and can extract or acquire the sulfur it needs by leaching it out of pyrite, or fool's gold;

• *Aquifex pyrophilus*, a similarly small organism that runs on the energy released when hydrogen is combined with trace amounts of oxygen to form water; and

• *Pyrodictium occultum*, a disk-shaped microorganism that sustains itself by converting sulfur and hydrogen to foul smelling hydrogen sulfide at temperatures as high as 234 degrees F. (112 degrees C.).

Had I described such physiologies on exams in the micro class I took in college, I would have failed the course. But it turns out that not only do such creatures live on earth, they occupy a privileged position. Hyperthermophilic organisms reside on the deepest known branches of the evolutionary tree, suggesting that the last common ancestor of all organisms alive today on earth may have lived in a hydrothermal environment.

The frontiers of life are expanding: the range of habitable environments now includes polar regions well below the freezing point, as well as scalding thermal springs. As biologists explore those frontiers, we often speak of such organisms, in our anthropocentric way, as "extreme" or "unusual." But in an evolutionary sense, it may be that we are the unusual ones. And those resilient hyperthermophilic lifeforms may provide the closest terrestrial analogy to any life that might have existed, and perhaps may persist, on Mars.

Even a cursory examination of the images transmitted last year by the Mars *Pathfinder* landing craft and its slow-moving robot *Sojourner* suggests that the ingredients required for life are in short supply on the surface of Mars. The *Pathfinder* analyses confirm and extend those made twenty years ago by *Pathfinder's* intellectual and technological predecessor, the Viking missions of 1976. The two Viking spacecraft found a landscape that is inhospitable even by the highly tolerant standards of bacteria. The Martian surface is highly oxidizing (hence its red color), essentially devoid of organic matter, and apparently without persistent bodies of water. It is subject to strong ultraviolet radiation, and it is extremely cold— the mean annual temperature at the Martian equator is about sixty degrees below zero F.

The Viking missions convinced many biologists that there is no life on Mars— although a more careful and correct interpretation of the data would be that life is unlikely to exist currently on the Martian surface. Such an interpretation leaves open the possibility of subsurface life. Perhaps—Just perhaps—life still thrives in subterranean oases where hydrothermal water circulates through volcanically heated crust—and if not now, then possibly in the past.

Although Viking failed to detect micro organisms, it did yield remarkable observations that have fueled continuing interest in Martian biology. Simply put, the Viking data suggest that whereas the surface of Mars today is very different from the surface of the earth, four billion years ago—when life was emerging on earth—the two planets had much more in common. Early in its history Mars had a thick atmosphere, abundant volcanism and, at least intermittently, liquid water. Under those conditions, the conclusion is inescapable that Mars harbored widespread hydrothermal activity both above and below ground. The Viking craft found what appear to be dry river channels on Mars, as well as evidence of flash flooding—large, striking geologic features that were almost certainly carved by running water.

Surprisingly, then, the scientific discipline best suited for finding evidence of life on Mars may be not microbial ecology but paleontology. Frequent encounters with dinosaurs in museums and movie theaters have assured the public that large creatures can leave an interpretable fossil record in the form of bones. But fossils of…bacteria? It turns out that some microorganisms form extracellular walls that, like mineralized skeletons, resist decay. When buried in fine-grained sediments, or entombed in precipitated minerals, such remains provide a remarkable record of the deep evolutionary history of life.

Not only can microscopic bacteria become fossils, but recently investigators have discovered molecular fossils: complex organic molecules derived from living organisms. Patterns in the relative amounts of various carbon isotopes present in limestones and sedimentary organic matter can point to a biological origin. Less direct but equally important traces of life are the stromatolites, which are laminated trace fossils of microbial mat communities.

I have spent half a lifetime scouting for fossils—in cliffs and rocky canyons in Australia, China, Panama, Siberia, Sweden and elsewhere. In South Africa I dug up the remains of single-celled organisms that lived 3.5 billion years ago, when life on earth was nothing but a bold and rather tenuous innovation. I have focused not on trilobites or mastodons, but rather on the earliest traces of life, and on the even deeper question such finds raise: How did life begin?

When, more than a decade ago, I joined NASA's research program in planetary biology, I realized that here lay the logical extension of my work. The idea of expanding the search for life not only back in time but to other planets is wildly exciting—but also quite reasonable. Mars, like the earth, has sedimentary rocks where the remains of organisms may have been buried. And that fossil record may actually be easier to read on Mars than on earth.

The reason is that Mars is smaller, and lacks the volatile internal heat sources that have given rise to the tectonic shifts of the continents of the earth. That constant shifting, the consequent smashup and erosion of the continents, and the continual upwellings of magma to form new terrestrial crust, have led to the destruction of much of the most ancient sedimentary record. On Mars, however, where the crust has remained more or less stationary, investigators can have access to a direct geological record of the earliest chemical and biological evolution.

Terrestrial experience suggests where and how to look for life on Mars. Both biology and paleontology indicate that investigators should seek evidence of water and hydrothermal mineral deposits. If life was ever present on Mars, it most likely left its calling card where muds dropped out of suspension in an ancient water body or where minerals such as calcium carbonate have accumulated at or near the surface of the planet.

But a landing site chosen according to the usual criteria—because it is flat or relatively easy to reach—probably will not lie within rover range of such prime targets for biological investigation. Like more conventional paleontologists, exopaleontologists who study Mars must begin by consulting maps that show the distribution of key features on the surface of the planet. Existing images show that biologically promising locations do exist—on Apollinaris Patera, for instance, an ancient volcano whose summit displays whitish patches that can be interpreted as precipitates formed from hot escaping gases; or at Dao Vallis, a large outflow channel on the flank of Hadriaca Patera, another ancient Martian volcano. Here water flowed out from the heated subsurface and could have sustained hydrothermal activity, leaving behind mineral deposits that could preserve microbial remains.

Careful mapping must be followed by meticulous field exploration. Early in the next century NASA is scheduled to launch another in its series of Mars lander missions. The payload will include an advanced technological descendant of *Sojourner*—a mechanical geologist able to make a wide range of maneuvers, images and analyses on command from earth. *Sojourner* looked at pretty much whatever rocks it stumbled across. The new ambulatory robot will travel about 100 meters a day (compared with a few meters for *Sojourner*); come equipped with more sophisticated photographic and chemical equipment; and, unlike Sojourner, have the ability to take core samples of rock.

The new rover will collect about 100 grams of such samples from Martian rocks and prepare them for retrieval by a later mission, scheduled for 2007.

When the new rover lands, I, along with a number of other scientists, will be stationed at the jet Propulsion Laboratory with the once-in-a-lifetime chance to see what the rover sees and instruct it to do what we want it to do. Fossil hunting is a tactile activity, and doing it remotely—with a screen for eyes and remote-controlled robotic limbs for hands—will not be easy. But until people can land on Mars, that seems the best way to go about the painstaking work of extraterrestrial paleontology.

Once rocks from Mars reach our earthbound laboratories what will we look for in them? There can be no more basic question, but the answer is far from simple. Paleontological studies of the early earth rely on knowledge of terrestrial biology. Presumably then, morphological, sedimentary or chemical patterns that organisms are known to form, and that do not derive from any known nonbiological processes, all count as evidence of ancient life. That second criterion is critical. It explains why fossils of dinosaurs, trilobites and even cyanobacteria are readily identified as biological, whereas the simple structures and chemical compounds in the Allan Hills meteorite are not.

Understanding the limits of physical and chemical processes is also important, because it is hard to know which features of terrestrial life are likely to prove general and which are the specific products of our own peculiar natural history. Most biologists agree that life elsewhere, if it exists, is likely to be based on the chemistry of carbon, and it may even rely on amino acids, nucleotides and sugars similar to the ones in our bodies. But photosynthesis, multicellularity and jointed legs are far from givens. And extraterrestrial paleontologists composing essays on word processors? Such things may be rare in the universe.

Whatever the outcome of continuing studies of Mars, David McKay and his colleagues have taken a bold leap, which has given us all a precious opportunity to road test our philosophical and analytical approaches to extraterrestrial samples, years in advance of the day when intelligently chosen chunks of Mars are rocketed back to the earth. More than that, McKay's work has resonated deep within the human spirit. It has crystallized our resolve to address one of humankind's oldest and most basic questions: Are we alone in the universe? It is our remarkable good fortune to be part of the first generation in history that has a chance of finding the answer.

## Questions:

1. What makes the new ambulatory robot (the more advanced descendant of *Sojourner*) more technologically advanced and better for geologists to use?

2. What are McKay's 4 lines of evidence for ancient life on Mars?

3. True or False. Mars at one time in its history had a thick atmosphere, abundant volcanism and, at least intermittently, liquid water.

Answers are at the back of the book.

## Activity:

Do some research and find out if there have been any new discoveries on this topic since this article was published in 1998 and share it with the rest of the class. Use any resource available. For starters you could try the following Website:
**http:// ispec.scibemet.com/lifemars.shtml** then select: NASA JSC-Evidence of Primitive Life from Mars.

# 13

Traditionally, since the discovery of Pluto, our solar system has contained nine planets. However, since Pluto is so small, only one five hundredth of earth's mass, many see it as an asteroidal object. The debate over Pluto's planetary status has been around since its discovery. For now, the International Astronomical Union will allow Pluto to remain the ninth planet.

# Pluto Reconsidered

By Kelly Beatty

In a 1995 film titled *The Englishman Who Went Up a Hill But Came Down a Mountain*, residents of a small Welsh town are shocked to learn that their beloved Ffynnon Garw—the first mountain in Wales—is 16 feet too short to be declared a mountain by government surveyors from England. Roused to action, the townspeople heap dirt upon the hill until it exceeds the 1,000-foot qualification for proper mountainhood.

Sometimes life imitates art, and in recent months a similar saga has played out among the world's astronomers. At issue is whether distant Pluto should properly be considered a planet, some lesser type of solar-system object, or both.

Last year Brian G. Marsden, who directs the Minor Planet Center in Cambridge, Massachusetts, suggested that Pluto be assigned the number 10,000 in the MPC's master list of asteroidal objects. But what began as a simple matter of celestial bookkeeping soon became rancorous. Widely interpreted as a move to diminish Pluto's status, the initiative drew howls of protest from various astronomical quarters and lots of publicity. By early February 1999 beleaguered officials of the International Astronomical Union (IAU) had reaffirmed that Pluto is a planet and shelved plans to add it to the minor-planet list.

Pluto's identity crisis is hardly new. Pointed questions about this little world's size and mass arose within weeks, of its discovery by Clyde Tombaugh in 1930. A significant adjustment followed the 1978 discovery of a companion moon, Charon, whose orbital parameters showed that Pluto had just 1/500 of Earth's mass. The issue of planetary worthiness again took center stage in 1996, when the Hubble Space Telescope provided the first clearly resolved features on Pluto's disk.

Even though Pluto has always been something of a misfit among its more massive siblings, the astronomical community at large has continued to count it among the major planets for more than six decades. However, the greatest challenge to Pluto's planethood came in 1992, when observers David Jewitt and Jane X. Luu spotted a sizable something orbiting out beyond Neptune. Their discovery of 1992 QB$_1$ did more than certify the existence of a trans-Neptunian cometary swarm, the Kuiper Belt, after decades of conjecture—it finally provided a raison d'être for Pluto. More and more, Pluto appeared to be simply the largest of an estimated 70,000 observable objects thought to be between 30 and 50 astronomical units of the Sun.

So is Pluto the ninth planet or not? Astronomers themselves have been sending a mixed message in recent years. As noted by Daniel W. E. Green, associate director of the IAU's Central Bureau for Astronomical Telegrams, dozens of professional papers, textbooks, and popular articles have questioned little Pluto's inclusion with the likes of its more

From "Pluto Reconsidered," by K. Beatty, Sky and Telescope, May 1999, pp. 48-52. Reprinted with permission.

massive neighbors. "It really is time that we stop teaching our children and students some old picture of the solar system from the 1940s and 1950s," Green wrote in 1997. A number of his colleagues agree. I firmly believe that if Pluto were discovered today, we wouldn't be calling it a planet," observes Harold F. Levison (Southwest Research Institute).

## An Offer of Dual Citizenship

Aside from an occasional pulse of news-media interest, however, the sticky issue of Pluto's planetary qualifications remained low key in the years following 1992 QB₁'s discovery. Meanwhile, a minor-planet milestone was approaching that would raise the stakes considerably and require the direct intervention of IAU higher ups. Asteroids were being discovered an cataloged so rapidly that the count of those with assigned numbers, a stamp of orbital reliability, would reach 10,000 in early 1999. Traditionally such "millennial" asteroids are given names of historic importance.

Why not give the coveted 10,000th assignment to Pluto, Marsden wondered, and follow it sequentially with numbers for some of the more secure Kuiper Belt finds? Discussions with his IAU bosses, Hans Rickman (Uppsala Astronomical Observatory) and Michael A'Hearn (University of Maryland), convinced Marsden that a kind of "dual citizenship" was viable. Pluto could remain a planet in the hearts and minds of astronomers everywhere, even as it (and other Kuiper Belt objects, were added to the master catalog of minor-planet orbits. He reasoned that keeping tabs on Pluto in this way might eliminate its inadvertent rediscovery by observers who occasionally mistake it for a faint new object.

Marsden has been trying to work Pluto into the minor planet catalog for a long time. In 1980 he made a pitch for reclassification at a meeting celebrating the golden anniversary of Pluto's discovery (with Tombaugh among the attendees). Noting the issue of too little mass, Marsden also pointed out that the planet-crossing orbit of Pluto was not unlike those of 2060 Chiron and 944 Hidalgo. "By demoting Pluto to minor-planet status, I do not mean to imply that it is lacking in significance," he later wrote in the journal *Icarus*, but he reiterated that continuing to call Pluto a major planet would be a mistake.

Momentum for the "10,000 Pluto" initiative built gradually and quietly throughout 1998. To become binding, it needed the approval of the IAUs Small Bodies Names Committee, whose 11 members include Marsden, Rickman, and A'Hearn. But word of this activity reached Richard P. Binzel (MIT), who has studied Pluto extensively. He objected to Marsden's notion of dual citizenship, arguing that it was "tantamount to officially demoting Pluto from its historical planetary status." Binzel proffered a counterproposal, suggesting that an entirely new catalog be created solely for Kuiper Belt objects. Pluto would remain a planet but be assigned "K/1," to acknowledge its status as king of the Kuiper Belt.

Marsden bristled at the charge that he was trying to lessen Pluto's status, retorting, "The people who claim it's a demotion are barking up the wrong tree." Moreover, he opposed creating of a new catalog, and a stalemate ensued when neither his proposal nor Binzel's won widespread acceptance. It fell to A'Hearn, as president of the IAU's Division III (Planetary Systems Sciences), to disentangle the Gordian knot of nomenclature. "There are no rules on how to deal with this kind of issue," he conceded. "We're making them up as we go along." Throughout October and November, the e-mail pathways sizzled with debate.

In December A'Hearn polled the division's 12-member executive committee, but a clear consensus was still lacking. Alerted to the ongoing deliberations by news stories and e-mail, other planetary scientists entered the fray. Even though none of the proposals before the IAU called for Pluto's demotion outright, some openly endorsed such action anyway. "Astronomers are admitting something they've known for a long time," David W. Hughes (University of Sheffield) told a British radio audience. Other researchers—particularly those involved in Pluto's study—were strongly opposed. "Jerking Pluto's planetary status is about as stupid as claiming it to be an escaped satellite of Neptune," countered Robert L. Marcialis (University of Arizona). Mostly, there was confusion over both the process and the participants. "The whole thing is a miasma," groaned Edward L. G. Bowell (Lowell Observatory).

While praising A'Hearn for his handling of the difficult situation, IAU officials found the strident opposition unsettling. "I have always been a supporter of numbering Pluto," Rickman says, "but right now it would be unwise for the IAU to support it." Even A'Hearn, who for years has thought and taught that Pluto doesn't fit in as a planet, eventually backed away from Marsden's proposal.

Ultimately Johannes Andersen, the union's General Secretary, stepped in to help resolve the four-month standoff. "The IAU is neither a parliament nor a court," he told *Sky & Telescope*. "For us to make an out-of-the-blue decision affecting the status of Pluto in the midst of a controversy would make no sense." On February 3rd, Andersen released a statement denying that the IAU had intended to change the status of Pluto and announcing that no minor-planet number would be assigned to it. Soon thereafter Binzel withdrew his plan for a Kuiper Belt list. The status quo had prevailed.

## Bigger Questions

Everyone involved at least agrees on one thing: the entire debate might have been avoided if astronomers knew a planet when they saw one. "The working definition at present," says A'Hearn, "is simply a cultural one: anything that we have called a planet for more than a few years. The recent fracas elicited such tongue-in-cheek reassessments as whether Mars is really an asteroid ("it's a source of meteorites") or Jupiter a star ("all that hydrogen and helium").

Notably, consensus is near on what defines the upper end of the planetary spectrum: an object massive enough to fuse deuterium in its core is widely considered to be "substellar," a brown dwarf. However, settling the lower limit is fraught with pitfalls—especially if the criteria are set so as to include Tombaugh's famous find.

As shown above, Pluto is hardly the massive perturber envisioned prior to its discovery. Instead, it is outsized by seven of the solar system's natural satellites, including the Moon. Yet Pluto is at least 10 times more massive than the largest asteroid, 1 Ceres, and it dwarfs all known Kuiper Belt finds. Moreover, it has an atmosphere and a substantial moon. "Pluto is demonstrably a planet by any reasonable criterion," argues S. Alan Stern (Southwest Research Institute). He adds that Pluto may once have been significantly larger, perhaps having several times its present mass, before a giant collision cleaved off Charon and sent sizable chunks flying away into similar orbits.

Some astronomers have backed a simple definition for planethood: anything more than 1,000 km across. By this standard, Pluto is "in" and Ceres, 933 km across, is "out" (barring any additions by future Welsh astronauts). However, many find the 1,000-km yardstick too arbitrary and lacking any physical basis. One suggestion with many adherents is that anything large enough to shape itself into a sphere should be a planet—a qualification satisfied by Pluto as well as big asteroids like Ceres. "Loose definitions serve us better sometimes, especially when we're finding so many objects that don't fit," observes Mark V. Sykes (University of Arizona).

Pluto aside, the conflicts over planetary status will likely arise again—even in our own solar system. Stern calculates that many yet-to-be-discovered objects in the Kuiper Belt should equal or even exceed the ninth planet in size. When and if they are found, another referendum on Pluto's planethood may occur. For now, however, IAU officials are siding with tradition.

---

### Questions:

1. What has been the greatest challenge to Pluto's planetary status?
2. What defines a planet at the upper end of the planetary spectrum?
3. What are some of the simple definitions of a planet?

### Activity:

Using viable internet sources write a page about the terrestrial and atmospheric characteristics of Pluto. Do you think it should remain a planet?

# PART 4
# Resources and Pollution

# 14

There are many different species of fish, mammals, birds and reptiles that may become extinct at the hands of humans. The leading factor contributing to the decline of these species is habitat loss, by the conversion and disruption of ecosystems such as forests, wetlands, deserts, riverways, grasslands and so on. In 1992 the Convention of Biological Diversity was begun and signed by 169 countries. These countries must develop a system, individually, to protect our lands. To ensure protection, environmental education programs must be employed and native peoples must be taught ways to make a living while protecting natural treasures. Humans have the power to change the fate of these environmentally important species if we act now.

# Sharing the Planet: Can Humans and Nature Coexist?

By John Tuxill and Christopher Bright

Like the dinosaurs 65,000,000 years ago, humanity finds itself in the midst of a mass extinction, a global evolutionary convulsion with few parallels in the entire history of life. Unlike the dinosaurs, though, humans are not simply the contemporaries of a mass extinction—they are the reason for it.

The loss of species touches everyone, no matter where or how they live. Earth's endowment of species provides humanity with food, fiber, and many other products and "natural services" for which there is no substitute.

About 25% of drugs prescribed in the U.S. include chemical compounds derived from wild organisms, and billions of people worldwide rely on plant- and animal-based traditional medicine for their primary health care. Biodiversity provides a wealth of genes essential for maintaining the vigor of crops and livestock. It provides pollination services, mostly in the form of insects, without which we could not feed ourselves. Frogs, fish, and birds provide natural pest control; mussels and other aquatic organisms cleanse our water supplies; and plants and microorganisms create our soils.

Birds, mammals, reptiles, amphibians, and fish constitute the vertebrate animals, distinguished from invertebrates by an internal skeleton and a spinal column—a type of anatomy that permits, among other things, complex neural development and high metabolic rates.

Vertebrates combined total about 50,000 species and can be found in virtually all environments on Earth, from the frozen expanses of Antarctica to scorching deserts and deep ocean abysses. By virtue of the attention they receive from researchers, vertebrates can serve as ecological bellwethers for the multitude of small, obscure organisms that remain undescribed and unknown. Since vertebrates tend to be relatively large and to occupy the top rungs in food chains, habitats healthy enough to maintain a full complement of native vertebrates will have a good chance of retaining the invertebrates, plants, fungi, and other small or more obscure organisms found there. Conversely, ecological degradation often can be read most clearly in native vertebrate population trends.

## Disappearing Birds

Estimates are that at least two out of every three bird species are in decline worldwide. Four percent—403 species—are endangered.

The most threatened major groups include rails and cranes, parrots, terrestrial game birds (pheasant, partridges, grouse, and guans), and pelagic seabirds (albatrosses, petrels, and shearwaters). About one-quarter of the species in each of these groups is endangered. While just nine percent of songbirds are threatened, they still contribute the single largest group of endangered species because they are far and away the most species-rich bird order.

The leading culprits in the decline of birds are a familiar set of interrelated factors, all linked to human activity: habitat alteration, overhunting, exotic species invasions, and chemical pollution of the environment. Habitat loss is by far the leading factor—at least three-quarters of all threatened bird species are in trouble because of the transformation and fragmentation of forests, wetlands, grasslands, and other unique habitats by human activities, including intensive agriculture, heavy livestock grazing, commercial forestry, and suburban sprawl.

In some cases, habitat alteration is intensive and large-scale, as when an internationally funded development project converts large areas of native forest to plantation crops, or a large dam drowns a unique river basin. In other instances, habitat is eroded gradually over time, as when a native grassland is fragmented into smaller and smaller patches by farming communities expanding under a growing population.

High concentrations of gravely endangered birds are found on oceanic islands worldwide. Birds endemic to insular habitats account for almost one-third of all threatened species and 84% of all historically known extinctions. Since island birds often are concentrated in just a handful of populations, if one such group is wiped out by a temporary catastrophe such as a drought, the birds usually have few population sources from which they can recolonize the formerly occupied habitat. Equally important is that many island birds have evolved in isolation for thousands or even millions of years. Such species are particularly vulnerable to human hunting, as well as predation and competition from non-native, invasive species. (Invasives are highly adaptable animals and plants that spread outside their native ecological ranges—usually with intentional or inadvertent human help—and thrive in human-disturbed habitats.)

No island birds have been more decimated than those of Hawaii. Virtually all of its original 90-odd bird species were found nowhere else in the world. Barely one-third of the species remain alive today, and two-thirds of these continue to be threatened with extinction. The degree of ecological disruption is so great that all lowland Hawaiian songbirds are non-native species introduced by humans.

An equally disturbing trend is population declines in more widespread species, particularly those that migrate seasonally between breeding and wintering grounds. Long-term population declines are tied to a host of contributing hazards. Habitat loss squeezes species on both breeding and wintering grounds, as well as at key stopping points—such as rich tidal estuaries for shorebirds—along their migratory routes. In North America, the loss of almost half of all wetlands has been a major factor behind a 30% drop in the population of the continent's 10 most abundant duck species.

Exposure to chemical pollution is another hazard many birds face. The greatest risk of pesticide and pollution exposure occurs in developing countries, where many chemicals banned from use in industrial nations continue to be applied or discharged indiscriminately.

The decline of migratory birds is sobering because it is a loss not just of individual species, but of an entire ecological phenomenon. Present-day migrants must negotiate their way across thousands of miles of tattered and frayed ecological landscapes. The fact that many birds continue to make this journey, despite the threats and obstacles, is cause for hope and inspiration. Yet, as long as bird diversity and numbers continue to spiral downward, there can be no rest in the effort to protect and restore breeding grounds, wintering areas, and key refueling sites that all birds—migratory and resident—can not live without.

## Mammals' Dark Future

About 25% of all mammal species are treading a path that, if followed unchecked, is likely to end in their disappearance from Earth. Out of almost 4,400 mammal species, about 11% already are endangered. Another 14% remain vulnerable to extinction, including the Siberian musk deer, whose populations in Russia have fallen 70% during this decade due to increased hunting to feed the booming trade in musk, used in perfumes and traditional Asian medicine. An additional 14% of mammal species tend to have larger population sizes or be relatively widespread, but nonetheless face pressures that have them on the fast track to threatened status in the not-too-distant future.

Among major mammalian groups, nearly half of primate species (lemurs, monkeys, and apes) are threatened with extinction. Also under severe pressure are hoofed mammals (deer, antelope, horses, rhinos, camels, and pigs), with 37% threatened; insectivores (shrews, hedgehogs, and moles), with 36%; and marsupials (opossums, wallabies, and wombats) and cetaceans (whales and porpoises), at 33% each. In slightly better shape are bats and carnivores (dogs, cats, weasels, bears, raccoons, hyenas, and mongooses), at 26% apiece. Rodents are the least-threatened mammalian group, at 17%, and the most diverse.

The biggest culprit in the loss of mammalian diversity in the late 20th century is the same as that for birds—habitat loss and degradation. As humankind converts forests, grasslands, riverways, wetlands, and deserts for intensive agriculture, tree plantations, industrial development, and transportation networks, many mammals are relegated to precarious existences in fragmented, remnant habitat patches that are more ecological shadows of their former selves.

The loss of habitat also afflicts marine mammals, though it usually proceeds as gradual, cumulative declines in habitat quality rather than wholesale conversion of ecosystems (as when a forest is replaced by a housing development). Marine mammals, particularly those that inhabit densely populated coastal areas, have to content with polluted water and food, physical hazards from fishing gear, heavy competition from humans for the fish stocks on which they feed, and dangerous, noisy boat traffic.

In addition to habitat loss, at least one in five threatened mammals faces direct overexploitation
—excessive hunting for meat, hides, tusks, and medicinal products—and persecution as predators of and competitors with fish and livestock. Overexploitation tends to affect larger mammals disproportionately over smaller ones,

and when strong market demand exists for a mammal's meat, hide, horns, tusks, or bones, species can decline on catastrophic scales.

While the drastic population crashes of great whales, elephants, and rhinos are well-known, the long shadow of overexploitation actually reaches much further. For instance, only the most remote or best-protected forests throughout Latin America have avoided significant loss of tapirs, white-lipped peccaries, jaguars, wooly and spider monkeys, and other large mammals that face heavy hunting pressure from rural residents. Much of this hunting is for home subsistence—wild game meat is an important source of protein in the diets of rural residents, particularly for indigenous people. One estimate pegs the annual mammal take in the Amazon Basin at more than 14,000,000.

Throughout south and east Asia, a major factor fueling excessive wildlife exploitation is the demand for animal parts in traditional medicine. Tigers—the largest of all cats—once ranged from Turkey to Bali and the Russian Far East, and have been the subject of organized conservation projects for more than two decades. At first, these projects appeared to be having some success—until the mid 1980s brought a burgeoning demand in east Asia for tiger parts as aphrodisiacs and medicinal products. With the body parts of a single tiger potentially worth as much as $5,000,000, illegal hunting skyrocketed, particularly in the tiger's stronghold—India. Wild tigers now total barely 3-5,000, many in small isolated populations that will be doomed without more intensive protection.

The loss of a region's top predators or dominant herbivores is particularly damaging because it can trigger a cascade of disruptions in the ecological relationships among species that maintain an ecosystem's diversity and function. Large mammals tend to exert inordinate influence within their ecological communities by consuming and dispersing seeds, creating unique microhabitats, and regulating populations of prey species. In Côte d'Ivoire, Ghana, Liberia, and Uganda, certain trees—including valuable timber species—have shown reduced regeneration after the crash of elephant populations, which the trees depend on for seed dispersal. Similarly, decades of excessive whaling reduced the number of whales that died natural deaths in the open oceans. This may have adversely affected unique deep-sea communities of worms and other invertebrates that decompose the remains of dead whales after they have sunk to the ocean floor.

Mammals in most regions have been less susceptible than birds to invasive species, but there is one big exception—the unique marsupial and rodent fauna of Australia, long isolated from other continents. The introduction of non-native rabbits, foxes, cats, rats, and other animals has combined with changing land use patterns during the past two centuries to give Australia the world's worst modern record of mammalian extinction.

## Reptiles in Retreat

Among reptiles, species are declining for reasons similar to those affecting birds and mammals. Habitat loss is again the leading factor, contributing to the decline of 68% of all threatened reptile species. In island regions, habitat degradation has combined with exotic species to fuel the decline of many unique reptiles. In Ecuador's Galápagos archipelago, the largest native herbivores are reptiles—long-isolated giant tortoises and land and marine iguanas found nowhere else in the world. Introduced goats are winning out over the native reptiles, however, and these interlopers have eliminated unique populations of tortoises on three of 14 islands within the Galápagos chain. At least two other tortoise populations are in imminent danger.

In addition, 31% of threatened reptiles are affected directly by hunting and capture by humans. This figure may be somewhat inflated since the reptile groups most thoroughly assessed—turtles and crocodilians—are among those most pursued by humans. Nevertheless, the high percentage is a clear indication of the heavy exploitation suffered by these species.

The plight of sea turtles has been studied and publicized since at least the 1960s, and all seven species are judged to be endangered, with many populations continuing to dwindle. Although there has been progress on protecting sea turtles at some of their best-known nesting grounds, illegal poaching for meat and eggs remains a widespread problem. Where beaches are lit at night with artificial lights, as at tourist resorts, hatchling turtles become disoriented and crawl toward the land rather than toward the sea. Moreover, sea turtles continue to suffer inadvertent, but significant, mortality from nets set for fish and shrimp.

Although less well-known than their seagoing relatives, tortoise and river turtle species are exploited intensively in certain regions, to the point where many populations are depleted greatly. Tortoises and river turtles throughout Southeast Asia long have been an important source of meat and eggs for local residents. There is a burgeoning international trade in these species to China, where they are used in traditional medicine.

Certain species of crocodilians still suffer from overhunting (such as black caimans in the Amazon Basin) and from pollution (such as the Indian gharial and the Chinese alligator), but this is one of the few taxonomic groups of animals whose over-all fate actually has improved over the past two decades. Since 1971, seven alligator and crocodile species have been taken off the endangered list, including Africa's Nile crocodile and Australia's huge estuarine crocodile. In part, these recoveries are due to the development of crocodile ranching operations that harvest the animals for their meat and hides. When combined with effective wildlife protection efforts, this can take hunting pressure off wild populations.

Habitat loss remains a serious issue, affecting some 58% of threatened amphibians. Much of this is due to the drainage, conversion, and contamination of wetland habitats. In addition, the spread of road networks and vehicular traffic leads to increased amphibian mortality that can decimate local populations.

In recent years, amphibians have captured worldwide attention due to the rapid and unexplained decline—and, in some cases, even extinction—of frog species in relatively pristine, intact ecosystems where habitat loss is not a factor. These mysterious decreases have been well-documented among frogs in little-disturbed mountain habitats in Central America and the western U.S., as well as in 14 species of rainforest-dwelling frogs in eastern Australia.

Researchers have advanced various explanations for these declines, including disease epidemics caused by invasive pathogens; increases in ultraviolet radiation, which inhibits egg development; introduced predators, particularly game fish like bass and trout; acid rain and other industrial pollutants; and unusual climatic fluctuations, such as extended drought.

Most likely, it is not a single factor, but, rather, synergistic combinations that best explain the declines. For instance, the presence of industrial pollutants may stress and weaken frogs and make them more susceptible to infectious diseases. It may be that frogs, with their highly permeable skins and life cycles dependent on both aquatic and terrestrial habitats, are signaling—more clearly than any other group of organisms—the gradual, but global, decline of the planet's environmental health.

## Fish Under Siege

The causes of fish endangerments—habitat alternation, exotic species, and direct exploitation
—are no different from those affecting other species, but they appear to be more pervasive in aquatic ecosystems. Freshwater hot-spots of fish endangerment tend to be large rivers heavily disturbed by human activity (such as the Missouri, Columbia, and Yangtze) and unique habitats that are host to endemic fish faunas, such as tropical peat swamps, semi-arid stream systems, and isolated large lakes. Saltwater hot-spots include estuaries, heavily disturbed coral reefs, and other shallow, near-shore habitats.

Although degradation of terrestrial habitats such as forests may be more obvious and get the most attention, freshwater aquatic habitats receive an even heavier blow from humanity. More that 40,000 large dams and hundreds of thousands of smaller barriers plug up the world's rivers—altering water temperatures, sediment loads, seasonal flow patterns, and other river characteristics to which native fish are adapted. Levees disconnect rivers from their floodplains, eliminating backwaters and wetlands that are important fish spawning grounds.

Agricultural and industrial pollution of waterways further reduces habitat for fish and other aquatic life. Agricultural runoff in the Mississippi River basin is so extensive that, when the river enters the Gulf of Mexico, the overfertilized brew of nutrients it carries sparks huge algal blooms, depleting the water of oxygen and creating a "dead zone" nearly the size of New Jersey.

Moreover, many fish species face a high degree of exploitation from commercial fisheries, particularly marine fish and species like salmon that migrate between salt and fresh water. About 68% of all threatened marine species suffer from overexploitation. The days when experts thought it impossible to deplete marine fish populations are long gone, and scientists now realize that overexploitation is a serious extinction threat for many ecologically sensitive species.

Seahorses, for example, are captured for use in aquariums, as curios, and in traditional Chinese medicine. Sharks are a second group of marine fish headed for trouble. They are valued for their skin, meat, cartilage (reputed to have anti-cancer properties), liver oil, and especially fins, which are among the highest-valued seafood commodities due to their popularity in east Asian cuisine.

Sturgeon, one of the most ancient fish lineages, occur in Europe, northern Asia, and North America, and long have been harvested for their eggs, famous as the world's premier caviar. Russia and Central Asia are home to 14 sturgeon species—tops in the world—and produce 90% of the world's caviar, mostly from the Black and Caspian Sea regions. The sturgeon fishery was relatively well-regulated during the Soviet era, but massive water projects and widespread water pollution led to sturgeon population crashes, so that all 14 species are highly endangered.

With the collapse of native fish faunas in many river basins and lake systems, and with growing awareness that many marine fish are in decline, the evidence suggests that biological diversity is faring no better underwater than on land. As noted, one out of every three fish species appears to be on the path to extinction. If this percentage holds up, it portends a grim future for other aquatic life on Earth.

If the trends evident in vertebrates hold for other organisms, extinction would appear to be a near-term possibility for about a quarter of the world's entire complement of species. If the current scientific consensus on the rate and scale of climate change proves accurate, over the next century, natural communities will face a set of unprecedented pressures. A warmer climate probably will mean changes in seasonal timing, rainfall patterns, ocean currents, and various other parts of the Earth's life-support system.

In the evolutionary past, the ecological effects of abrupt climate shifts were cushioned somewhat by the possibility of movement. One part of a plant's or animal's range might dry out and become uninhabitable, but another area might grow more moist and become available for colonization. Today, with more and more species confined to fragmented remnants of their former range, this kind of compensatory migration is less and less likely.

In the face of current and expected declines, the world's governments have clear moral and practical reasons to act. One course of action should involve pursuit of the process begun at the 1992 Earth Summit in Rio de Janeiro, which resulted in the Convention on Biological Diversity (CBD), signed by 169 countries. This and other environmental treaties provide important forums for coordinating international responses to biodiversity issues. To some degree, they can function as a sort of international mechanism for self-policing.

Obviously, treaties are only as effective as the will and competence of signatory countries permit. The CBD requires all participant countries to prepare national strategies for conserving their biodiversity. Because of its comprehensiveness, it represents the most thorough test to date of the international community's will to face up to the biodiversity crisis. Nevertheless, the primary cause of that crisis—habitat loss—is likely to escape the CBD in large measure, as it has most other treaties. Habitat loss is an issue that must be solved mainly on a national and local level.

The main approach that countries have taken to safeguard habitat has been to establish systems of national parks, wildlife refuges, forest reserves, marine sanctuaries, and other formally protected areas. Protected lands safeguard some of the Earth's greatest natural treasures and have made a big difference for some "conservation-dependent" vertebrates that otherwise almost certainly would be sliding into extinction. These include about 40 species of eastern and southern Africa, such as giraffes, hyenas, wildebeest, and impala. The populations of these animals presently are out of danger, in large part because of an extensive reserve system in their home countries. Yet, despite these notable successes, current networks of protected areas are nowhere near capable of saving most biodiversity.

A major shortcoming of the reserve system is a lack of implementation. Many parks exist on paper, but completely are unprotected on the ground. These "paper parks" are most common in developing countries, which hold the bulk of the world's biodiversity, but have the least in the way of money or expertise to devote to managing protected areas. As a result, many officially designated reserves are subject to agricultural development, mining, extensive poaching, and other forms of degradation.

The "paper park" syndrome reflects the lack of a wider social commitment to protect biodiversity and wildlands. Without such a commitment—or a viable plan for generating it—more funding alone is unlikely to improve matters significantly. The tactics for building that commitment will require two basic strategies:

• Environmental education programs must be built into school curricula (preferably beginning at an early age) to help people understand the complexity and intrinsic value of natural communities.

• Practical, culturally sensitive development initiatives are needed that can help local people make a living from nature without permanently damaging it. Well-planned ecotourism projects can play such a role, for example, as can "biodiversity prospecting"—the search for species that might yield new chemicals, drug precursors, genes, or other beneficial products.

The biggest opportunities from this dual strategy perhaps can be seen where biological diversity meets social diversity. A great deal of the natural wealth that conservationists seek to protect is on land and under waters long managed by local people. Communities throughout Asia and Africa, as well as the indigenous cultures of the Americas, traditionally have protected many forests, mountains, and rivers as sacred sites and ceremonial centers. Such peoples often have a great fund of pragmatic knowledge, too. They know how the local weather works, which organisms produce powerful chemicals, and what grows where.

Some of the biggest mistakes in natural areas conservation have involved the forcible removal of long-term residents from newly designated parks. Relocating such individuals or denying them access to traditional plant and animal resources has generated a great deal of ill will toward protected areas worldwide. In some cases, local people have reacted by purposefully neglecting plants and animals they previously had managed wisely for generations. Even in cases where communities have expressed a willingness to move out of a protected area voluntarily—say, to obtain better schooling for their children or improved medical care—governments often have not kept their promises to provide land and housing equal to what the relocated residents left behind.

To give biodiversity and wildlands breathing space, humans must find ways to reduce the size of the imprint on the planet. That means stabilizing and ultimately reducing the human population. It means far greater efficiency in materials and energy use. It means intelligently planned communities. It means educational standards that build an awareness of our responsibility in managing 3,200,000,000 years' worth of biological wealth. Ultimately, it means replacing our consumer culture with a less materialist and far more environmentally literate way of life.

Humans, after all, are not dinosaurs. We can change. Even in the midst of this mass extinction, we still largely control our destiny, but only if we act now. The fate of untold numbers of species depends on it. So does the fate of our children, in ways we barely can begin to conceive.

## Questions:

1. What is a 'paper park?'
2. What is meant by invasive species?
3. What consequences occur due to the loss of top predators or dominant herbivores?

Answers are at the back of the book.

## Activity:

We will use the Internet for this activity. Go to the 'Earth Wise Journeys' Website (Earth friendly Travel for Discovery of our Global Community) at:
**http://www.greenmoney.com/gmg/ecotour.htm**
Discover some of the places/countries that provide ecotours and the diverse array of plants and animals that can be observed there. Discuss these with the rest of the class.

15

In the past, economists and environmentalists have usually been viewed as having different perspectives on environmental issues. However, environmentalists have begun to appreciate the role economics can play in social solutions to these issues. Today, the U.S. economy fails to include all the costs of production, especially the environmental costs. These types of market failures are evident in the subsidization of the timber and oil industries, and cost the taxpayers millions of dollars. Green taxes offer a strategy to include these costs and to transfer the burden to those activities that incur environmental costs.

# Sharing the Wealth

By Brian Dunkiel, M. Jeff Hamond and Jim Motavalli

When the *Exxon Valdez* went aground in Prince William Sound in 1989, spilling millions of gallons of oil, it caused grievous environmental damage which will never be fully erased. But Exxon absolved itself of future responsibility with a $1 billion settlement. And because of the current tax laws, Exxon could deduct that settlement as a business expense, sticking taxpayers with $250 million of the cleanup.

All too often, the nation's tax policy is in direct conflict with environmental goals, including efforts to protect habitat and biodiversity. Few environmentalists give tax policy much attention, yet the tax code and budget policy in general may be the largest influences on conservation efforts. One tax break to the oil industry can create the opportunity and financial incentive to launch drilling expeditions in several sensitive habitats.

Our current internal revenue tax code dates back to colonial times and reflects colonial attitudes. It's based on the philosophy of people like Louis XIV's financial advisor, Jean Baptiste Colbert, who once said, "The art of taxation consists in so plucking the goose as to obtain the largest possible amount of feathers with the smallest possible amount of hissing." The tax code has since increased only in complexity, not

in outlook. The original income tax law was 14 pages long; today, the code encompasses 7.5 million words, in 9,000 individual sections. Only four taxes are explicitly identified as "environmental," but many others have significant effects on the natural world.

When our nation was young, the emphasis was on opening up what seemed like a limitless wilderness, and little thought was given to natural resources. Despite our growing awareness of the long-term costs of environmental degradation, tax priorities haven't changed. In 1998, according to a Friends of the Earth report called *Dirty Little Secrets: Polluters Save While People Pay*, anti-environmental tax breaks, like those that subsidize oil exploration and logging in national forests, cost the nation $20 billion in a five-year period.

Pollution, for the most part, is a business write-off. Eliminate these incentives and the resulting revenue would equal the federal income tax paid by 12 million low-income Americans, or the populations of both Arkansas and Montana.

Some tax shelters hit close to home. Professor Oliver Houck of Tulane University points out that the tax provision that allows Americans to deduct mortgage interest paid on second homes is a major impediment to the protection of threatened and endangered species. The subsidy, the largest U.S. tax break for development, cost the U.S. Treasury $43 billion in lost revenue in 1998. It is an economic motivator for vacation home construction, encouraging more of them (along with roads and related amenities) to be built in pristine or environmentally-sensitive regions. Such favorable circumstances also encourage the construction of larger homes, on larger pieces of land, with consequently longer access roads.

Oil and gas tax subsidies save the industries $1.3 billion per year, with oil companies getting subsidies for drilling on federal land and for exploring in deep oceans. One tax break allows large oil and gas producers to immediately deduct 70 percent of their "intangible" drilling and development costs—including expenses for wages, fuel, repairs, hauling, supplies and site preparation. The remaining 30 percent of costs can be deducted over five years. These quick tax deductions allow oil and gas companies to depreciate their assets much faster than they actually wear out, making new drilling projects very attractive and providing a $2.6 billion subsidy over five years.

Further incentives for the industry allow oil and gas investors to deduct losses even when they weren't substantially involved in the actual operation—a so-called "passive loss" deduction, which was eliminated for all other industries as part of the 1986 Tax Reform Act. This special loophole cost Americans $295 million in the last half decade.

In the real world, lucrative subsidies like these persuade oil and gas companies to drill in pristine areas they would otherwise leave alone, like the St. George Basin, off the west coast of Alaska in the Bering Sea. The region, a gateway for virtually every marine mammal, bird and fish species migrating through to the North Pacific, was estimated to have only a 28 percent chance of harboring commercially-viable offshore oil. Yet oil companies were willing to pay almost $500 million for exploration there, a dubious business proposition that makes sense only in light of the generous tax advantages.

The timber industry, like the oil and gas industries, enjoys its own special tax breaks that will cost ordinary taxpayers about $1.1 billion over the next five years. But many tax-shifting advocates believe that, rather than eliminating these tax breaks altogether, they should be reformed to reward sustainable timber practices.

But not all perverse polluter subsidies go to big corporations. Soccer moms and other sport-utility vehicle (SUV) owners are among the beneficiaries of the SUV exemption from the gas guzzler tax, which would run as high as $7,700 for a Lincoln Navigator. The tax code encourages the production and purchase of trucks and SUVs, now 51 percent of all vehicles sold, even though they spew 30 percent more carbon monoxide and hydrocarbons, and 75 percent more nitrous oxides, than ordinary cars. SUV sales are, of course, encouraged by extremely low gasoline prices, resulting in part from our timidity in imposing energy taxes.

## Toxic Taxes

Mining companies also share the wealth; they are allowed to deduct exploration costs in the year they were incurred, rather than spreading them over the lifetime of the property (the usual practice for business investments). The practice encourages environmentally-destructive mining activity that would otherwise be dismissed as not economically viable. In recent years, mining companies have claimed more than $100 million annually in such deductions.

Mining is hardly an environmental activity; it irreparably scars the landscape and pollutes surface and ground water, destroying the habitats of many species of plants and animals, including those listed as endangered. There are more than 550,000 abandoned mines spread over 32 states, and many of them are listed on the Superfund National Priority List, with estimated cleanup costs in the billions, burdening taxpayers.

Another tax break for mining companies allows them to automatically deduct a certain percentage from their gross income to reflect their mines' dwindling value over time. This fixed depreciation, known as the percentage depletion allowance, results in a five to 22 percent reduction in annual taxable value (depending on the substance mined). The highest deductions are actually given to the most dangerous, toxic substances, including uranium, lead, mercury and asbestos, creating absurd contradictions in governmental environmental policy. Mining companies often recoup more money through this tax loophole than they actually invest in the mine! Government estimates show that taxpayers have, in effect, paid $1.5 billion in subsidies for mining operations through this deduction.

While taxpayers are subsidizing lead mining, local public health and environmental agencies are struggling with a vast children's health crisis caused by pervasive lead poisoning. Such poisoning, declining now because lead was removed from motor fuels, still affects nearly nine percent of U.S. preschoolers, 1.7 million kids. Federal agencies spend $200 million a year on prevention and testing programs. Mercury-contaminated fish have sparked another health crisis, even as the federal government subsidizes mercury mining.

In contrast to the tax breaks for extractive industries, solar and geothermal research and development received only $58 million in subsidies, $22.4 for pollution to every $1 for technological research and development. Other big polluters also get a free ride—the trucking industry, for instance, pays only 65 percent of what it should in taxes, according to some environmental groups. Agribusiness is favored with tax provisions that hurt small farmers and discourage sustainable and organic agriculture.

## The Tax Shift

Environmentalists are proposing a "tax shift" to redirect the incentives in the tax code. The goal, as *The Ecology of Commerce* author Paul Hawken puts it, is to give people and companies positive incentives to avoid taxation. The green economists would purge the tax code of regulations and loopholes that clearly encourage environmental degradation, such as the $17 billion cost of tax-free parking. New levies would be applied on pollution-generators like products containing lead, gas guzzling cars, ozone-depleting chemicals and the burning of fossil fuels. Taxes would be judged on their real contribution to the economy, in terms of job creation and productivity growth, equity for the people paying them, and resource conservation.

New and progressively-graduated taxes could shift 10 percent of the federal tax burden in the next 10 to 20 years. As defined by Alan Thein Durning and Yoram Bauman in their book *Tax Shift*, and by Redefining Progress in *Tax Waste, Not Work*, the levies could include:

• **Carbon taxes** to decrease the generation of greenhouse gases threatening worldwide climatic change. Governments could impose a tax—say, $50 per ton of carbon emissions—or combine a smaller tax with user fees or revenues from the sale of pollution credits;

• **Pollution taxes** to reduce the contaminants flowing into our rivers and streams, filling our landfills and eroding the quality of our soil. There are an estimated 250 human-made chemicals harbored in the living tissue of the average American;

• **Point source taxes** to reduce pollutants pouring forth from the outflow pipes and smokestacks of sewage treatment plants, factories and incinerators;

• **Traffic taxes** in the form of tolls imposed strictly during rush-hour congestion periods, could promote the use of carpools and mass-transit, as well as flextime work hours.

Tax incentives would be offered to invest in energy efficiency and technological improvements. For example, a system of new taxes and permits may dramatically reduce global warming gases and nonpoint source pollution, which is toxic runoff into rivers and streams. Taxes could be levied on development resulting in loss of biodiversity in wilderness areas. Other new taxes, some of which have already been proposed on the state level, are levies on carbon dioxide emissions and gasoline, taxes on pollutants, taxes on virgin materials and increased fees for using public resources.

Reforming tax laws is an important tool for conservation activists because tax policy is a blunt instrument and its influence on behavior sweeps broadly. Tax policy does not tell people or corporations how to behave; it merely creates incentives or disincentives, and it's a tool to complement environmental laws. Tax-based policies shift green reforms from "end of pipe" penalty solutions to economic incentives, a move that businesses should support because it reduces their regulatory burden.

## Tax Confusion

Despite this, the public is confused about green taxes. Though a 1998 public opinion poll shows that an impressive 71 percent of American voters favor tax shifting as a way to reform the system (a sentiment that crosses party lines), voters have often been confused by well-financed media campaigns. In 1993, President Clinton proposed an energy tax that could have helped foster wind and solar power while reducing the budget deficit. The tax was defeated by two votes, however, because voters were convinced it was a bad idea by a $6 million oil and gas industry campaign. "Back then, we were novices at this," admits Gawain Kripki of Friends of the Earth.

"But now the fine tuning and public education is ready."

One part of that public education effort is to demonstrate to Americans that they're getting unfairly taxed for working. More than 70 percent of U.S. families pay more in payroll taxes to support Social Security and Medicare than they do in federal income taxes. "Instead of taxing payrolls, America should tax pollution," says Redefining Progress founder Ted Halstead, who reckons that new levies on polluting industries could yield hundreds of billions of dollars in new revenues that could "strengthen our economy, boost wages and job creation, fix our troubled tax system and protect the environment, all without raising the deficit. What more could Americans want from a tax plan?"

The payroll tax is strikingly regressive. A worker making $30,000 a year is taxed at 15.3 percent, while his $300,000 a year boss pays only 5.7 percent of his income.

Some conservatives, who largely support flat taxes and other "reforms" championed by the rich, are cool to environmental tax shifting. Dan Mitchell, an economist with the Washington, D.C.-based Heritage Foundation, argues that "environmentalists think all energy consumption is bad. But I have no shame or embarrassment about energy consumption because we need energy to help our economy grow." Like most conservatives, Mitchell inherently opposes any new taxation, but even he begrudgingly admits that green taxes have some validity. "If I were held down with a gun to my head and my left arm about to be cut off and I had to raise taxes, there's little question that a tax on pollution or emissions would be the way to go," he says.

Though most politicians pay lip service to the environment, getting a sweeping tax shift through Congress is no easy task. Treasury Secretary Robert Reich, for instance, likes the concept of green taxes, but doesn't see it becoming policy soon. "I wish I could be optimistic, but politically, it's a very hard sell," he says. "Energy states are very powerful in Congress."

A congressional economist, who asked not to be identified, says the "clever" thing about tax shifting is that it doesn't shake out as a new burden on middle-class taxpayers. "You can make the average person break even, or come out ahead," she says. But she's skeptical about a tax shift's effect on the overall economy. "Can you reduce pollution without encouraging less production, and therefore less labor?" she asks, perhaps with the fate of laid-off coal miners in mind. Nada Eissa, a professor of economics at the University of California at Berkeley, thinks we can achieve that desirable end—if the right reforms are put in place. "An environmental tax where revenues are recycled to reduce payroll or income taxes could increase labor supply," she writes in a paper published by Redefining Progress.

Andrew Hoerner of the Washington, D.C.-based Center for a Sustainable Economy thinks that shifting the tax burden from work to pollution will boost the job prospects of working families all over the world, particularly in Europe, where environmental tax reform is becoming popular. "In the past," he says, "environmental, economic and social justice concerns have been seen as competitors for a limited pool of resources. Environmental tax reform is a member of an emerging family of policy approaches that harmonize these concerns by simultaneously promoting all of them."

## The View From Europe

As the newly elected German government—a coalition of Social Democrats and Greens—takes control, another European nation will rev up its economy and protect the environment by shifting taxes onto pollution and taking pressure off existing taxes. Burdened by high social security taxes and unemployment, several European countries either have or are discussing increasing taxes on polluting fuels (like coal and gasoline), and using these revenues to reduce taxes on labor and employment to stimulate job creation.

"Germany is finally catching up with ecological tax reformers like the Netherlands, Denmark, Sweden and Norway," says Kai Schlegelmilch, project manager of the Climate Policy Division at Germany's Wuppertal Institute for Climate, Environment and Energy. Will the U.S. follow Germany's lead? Schlegelmilch fears that increasing taxes in the U.S. is almost like committing political suicide." He suggests that "a very important first step is to phase out all environmentally-damaging subsidies, such as those for fossil fuels. But it has to be accepted that higher energy prices must follow." That may doom the reform right there.

The powerful chairman of the House Ways and Means Committee, Bill Archer (R-TX), comes from oil country and views eliminating the billions of dollars in tax breaks enjoyed by oil and gas companies as a tax hike.

There are few Europe-wide green taxes, but many individual countries have successfully introduced them. Four main categories of green taxes are already in use:

• **Fiscal environmental taxes** are levied on waste and emissions. Sweden has placed taxes on carbon-based fuels and carbon dioxide emissions, as well as emissions from domestic airline flights. Denmark has instituted taxes to reduce waste generation and increase recycling and reuse.

• **Incentive changes** encourage less polluting actions by taxing bad ones. Sweden taxes leaded gas and polluting diesel fuels, and synthetic fertilizer use. Nitrate emissions are charged at a rate four times that of regular emissions. France and Germany tax water pollution; revenues then are used to build new and better wastewater treatment plants.

• **Cost-covering charges** are levied on users for general waste and pollution. Germany has a tax on items as varied as hazardous wastes and disposable fast-food packaging. The Netherlands has a water pollution user tax, revenues from which build water treatment plants. It also taxes household wastes. Great Britain has a landfill tax, and uses the revenue to reduce assessments on payrolls.

• **Specific cost-covering charges** may also be levied, on everything from batteries to aircraft noise. France has implemented a sulfur dioxide tax and landfill fees, with the funds flowing into environmental investments. Denmark taxes pesticides, herbicides and fungicides, and is proposing taxes on the use of toxic heavy metals and chlorinated solvents. In January, Switzerland began taxing volatile organic compounds (VOCs) in order to reduce ground-level ozone; and high-sulfur heating oil will be taxed beginning July 1, with revenues going into the national health insurance fund. Other countries adopting green tax approaches include Korea, Taiwan and Singapore.

## A Growing Consensus

The growing international agreement to shift the tax burden to polluting industries has as yet had little impact on the U.S., though green seeds have been planted in the states. The federal government raises $1.5 trillion in revenue each year, almost 90 percent of which comes from taxes on payroll, personal and corporate income. Through the incentives it creates, the tax code necessarily exerts a strong influence over the behavior of consumers, investors and businesses. But, all too often, these incentives lead to environmental harm.

Reforming the tax system now would have two benefits: The tax system would complement environmental regulation rather than frustrating it (as it now does); and the tax code could harness market forces so that they work for the environment, not against it. Given the large impact that taxes have, they could be a powerful tool for promoting sustainable development, while actually helping the economy and supporting labor. As the Worldwatch Institute puts it, "For progressives, [tax shifting] has the appeal of protecting the environment by making the polluter pay and reducing unemployment. For conservatives, it offers the advantage of using the market, rather than regulatory agencies, to protect the environment, and allows for cuts in much-resented income or sales taxes that may inhibit constructive economic activity." And who could have a problem with that? CONTACT: Center for a Sustainable Economy, 1731 Connecticut Avenue NW, Suite 500, Washington, DC 20009/(202)234-9665; Friends of the Earth, 1025 Vermont Avenue NW, Washington, DC 20005-6303/(202)783-7400; Redefining Progress, 1 Kearney Street, Fourth Floor, San Francisco, CA 94108/(415)781-1191.

**Questions:**

1. What are some of the polluting industries and activities that are subsidized by our economy?
2. Why was Clinton's proposed energy tax defeated in 1993?
3. What are the four main categories of taxes already in use in many European countries? List and explain.

Answers are at the back of the book.

**Activity:**

Go to your local gas station and find out how much of the price per gallon for gas goes to taxes.

**16**

It has been suggested that lack of water is already constraining agricultural output in many parts of the world and could be a serious problem by 2025. Rain-fed land will become more important to global food security and improving the efficiency of irrigation can increase agricultural water productivity. By using efficient sprinklers, drip systems and other methods of irrigating water more directly to the crops roots can decrease evaporation. Efforts are necessary to raise water productivity of the global crop base, both rain-fed and irrigated.

# Water for Food Production: Will There Be Enough in 2025?

By Sandra L. Postel

This year marks the 200th anniversary of the publication of Thomas Malthus's famous essay postulating that human population growth would outstrip the earth's food-producing capabilities. His writing sparked a debate that has waxed and waned over the last two centuries but has never disappeared completely. Stated simply, Malthus's proposition was that because population grows exponentially while food supplies expand linearly, the former would eventually outpace the latter. He predicted that hunger, disease, and famine would result, leading to higher death rates.

One of the missing pieces in Malthus's analysis was the power of science and technology to boost land productivity and thereby push back the limits imposed by a finite amount of cropland. It was only in the twentieth century that scientific research led to marked increases in agricultural productivity. Major advances, such as the large-scale production of nitrogen fertilizers and the breeding of high-yield wheat and rice varieties, have boosted crop yields and enabled food production to rise along with the world population (Dyson 1996). Between 1950 and 1995, human numbers increased by 122% (US Bureau of the Census 1996), while the area planted in grain expanded

by only 17% (USDA 1996, 1997c). It was a 141% increase in grainland productivity, supplemented with greater fish harvests and larger livestock herds, that allowed food supplies to keep pace with population and diets for a significant portion of humanity to improve.

Despite this remarkable success, concern about future food prospects has risen in recent years because of a marked slowdown in the growth of world grain yields, combined with an anticipated doubling of global food demand between 1995 and 2025 (McCalla 1994, FAO 1996). Whereas annual grain yields (expressed as three-year averages) rose 2-2.5% per year during every decade since 1950, they registered growth of only 0.7% per year during the first half of the 1990s (Brown 1997, USDA 1997a, 1997b). Excluding the former Soviet Union, where the political breakup and economic reforms led to large drops in productivity, global grain yields increased an average of 1.1% per year from 1990 to 1995, approximately one-half the rate of the previous four decades (Brown 1997). Today, the principal difference between those analysts projecting adequate food supplies in 2025 and those anticipating significant shortfalls is the assumed level of productivity growth—specifically, whether annual productivity over the next three decades is likely to grow at closer to the 1% rate of the 1990s or the 2-2.5% rate of the previous four decades.

Water—along with climate, soil fertility, the choice of crops grown, and the genetic potential of those crops—is a key determinant of land productivity. Adequate moisture in the root zone of crops is essential to achieving both maximum yield and production stability from season to season. A growing body of evidence suggests that lack of water is already constraining agricultural output in many parts of the world (Postel 1996, UNCSD 1997). Yet to date, I am aware of no global food assessment that systematically addresses how much water will be required to produce the food supplies of 2025 and whether that water will be available where and when it is needed. As a result, the nature and severity of water constraints remain ill defined, which, in turn, is hampering the development of appropriate water and agricultural strategies.

In this article, I estimate the volume of water currently consumed in producing the world's food, how much additional water it will take to satisfy new food demands in 2025, and how much of this water will likely need to come from irrigation. I then place this expected irrigation demand in the context of global and regional water availability and trends. Finally, I discuss the policy and investment implications that emerge from the analysis.

## Total Water Consumed in Food Production

The volume of water consumed in producing current food supplies is much larger than estimates of agricultural water use typically suggest. These estimates have focused almost exclusively from the volume of water removed from rivers, lakes, and underground aquifers for irrigation. They typically neglect the soil moisture derived directly from rainfall that is consumed by agricultural cropping systems, pastures, and grazing lands. This omission is perhaps understandable, given that such rainfed lands do not require investments in dams, canals, and other water infrastructure and do not figure into projected demands on regional water supplies. Yet it results in an incomplete and misleading picture of the volume of water actually used to produce the world's food—and, by extension, of future water requirements for food production.

**Water consumed by crops and croplands.** In general, there is a linear relationship between a crop's water consumption and its dry matter yield up to the point at which water is no longer limiting (Sinclair et al. 1984). The amount of dry matter produced per unit of water transpired—which is known variously as a crop's water use efficiency or transpiration ratio—is the slope of this linear relationship, and it varies by crop, climate, and other factors. For example, climatic and other conditions being equal, $C_4$ crops, such as maize, tend to use water more efficiently than other grains because of their special anatomical and biochemical characteristics. A crop grown in a drier climate will transpire faster than the same crop grown in a more humid climate because of the larger vapor pressure gradient between the plant's stomata and the atmosphere. Thus, the volume of water a given crop uses will vary by crop type, climate, season, and other factors, but the basic linear relationship between dry matter production and transpiration generally holds for all crops and growing environments (Kramer and Boyer 1995).

In determining the amount of water consumed in producing the global food supply, several additional factors must be taken into account. Water is consumed not only through transpiration but also through evaporation from the soil and leaf surfaces. Under field conditions, evaporation is difficult to measure separately from transpiration, so the processes are typically referred to jointly as evapotranspiration. In addition, because only the edible portion of a crop contributes to food supplies, the portion of a crop's dry matter that is actually harvested (known as the harvest index) must also be taken into account. The water-use efficiency of the harvested yield is expressed as the harvested crop yield per unit of water evapotranspired and is often denoted by Ey. These values are shown in Table 1, along with the total 1995 production of each crop or crop category. Lacking detailed regional data, the estimated global crop water requirements shown in Table 1 were derived by multiplying the inverse of the midpoint of the Ey value for each crop or crop category by the 1995 global production of that crop. This calculation results in an estimated minimum water requirement for the 1995 global harvest of crops of approximately 3200 km³ (3200 billion cubic meters).

Not surprisingly, wheat, rice, maize, and other grains—the staples of the human diet and also sources of feed for livestock—account for more than 60% of the total crop evapotranspiration requirement. Soybeans and other oilseed crops account for 17% of this requirement, and sugar cane alone accounts for approximately 6%. It is important to emphasize that the values in Table 1 do not reflect how much water is *actually* consumed in crop production but rather the *minimum required* for that production. Inefficiencies in irrigation that result in evaporative losses, for example, are not taken into account; I address such additional consumptive uses of water in a later section.

The plants from which the world's food commodities are harvested represent only a portion of total cropland biomass. The net photosynthetic product of the world's croplands has been estimated at $15 \times 10^9$ t/yr (Ajtay *et al.* 1979, Vitousek *et al.* 1986). Assuming that an average of 2 g biomass is produced per 1 L of water evapotranspired (Monteith 1990, Postel *et al.* 1996), a total of 7500 km$^3$ would be consumed through evapotranspiration in cropland ecosystems—more than twice the estimated evapotranspiration of the crop plants themselves. Because crop production depends on the productivity of the supporting ecosystem, this higher figure may more accurately reflect the total amount of water consumed through evapotranspiration on the world's croplands.

**Water consumed by converted pasture and grazing land.** The world's domesticated animals—including 1.3 billion cattle, 900 million pigs, and more than 12 billion chickens (FAO 1996)—contribute meat, milk, eggs, and other items to the human diet. Of the 2700 kilocalories available per capita per day on average worldwide (FAO 1995), approximately 16% comes from animal products. However, this share varies greatly by country and region: For example, 32% of the estimated 3410 calories per capita per day available in Europe comes from animal products, compared with just 7% of the average 2282 kilocalories per capita per day available in Africa (FAO 1995).

Livestock variously eat grass, hay, feed grain, and food waste. Although the feed grain and food waste are included in the crop production figures in Table 1, a separate calculation needs to be made to account for evapotranspiration on converted pasture and grazing land. Again, assuming an average biomass production rate of 2 g/L of water, the estimated water consumption occurring on pasture and rangeland totals 5800 km$^3$/yr.

**Non-beneficial evapotranspiration of irrigation water and from aquaculture ponds.** Irrigated lands—those receiving artificial water applications to supplement natural rainfall—totaled $249.5 \times 10^6$ ha in 1994, the most recent year for which data are available (FAO 1996). Because irrigation makes possible more than one harvest a year on the same parcel of land and allows farmers greater control over the watering of their crops, these lands are disproportionately important in global food production; they represent just 17% of the world's total cropland area but yield on the order of 40% of the world's food (Rangeley 1987, Yudelman 1994).

Shiklomanov (1996) estimated that in 1995 a total of approximately 2500 km$^3$ was withdrawn from rivers, lakes, and aquifers for irrigation. However, a portion of this water never benefits a crop. Some of it is lost to evapotranspiration as the water is stored in ponds or reservoirs, transported by canals, and applied to farmers' fields. Water percolating into the soil through unlined canals or running off the end of a farmer's field also represents inefficiency and can degrade both land and water quality. But because this water is not evapotranspired, it is theoretically available to be used again and so is not counted as a loss. No good global estimate of nonbeneficial irrigation water losses exists, but they may amount to approximately 20% of the volume withdrawn (Perry 1996). Applying this figure to the 1995 estimate of irrigation withdrawals suggests unproductive evapotranspiration losses of 500 km$^3$.

Water also evaporates from ponds used in fish farming, an increasing source of protein worldwide. These evaporation losses are difficult to estimate because aquaculture production can occur in coastal bays or estuaries, indoor tanks, or artificial ponds. Currently, evaporation from ponds is negligible relative to the total water consumed in food production. Yet fish farming is growing rapidly: Aquaculture production tripled between 1984 and 1995, from $7 \times 10^6$ t/yr to $21 \times 10^6$ t/yr, and in 1995 it accounted for 19% of the global fish harvest (McGinn 1997). As aquaculture expands, pond evaporation will increase and may factor significantly into the water budgets of water-short areas.

Summing the estimated volumes of water consumed by cropping systems, grasslands and pasture, and non-beneficial evaporation of irrigation supplies yields an estimate of total water consumption for food production in 1995 of 13,800 km$^3$/yr—or nearly 20% of the total annual evapotranspiration occurring on the earth's land surface. For the 1995 population of 5.7 billion (PRB 1995), this global total translates to an annual average of approximately 2420 m$^3$ per capita.

## Changing Structure of Global Food Sources

The structure and sources of the global food supply in 2025 will not be simply an extrapolation of past trends. Serious constraints exist on the expansion of grazing land, fisheries, and cropland, which suggests that most of the additional food required in the future will need to come from higher productivity on existing cropland. This shift has important implications for the volume and sources of water that will be required to satisfy future food needs.

**Rangeland constraints.** According to a global assessment of soil degradation (Oldeman *et al.* 1991), overgrazing has degraded some 680 x 10$^6$ ha of the world's rangelands since midcentury. This finding suggests that 20% of the world's pasture and range is losing productivity and will continue to do so unless herd sizes are reduced or more sustainable livestock practices are put into place. With the global ruminant livestock herd, now numbering about 3.3 billion, unlikely to increase appreciably, most of the increase in meat production will need to come from grain-fed livestock.

**Fisheries constraints.** The wild fish catch from marine and inland waters totaled 91 x 10$^6$ t in 1995, little more than in the late 1980s. On a per capita basis, the 1995 global fish catch was down nearly 8% from the 1988 peak (McGinn 1997). With the United Nations Food and Agriculture Organization (FAO 1993) reporting that all 17 of the world's major fishing areas have either reached or exceeded their natural limits, no growth can be expected in the oceanic catch. Aquaculture, the most rapidly growing source of fish, now accounts for one of every five fish consumed, a share that is expected to increase (McGinn 1997). Although fish is a more water efficient source of animal protein than virtually any other grain-fed source, the expansion of aquaculture will increase pressures on both cropland and water supplies in the future.

**Cropland constraints.** With production from both rangelands and fisheries reaching natural limits, most of the increased food supply in 2025 will need to come from cropland. However, on a net basis, cropland area is unlikely to increase appreciably. As much as 10$^7$ ha may be lost each year due to erosion, other forms of degradation, or conversion to nonfarm uses (Leach 1995, Pimentel *et al.* 1995). Because such losses are often not fully counted in official statistics—which show that cropland expanded an average of 1.6 x 10$^6$ ha/yr between 1979 and 1994 (FAO 1996)—net cropland expansion could well be close to zero or even negative. Moreover, possibilities for opening up new cropland are mostly in areas in which the long-term crop production potential is relatively low and the biodiversity and other ecological costs are very high, such as in Brazil and central Africa.

**Implications for future water requirements.** By definition, the water requirements of rain-fed crops are met by rainfall, which is supplied freely by nature and rarely counted in estimates of global agricultural water use. With net cropland area unlikely to expand much if at all, the potential for increased use of direct rainfall to meet crop evapotranspiration requirements is limited largely to improving the productivity of rainwater on existing croplands, both irrigated and rain-fed. Terracing, mulching, contour bunding (placing stones or vegetation along contours), and other methods of capturing rainwater to enhance soil moisture have proven effective at increasing yields of rain-fed crops (Unger and Stewart 1983, Critchley 1991, Reij 1991). Rain-fed production may also benefit from greater focus on boosting total crop output from the land—for example, through agroforestry and synergistic inter-cropping—as opposed to boosting the yields of single crops.

Globally, the volume of water available for crop evapotranspiration will need to roughly double by 2025 if total crop production is to double. Although actual crop water requirements in 2025 will depend on the crop mix, the climate under which crops are grown, changes in the harvest index, and other factors, a doubling is a reasonable assumption. Because net cropland area is likely to expand minimally if at all, I assume no increase in the water use of related cropland biomass and focus on the direct evapotranspiration requirements of crops, an estimated 6400 km$^3$ in 2025.

How this additional water for crop evapotranspiration will be partitioned between rainfall and irrigation is impossible to project, especially given that the current partitioning of the crop water supply can be approximated only roughly. However, if 40% of the global harvest currently comes from irrigated land and if, on average, 70% of the soil moisture on this irrigated land comes from irrigation water (the other 30% comes directly from rainfall), then irrigation water would account for about 900 km$^3$ of the 3200 km$^3$ required for crop evapotranspiration in 1995; the other 2300 km$^3$ would have been supplied directly from rainfall. It seems reasonable to assume that modest cropland expansion and enhanced rainwater productivity might allow productive use of rainfall for crop evapotranspiration to increase by 50% between 1995 and 2025. To satisfy the global crop water requirement in 2025, the volume of irrigation water consumed by crops would thus need to more than triple—from an estimated 900 km$^3$ in 1995 to 2950 km$^3$—and irrigation's share of total crop water consumption would rise from 28% to 46%. The volume of irrigation water annually available to crops as soil moisture would need to expand by 2050 km$^3$ equivalent to the annual flow of 24 Nile Rivers or 110 Colorado Rivers.

## Prospects for Supplying the Needed Irrigation Water

Current trends in water use and availability strongly suggest that supplying an *additional* 2050 km$^3$ per year for consumptive agricultural use on a sustainable basis will be extremely difficult. A variety of trends and indicators signal that water constraints on agriculture are already emerging, both globally and regionally.

**The global demand-supply outlook.** Of the 40,700 km$^3$ that run to the sea each year in rivers and aquifers, only an estimated 12,500 km$^3$ are actually accessible for human use, of which human activities already appropriate an estimated 54% (Postel *et al.* 1996). By 2025, water withdrawals for irrigation could approach 4600 km$^3$/yr, assuming 3500 km$^3$/yr of consumptive use (both beneficial and nonbeneficial) and somewhat higher irrigation efficiency than at present. In addition, estimates by the Russian hydrologist Igor Shiklomanov (1993) suggest that worldwide household,

municipal, and industrial water uses currently average approximately 240 m$^3$/yr per capita. Greater use of more efficient household and industrial technologies could reduce this per capita requirement substantially (Postel 1992), but the resulting savings would be partially offset by the water needed to meet minimum drinking and household requirements of the more than 1 billion people now lacking them (Gleick 1996).

Assuming an average global per capita household, municipal, and industrial water use of 200 m$^3$/yr, the combined demand in these sectors would total some 1640 km$^3$ in 2025. Adding this amount to estimated irrigation withdrawals and reservoir losses suggests that global withdrawals in 2025 could total 6515 km$^3$. This estimate exceeds by 26% that of Shiklomanov (1996), in large part because of the higher global irrigation water requirement that emerges from the more detailed crop-water analysis carried out in this study.

Adding in greater instream flow needs to dilute pollution, human appropriation of accessible runoff in 2025 could exceed 70%, up from just over 50% at present, even with fairly optimistic assumptions about supply expansion (Postel *et al.* 1996). Both the dams and other infrastructure built to meet the higher demand, as well as the high level of human co-option of the supplies available, would cause much greater loss of valuable freshwater ecosystem services (Postel and Carpenter 1997), further decline of fisheries, and more rapid extinction of species that depend on aquatic ecosystems.

**Global irrigation trends.** Worldwide growth of irrigated area has dropped from an average of 2% per year between 1970 and 1982 to 1.3% per year between 1982 and 1994 and shows no sign of picking up speed. Rising construction costs for new irrigation projects and the declining number of ecologically and socially sound sites for the construction of dams and river diversions have led international donor institutions and governments to reduce irrigation investments. Irrigation lending by the four major donors—the World Bank, the Asian Development Bank, the US Agency for International Development, and the Japanese Overseas Economic Cooperation Fund—peaked in the late 1970s and dropped by nearly half over the next decade (Rosegrant 1997). Governments in many Asian countries—including China, the Philippines, Bangladesh,

India, Indonesia, and Thailand—also cut back irrigation investments substantially during the 1980s. Although private investment has countered this trend somewhat, irrigation worldwide has been growing at a slower pace than population: Per capita irrigated area peaked in 1978 and fell 7% by 1994, the latest year for which data are available (Gardner 1997).

At the same time, the steady buildup of salts in irrigated soils is leading to a decline in the productivity of a portion of the existing irrigation base. Estimates suggest that salinization affects 20% of irrigated lands worldwide (Ghassemi et al. 1995) and may be severe enough on 10% of these lands to be reducing crop yields. Spreading at a rate of up to $2 \times 10^6$ ha annually (Umali 1993), salinization is offsetting a portion of the gains achieved by bringing new lands under irrigation. Together, spreading soil salinization and the declining rate in the expansion of irrigation have contributed significantly to the decline in grain yield growth witnessed during the first half of the 1990s.

**Regional signs of water depletion and unsustainable use.** Groundwater overpumping and aquifer depletion now plague many of the world's most important food-producing regions, including the north plain of China, the Punjab of India, portions of Southeast Asia, large areas of north Africa and the Middle East, and much of the western United States (Postel 1996). Falling water tables not only signal limits on the ability to expand future groundwater use but also indicate that a portion of the world's current food supply depends on water that is used unsustainably—and therefore cannot be counted as a reliable portion of the world's longterm food supply. Saudi Arabia, which as recently as 1994 was producing nearly $5 \times 10^6$ t of wheat by mining nonrenewable groundwater, illustrates this point well: When fiscal problems led the government to reduce the subsidies that had propped up this unsustainable wheat production, Saudi grain output plummeted 62% in two years, falling to $1.9 \times 10^6$ t in 1996 (USDA 1997a).

Many of the planet's major rivers are showing signs of overexploitation as well, adding to the evidence that it will be difficult to greatly increase agricultural water supplies. In Asia, where the majority of world population growth and additional food needs will be centered, many rivers are completely tapped out during the drier part of the year, when irrigation is so essential. According to a World Bank study (Frederiksen et al. 1993), essentially no water is released to the sea during a large portion of the dry season in many basins in Asia. These include the Ganges and most rivers in India, China's Huang He (Yellow River), Thailand's Chao Phraya, and the Amu Dar'ya and Syr Dar'ya in central Asia's Aral Sea basin. The Nile River in northeast Africa and the Colorado River in southwestern North America discharge little or no freshwater to the sea in most years (Postel 1996).

**Increasing competition for water.** Even as limits to tapping additional water supplies are appearing, agriculture is losing some of its existing water supplies to cities as population growth and urbanization push up urban water demands. The number of urban dwellers worldwide is likely to double to 5 billion by 2025. This trend will increase pressure to shift water out of agriculture to supply drinking water to growing cities, as is already happening in China, the western United States, parts of India, and other water-short areas.

In addition, rising public concern about the loss of fisheries, the extinction of aquatic species, and the overall decline of freshwater ecosystems is generating political pressure to shift water from agriculture to the natural environment, particularly in wealthier countries. In the United States, for example, the US Congress passed legislation in 1992 that dedicates $987 \times 10^6$ ml of water annually from the Central Valley Project in California, one of the nation's largest federal irrigation projects, to maintaining fish and wildlife habitat and other ecosystem functions. Among the objectives of the Central Valley Project Improvement Act is restoring the natural production of salmon and other anadromous fish to twice their average levels over the past 25 years (Gray 1994).

Further evidence of heightened competition for irrigation water comes from a county-level analysis of the 17 western US states (Moore et al. 1996), which found agricultural activities to be a factor in the decline of 50 fish species listed under the Endangered Species Act (ESA). This analysis also found that 235 counties contained irrigated land that drew water supplies from rivers harboring ESA-listed fish species. These findings suggest that US irrigated agriculture may face more widespread water losses because of legal obligations to protect species at risk.

## Water, Population, and the Global Grain Trade

Finally, a growing imbalance between population size and available water supplies is eliminating the option of food self-sufficiency in more and more countries. As annual runoff levels drop below 1700 $m^3$ per person, food self-sufficiency becomes difficult, if not impossible, in most countries. Below this level, there is typically not enough water available to meet the demands of industries, cities, and households; to dilute pollution; to satisfy other ecological functions; and to grow sufficient food for the entire population. Thus, countries begin to import water indirectly, in the form of grain.

Of the 34 countries in Africa, Asia, and the Middle East that have annual per capita runoff levels below 1700 $m^3$, all but two (South Africa and Syria) are net grain importers; 24 (70%) of these countries already import at least 20% of their grain. Collectively, their annual net grain imports, averaged over 1994-1996, totaled 48 x $10^6$ t, which suggests that water scarcity is to some degree driving about one-fourth of the global grain trade. With approximately 1500 ml of water required to grow 1 t of grain in these countries (higher than the global average because of the higher evapotranspiration rates in drier climates; FAO 1997), these annual grain imports represent 72 $km^3$ of water.

As populations grow, per capita water supplies will drop below 1700 $m^3$ per year in more countries, and countries that are already on the list of so-called water-stressed countries will acquire more people. By 2025, 10 more African countries will join the list, as will India, Pakistan, and several other Asian nations; China will only narrowly miss doing so. Given current population projections (PRB 1997), the total number of people living in water-stressed African, Asian, and Middle Eastern countries will climb 6.5-fold by 2025, from approximately 470 million to more than 3 billion. With nearly 40% of the projected 2025 population living in countries whose water supplies are too limited for food self-sufficiency, dependence on grain imports is bound to deepen and spread.

## Conclusions and Implications

Water availability will be a serious constraint to achieving the food requirements projected for 2025. The need for irrigation water is likely to be greater than currently anticipated, and the available supply of it less than anticipated. Groundwater overdrafting, salinization of soils, and re-allocation of water from agriculture to cities and aquatic ecosystems will combine to limit irrigated crop production in many important food-producing regions. At the same time, more and more countries will see their populations exceed the level that can be fully sustained by available water supplies.

The common presumption that international trade will fill emerging food gaps deserves more careful scrutiny. With each 1 t of grain representing approximately 1000 t of water, water-stressed countries will increasingly turn to grain imports to balance their water budgets. The majority of people living in water-stressed countries in 2025 will be in Africa and South Asia, home to most of the 1 billion people who are currently living in acute poverty (UNDP 1996) and the 840 million people who are currently malnourished (FAO 1996). It is questionable whether exportable food surpluses will be both sufficient and affordable for poor food-importing countries.

Given the limited potential for sustainable increases in cropland area and the mounting barriers to expanding irrigated area, measures are urgently needed to ensure that the best rain-fed land now in production remains in production. Rain-fed land does not compete directly with urban and industrial uses for water in the way that irrigated land does. In a world of deepening water scarcity, rain-fed land will thus become increasingly important to global food security. Whether through land-use zoning or other means, it deserves premium protection.

Clearly, greater efforts are needed to raise the water productivity of the global crop base, both rain fed and irrigated. Boosting by half the productive use of rainwater for crop evapotranspiration, as assumed in this analysis, will be difficult. Smallscale water harvesting, terracing, bunding, and other means of channeling and storing rainwater to increase soil moisture will be crucial. Successful examples of these types of projects in Africa (Critchley 1991), India (Centre for Science and Environment 1997), and elsewhere suggest greater potential for drought-proofing and increased rain-fed production than has been realized to date.

Improving irrigation efficiency can also increase agricultural water productivity. The estimated 500 $km^3$ of unproductive evaporation of irrigation water theoretically represents potential water savings sufficient to grow 450 x $10^6$ t of wheat, although only a portion of these losses could realistically and economically be captured. These savings increase the effective water supply without the need to build additional reservoirs or extract more groundwater. For example, researchers at the Sri Lanka-based International Irrigation Management Institute found that eliminating the flooding of rice fields prior to planting reduced water use by 25% (Seckler 1996). The portion of this reduction resulting from lower evaporative losses represents true water savings and effectively increases the available supply.

Efficient sprinklers, drip systems, and other methods of delivering irrigation water more directly to the roots of crops can also reduce unproductive evaporation. Research in the Texas High Plains has shown substantial water savings with low-pressure sprinklers that deliver water close to the soil surface rather than in a high-pressure spray (High Plains Underground Water Conservation District 1996). Water productivity gains of 20-30% or more are not uncommon when farmers shift to more efficient irrigation practices. Worldwide, however, such efficiency measures have spread slowly relative to their potential because of high upfront capital costs, relatively low crop prices, and heavy government subsidies that artificially lower irrigation water prices.

Improving the water-use efficiency of crops, shifting the mix of crops, and breeding crop varieties that are more salt tolerant and drought resistant may also increase agricultural water productivity. These gains do not come easily, however, because drawbacks can negate the potential benefits. For example, crop varieties that perform well under cooler temperatures may produce higher yields per unit of water consumed but have a lower harvest-index potential (Sinclair *et al.* 1984). Moreover, a good portion of the potential for improving crop water-use efficiency may already have been exploited. For example, breeders have already shortened the maturation time for irrigated rice varieties from 150 days to 110 days, substantially increasing that crop's water efficiency (IRRI 1995).

Finally, more equitable distribution of food may be necessary to satisfy the basic nutritional needs of all people as water constraints on agriculture increase. For the past three decades, the share of the world's grain supply fed to livestock has consistently ranged between 38% and 40% (Brown 1996). This large amount of grain—and, indirectly, water—could be used more productively to satisfy human nutritional requirements. For example, the diet of a typical US adult, with a relatively high percentage of calories derived from grain-fed livestock, includes enough grain to support the diets of four typical Indians.

Although it may be tempting to assert that the prospective shortage of water for crop production calls for stepped-up construction of large dams and river diversions to increase supplies, this conclusion is not sound. The aquatic environment is showing numerous signs of declining health, even at today's level of water use. Large dams and river diversions have proven to be primary destroyers of aquatic habitat, contributing substantially to the destruction of fisheries, the extinction of species, and the overall loss of the ecosystem services on which the human economy depends. Their social and economic costs have also risen markedly over the past two decades. Along with efforts to slow both population and consumption growth, measures to use rainwater and irrigation water more productively, to use food supplies more efficiently, and to alter the crop mix to better match the quantity and quality of water available offer more ecologically sound and sustainable ways of satisfying the nutritional needs of the global population.

## Acknowledgments

I am grateful to the Pew Fellows Program in Conservation and the Environment for the financial support to undertake this work. For helpful comments on an earlier draft of the manuscript, I thank Lester Brown, Harald Frederiksen, Johan Rockstrom, Jan Lundqvist, three anonymous reviewers, as well as Rebecca Chasan, editor-in-chief of *BioScience*.

## References Cited

Ajtay GL, Ketner P, Duvigneaud P. 1979. Terrestrial primary production and phytomass. Pages 129-182 in Bolin B, Degens ET, Kempe S, Ketner P, eds. The Global Carbon Cycle. New York: John Wiley & Sons.

Brown LR. 1996. Worldwide feedgrain use drops. Pages 34-35 in Brown LR, Flavin C, Kane H. Vital Signs 1996. New York: W. W. Norton.

____. 1997. The agricultural link: How environmental deterioration could disrupt economic progress. Washington (DC): Worldwatch Institute.

Centre for Science and Environment. 1997. Dying Wisdom. New Delhi (India): Centre for Science and Environment.

Critchley W. 1991. Looking After Our Land: Soil and Water Conservation in Dryland Africa. Oxford (UK): Oxfam.

Doorenbos J, Kassam AH. 1979. Yield Response to Water. Rome: Food and Agriculture Organization of the United Nations.

Dyson T. 1996. Population and Food: Global Trends and Future Prospects. London: Routledge.

[FAO] Food and Agriculture Organization of the United Nations. 1993. World review of high seas and highly migratory fish species and straddling stocks. Rome: Food and Agriculture Organization. Fisheries Circular no. 858.

____. 1995a. Irrigation in Africa in figures. Rome: Food and Agriculture Organization. Water Report no. 7.

____. 1995b. Production Yearbook 1994. Rome: Food and Agriculture Organization.

____. 1996a. Production Yearbook 1995. Rome: Food and Agriculture Organization.

____. 1996b. Food for All. Rome: Food and Agriculture Organization.

____. 1997. Water Resources of the Near East Region: A Review. Rome: Food and Agriculture Organization.

Frederiksen H, Berkoff J, Barber W. 1993. Water Resources Management in Asia. Washington (DC): World Bank.

Gardner G. 1997. Irrigated area up slightly. Pages 42-43 in Brown LR, Renner M, Flavin C. Vital Signs 1997. New York: W.W. Norton.

Ghassemi F, Jakeman AJ, Nix, HA. 1995. Salinisation of Land and Water Resources. Sydney (Australia): University of New South Wales Press.

Gleick PH. 1996. Basic water requirements for human activities: Meeting basic needs. Water International 21: 83-92.

Gray B. 1994. The modern era in California water law. Hastings Law Journal January: 249-308.

High Plains Underground Water Conservation District. 1996. November: The Cross Section. Lubbock (TX): High Plains Underground Water Conservation District.

[IRRI] International Rice Research Institute. 1995. Water: A Looming Crisis. Manila (The Philippines): International Rice Research Institute.

Kramer PJ, Boyer JS. 1995. Water Relations of Plants and Soils. San Diego: Academic Press.

Leach G. 1995. Global Land and Food in the 21st Century: Trends & Issues for Sustainability. Stockholm (Sweden): Stockholm Environment Institute.

McCalla AF. 1994. Agriculture and Food Needs to 2025: Why We Should Be Concerned. Washington (DC): Consultative Group on International Agricultural Research.

McGinn A. 1997. Global fish catch remains steady. Pages 32-33 in Brown LR, Renner M, Flavin C. 1997. Vital Signs 1997. New York: W.W. Norton.

Monteith JL. 1990. Conservative behaviour in the response of crops to water and light. Pages 3-16 in Rabbinge R, Goudriaan J, Van Keulen H, Penning de Vries FWT, Van Laar HH, eds. Theoretical Production Ecology: Reflections and Prospects. Wageningen (The Netherlands): Simulation Monographs 34, Pudoc.

Moore MR, Mulville A, Weinberg M. 1996. Water allocation in the American West: Endangered fish versus irrigated agriculture. Natural Resources Journal 36: 319-357.

Oldeman LR, van Engelen VWP, Pulles JHM. 1991. The extent of human-induced soil degradation. Annex 5 of Oldeman LR, Hakkeling RTA, Sombroek WG. World Map of the Status of Human-Induced Soil Degradation: An Explanatory Note. Wageningen (The Netherlands): International Soil Reference and Information Centre.

Perry CJ. 1996. The IIMI water balance framework: A model for project level analysis. Colombo (Sri Lanka): International Irrigation Management Institute. Research Report no. 5.

Pimentel D, et al. 1995. Environmental and economic costs of soil erosion and conservation benefits. Science 267: 1117-1123.

Postel S. 1992. Last Oasis: Facing Water Scarcity. New York: W.W. Norton.

____. 1996. Dividing the Waters: Food Security, Ecosystem Health, and the New Politics of Scarcity. Washington (DC): Worldwatch Institute.

Postel S, Carpenter S. 1997. Freshwater ecosystem services. Pages 195-214 in Daily GC, ed. Nature's Services: Societal Dependence on Natural Ecosystems. Washington (DC): Island Press.

Postel SL, Daily GC, Ehrlich PR. 1996. Human appropriation of renewable freshwater. Science 271: 785-788.

[PRB] Population Reference Bureau. 1995. 1995 world population data sheet. Washington (DC): Population Reference Bureau.

____. 1997. 1997 world population data sheet. Washington (DC): Population Reference Bureau.

Rangeley WR. 1987. Irrigation and drainage in the world. Pages 29-35 in Jordan WR, ed. Water and Water Policy in World Food Supplies. College Station (TX): Texas A&M Universityty Press.

Reij C. 1991. Indigenous Soil and Water Conservation in Africa. London: International Institute for Environment and Development.

Rosegrant MW. 1997. Water Resources in the Twenty-First Century: Challenges and Implications for Action. Washington (DC): International Food Policy Research Institute.

Seckler D. 1996. The New Era of Water Resources Management: From "Dry" to "Wet" Water Savings. Washington (DC): Consultative Group on International Agricultural Research.

Shiklomanov IA. 1993. World fresh water resources. Pages 13-24 in Gleick PH, ed. Water in Crisis. New York: Oxford University Press.

____. 1996. Assessment of Water Resources and Water Availability in the World. St. Petersburg (Russia): State Hydrological Institute.

Sinclair TR, Tanner CB, Bennett JM. 1984. Water-use efficiency in crop production. BioScience 34: 36-40.

Umali DL. 1993. Irrigation-Induced Salinity. Washington (DC): World Bank.

[UNCSD] United Nations Commission on Sustainable Deveopment. 1997. Comprehensive assessment of the freshwater resources of the world. New York: United Nations Economic and Social Council.

[UNDP] United Nations Development Programme. 1996. Human Development Report 1996. New York: Oxford University Press.

Unger PW, Stewart BA. 1983. Soil management for efficient water use: An overview. Pages 419-460 in Taylor HM, Jordan WR, Sinclair TR, eds. Limitations to Efficient Water Use in Crop Production. Madison (WI): American Society of Agronomy, Crop Science Society of America, Soil Science Society of America.

US Bureau of the Census. 1996. International electronic database. Suitland (MD): US Bureau of the Census. <http://www.census.gov/>

[USDA] United States Department of Agriculture. 1996, 1997a. Production, supply, and distribution (electronic database). Washington (DC): United States Department of Agriculture. <http://mann77.mannlib.cornell.edu/data-sets/international/93002/>

____. 1997b. World agricultural production. Washington (DC): United Stares Department of Agriculture.

____. 1997c. Grain: World markets and trade. Washington (DC): United States Department of Agriculture.

Vitousek PM, Ehrlich PR, Ehrlich AH, Matson PA. 1986. Human appropriation of the products of photosynthesis. BioScience 36: 368-373.

[WRI] World Resources Institute. 1994. World Resources 1994-95. New York: Oxford University Press.

Yudelman M. 1994. The future role of irrigation in meeting the world's food supply. In Soil and Water Science: Key to Understanding Our Global Environment. Madison (WI): Soil Science Society of America.

## Table 1. Estimated water consumption by crops worldwide, 1995.[a]

| Crop | Global production (x 1000 t) | Water-use efficiency of harvested yield[b] (kg/m$^3$) | Estimated water requirement (km$^3$/yr) |
|---|---|---|---|
| Wheat | 541,120 | 0.8–1.0 | 601 |
| Rice | 550,193c | 0.7–1.1 | 611 |
| Maize | 514,506 | 0.8–1.6 | 429 |
| Other grains | 290,236 | ~ 0.6–1.2 | 323 |
| Roots and tubers | 609,488 | ~ 4.0–7.0 | 111 |
| Pulses | 55,997 | ~ 0.2–0.6 | 140 |
| Soybeans | 125,930 | 0.4–0.7 | 229 |
| Other oilseeds | 125,749 | ~ 0.2–0.6 | 314 |
| Ground nuts | 27,990 | 0.6–0.8 | 40 |
| Vegetables and melons[d] | 487,287 | ~ 10.0 | 49 |
| Fruits (except melons)[d] | 396,873 | ~ 3.5 | 113 |
| Sugar cane[e] | 1,147,992 | 5.0–8.0 | 177 |
| Sugar beets[e] | 265,963 | 6.0–9.0 | 36 |
| Tobacco | 6,447 | 0.4–0.6 | 13 |
| Other[f] | | | 21 |
| Total | | | 3207 |

[a]Data from FAO (1996) and Doorenbos and Kassam (1979).

[b]The midpoints of these ranges are used to calculate the crop water requirement. Water-use efficiency values were not available for all crops, so where necessary I have attempted to make a reasonable assumption based on available information; these assumed values are denoted by a tilde (~).

[c]Rough rice; to calculate milled-rice production, multiply by 0.7.

[d]Statistics on fruit and vegetable production in many countries are unavailable, and much of the reported data excludes production from small household and community gardens, which can be substantial in some countries. The United Nations Food and Agriculture Organization (FAO 1996) has attempted to estimate total production of fruits and vegetables but does not break down these estimates by crop type. Thus, I have applied a reasonable water-use efficiency value based on known values for crops in these categories. Nevertheless, the margin of error for the estimated water requirements of fruits and vegetables is larger than that for the other crops.

[e]Values are for production of cane and beets, not for the raw sugar derived from them; per unit of sugar, beets are significantly more water efficient than cane.

[f]Coconuts, olives, tree nuts, coffee, cocoa beans, tea, and hops; the water requirements for these crops are little better than informed guesses, but this high margin of error does not significantly affect the global total.

## Questions:

1. What's evapotranspiration?

2. What's the harvest index?

3. Estimates suggest that salinization affects _____% of irrigated lands worldwide.

Answers are at the back of the book.

## Activity:

Go to the Internet Website:
**http://www.fao.org/wfs/fs/e/WatHm-e.htm** or **http://www.fao.org/wfs/fs/e/FSHm-e.htm**
to find examples of smale-scale, affordable irrigation techniques that poorer farmers can use and list some of them.

17

When considering land preservation, aesthetically pleasing areas are usually the first places to be taken into account. However, wetlands that provide purification of water, detention of water, and habitat to a multitude of species are also in need of preservation. Since 1780, approximately 53 percent of the wetlands in the United States have been lost to development and agriculture. The new "quick permits" under the U.S. Army Corps of Engineers will be replacing a more protective and stringent permit pertaining to the development of wetlands. To counteract these losses, mitigation efforts aim to increase the net acreage of wetlands by 100,000 acres. The struggle to meet the needs of society and to conserve ecologically important wetlands will continue and eventually come face to face with a population increase in coastal areas, spelling bad news for coastal wetlands and the many fish that spend part of their lives in these estuaries.

# Paradise Lost: America's Disappearing Wetlands

By Sally Deneen

She-crab soup arrives at restaurant tables on North Carolina's Outer Banks as a rich, sweet concoction, delighting tourists and new residents whose cars still boast license plates from their old states: Florida, Ohio, New York. As the ocean breezes sweep away the day-to-day worries of beach-bound visitors, Environmental Defense Fund (EDF) scientist Doug Rader realizes the days of the regional soup may be numbered. It's a simple axiom: No wetlands, no seafood.

Across San Francisco Bay from the Golden Gate bridge, the salty bay waters mingle with the melting snowcaps of the Sierra Mountains to form the largest estuary on the west coast of North and South Amercia. Yet, almost all of the freshwater marshes in this Caifornia delta are gone. Half of the tidal marshes have been destroyed, while others have been transformed into surreal, sunken farmlands. From the Gulf of Mexico's salt marshes to North Dakota's "prairie potholes," America's wetlands are disappearing rapidly, according to U.S. Fish and

Wildlife Service statistics comparing the colonial 1780s to the 1980s.

The rate: An acre a minute.

California has lost the greatest percentage (91 percent), but 21 other states have paved over or tilled at least half of their original wetlands. Fast-growing Florida has filled in the most acreage—a land size bigger than all of Massachusetts, Delaware and Rhode Island combined. Add the entire land size of California to that, and you can mentally picture the amount of wetlands lost since the Revolutionary War.

In cold, hard, economic terms, each acre of wetland is worth 58 times more money than an acre of ocean in the benefits it provides, according to *Science*. Wetlands act like sponges: The porous, jet-black peat helps soak up heavy rains and melting snow that otherwise may flood suburban yards. They also function like kidneys, filtering out dirt, pesticides and fertilizers before the unwanted run-off reaches lakes and streams. Without wetlands, excessive sediment can smother fish spawning areas and fertilizers can kill the prized fish sought by anglers.

Some of these soggy lands also serve as broad water-storage areas, allowing people to later enjoy these waters for iced tea and showers. And wetlands are a smorgasbord for frogs and migratory birds, and home to America's ducks. According to the National Audubon Society, wetlands compare to tropical rainforests in the diversity of species they support.

Yet which is more valuable to humans? According to Science, an acre of tropical forest is worth $817 for its ecosystem benefits. An acre of open ocean is worth $103. An acre of wetlands: $6,017.

Yet they continue to vanish.

Right now, Vice President Al Gore's office is fielding phone calls from concerned environmentalists and wildlife lovers who hope he will stave off "the biggest challenge to wetlands protection," says Robin Mann, an outraged member of the Sierra Club's Wetlands and Clean Water Campaign Steering Committee.

Shopping centers and riverfront homes conceivably could sprout up on soggy land without the usual requirements: notifying the public or asking for permission from the U.S. Army Corps of Engineers, the agency in charge of regulating the use or destruction of wetlands. The new "quick permits" would allow up to three acres of non-tidal wetlands to be developed or farmed, and up to 10 acres of any non-tidal wetlands to be destroyed as part of a "master planned development," notes Julie Sibbing, assistant director for wetlands and wildlife refuge policy at the National Audubon Society.

In some cases, a builder wouldn't have to notify the Corps at all. And the traditional requirement that wetlands be avoided where possible wouldn't apply—a crucial failing, say environmentalists and wildlife specialists. Don't like what's being built next door? Sorry. No public input would be allowed either, Sibbing adds.

Ironically, these "rubber-stamp permits," as Clean Water Network's Kathy Nemick calls them, are meant to quell public outcry, not rekindle it. They would replace the controversial and apparently more protective Nationwide Permit 26 (NWP 26), which allows up to three acres of isolated or headwater wetlands to be destroyed. The Corps has promised to ditch the more stringent permit by year's end.

It's no surprise the oil and gas industry wants the current permitting system changed. But this would've been a welcome innovation for retirees Bob and Mary McMacken, too. Their case is an example of how the old wetlands law was used badly: They received permits to build a house on less than an acre in a still-developing subdivision in Pennsylvania's Poconos, and lived there four years before a letter arrived in the mailbox telling them to cease and desist. Their property was a wetland, the Corps wrote. The message: Get out.

"This was a real emotional process to through," says Nancie G. Marzulla, president of Defenders of Property Rights, the nation's only public-interest legal foundation dedicated exclusively to protecting property rights. "'It took us two years to work with the Corps to get them absolved of all liability. . . They had already built there, for goodness sakes."

Trouble is, government scientists say the Corps' new proposal would destroy more wetlands and streams than the current dredge-and-fill permits. It also expands the scope of waters that could be filled in, and the Corps hasn't gathered data on the resulting environmental impacts either, writes a concerned Jamie Clark, director of the U.S. Fish and Wildlife Service. And what about endangered species? It could take two to three years to consult with Clark's agency and the National Marine Fisheries Service to hash out the possible impact. But that's too late: Some of the 16 new permits could be the law of the land as early as March.

And, the proposal "may not be consistent with" the Clean Water Act, which requires only "minimal adverse environmental impact," Clark wrote in a letter to Michael Davis, deputy assistant secretary of the Army, representing the Corps of Engineers.

"What we're demanding is that they withdraw the package," says environmentalist Mann, who is encouraging the public to write to Vice President Gore.

"We're losing 117,000 acres a year right now—so we're permitting a lot to be destroyed," says Nemsick, national coordinator for the Clean Water Network in Washington, D.C.

## On the Frontlines

Rader, the EDF scientist, speaks quickly, matter-of-factly, like a guy who has much to say and little time to say it. His immediate concern isn't the Corps' proposed new permit plan. He has other worries.

Rader can mentally connect the dots between the tasty sea creatures on dinner tables—softshell crab, blue crab and flounder—and the health of local wetlands. "All of those fish are directly linked to brackish-water estuaries that are girdled by wetlands," notes Rader.

Only four states have more wetlands than the popular resort destination of North Carolina, which has lost about half of its original soggy lands transformed into homes for new retirees, developments and farms. Time was when the state's two-legged population doubled just every 50 years. But as resort towns and cities grow, residents in some counties may quadruple in 50 years, Rader says."We're looking at a huge increase—particularly in the northern Outer Banks," he notes. "It means all bets are off in terms of estuarine environments."

"In 20 years, will all the fish here come from fish farms and foreign waters? I think that's a possibility," Rader says matter-of-factly.

That may be surprising, since some of the nation's largest fish nurseries are found along North Carolina's Pamlico Sound. The estuaries also have been rocked by headline-grabbing outbreaks of a fish-killing neurotoxin called *Pfiesteria piscicida*, believed to be caused by a chain reaction that occurs when waste draining off farms enters the rivers. The puzzling toxin causes a variety of symptoms in anglers, including wheezing, and nervous and respiratory system ailments. So people are advised not to eat fish when outbreaks occur.

Such suggestions aren't good for business: Commercial and sport fishing each year add at least $152 billion to the U.S. economy and provide about two million jobs, according to The Izaak Walton League of America. Three-fourths of the nation's fish production depends on marshes, estuaries and other wetlands, the league adds.

Though Rader feels a sense of optimism after the August announcement that about $221 million in federal money is on the way to restore local watersheds, and a 1997 state Marine Fisheries Reform Act now requires "no net loss" of wetlands, that doesn't mean all is well. A one-two punch diminishes the good news. For one thing, pigs outnumber people in North Carolina, and some of the fecal waste of the 10 to 11 million swine end up in rivers. Meanwhile,

farms and other development continue to eliminate wetlands and riparian buffer vegetation. So "the kidneys of these landscapes are being eliminated," Rader explains.

In trendy Portland, Oregon, about 40 percent of area wetlands have vanished in a decade, even though protective regulations were in place, according to wetland ecologist Mary Kentula of Oregon State University. The lesson, Kentula determined, was the need for better monitoring and protection in fast-growing areas around the U.S.

Down south, almost three-quarters of Louisiana's bottomland hardwood swamps have vanished as farmers till land drained long ago by the U.S. Army Corps of Engineers. Such swamps have always been the most common type of wetland in the U.S., claims the EDF. They're in the floodplains of rivers, such as the Mississippi, and they're found along slow-moving southern streams.

Draining the swamps of Louisiana has left the state's estimated 80 remaining black bears stranded in carved-up patches of land too small to support significant populations, and is linked to the decline of at least 80 other threatened or endangered species, according to an EDF and World Wildlife Fund joint study. Residents took the unusual step of passing a constitutional amendment to start a wetlands conservation fund a decade ago, and other anecdotal successes can be pointed out. Still, the EDF claims, "The expectation that public funds will become available for drainage continues to encourage destruction of bottomland hardwoods today."

In the willow wetlands of the sky-high Rocky Mountains, where moose delight hikers and 51 percent of the Southwest's birds depend on plants for some meals, estimates place wetland loss at 90 to 95 percent. The reasons: Cattle grazing, housing developments, ski resorts and conversion to agriculture.

That's not good news for anglers in what may be the nation's best trout fishery. "These streamside wetlands play a vital role by trapping and detaining large quantities of sediment, keeping it out of streams where it could otherwise obstruct spawning," reports the EDF.

Plus, for the anglers to eat trout, the trout need to eat invertebrates, which need to eat leaves. And those leaves drop from the wetlands' alder and willow trees around this time of year.

The Clinton Administration aims for a net *increase* of 100,000 acres of wetlands per year by encouraging the building of artificial wetlands. Yet, studies have shown that artificially created wetlands often dry up or die because scientists don't fully understand how to recreate the original soggy lands. In some cases, homeowners associations or commercial developers are left to tend the puzzling marshes, with decidedly checkered results.

That hasn't stopped a new trend toward "mitigation banking," which allows developers to destroy wetlands if they, in turn, give money to a mitigation bank such as Fort Lauderdale-based Florida Wetlandsbank The banks use the money to restore wetlands elsewhere—measures like restoring drainage or killing invasive exotic plants. The banks promise to maintain the restored wetlands forever. The value is, instead of having postage-stamp-sized wetlands dotting the landscape, you'll end up with a bigger stand of wetlands in an ecologically sound place, such as at the edge of the Everglades. The problem is, original wetlands function better.

"We still understand wetland functions relatively poorly. This hampers our ability to properly restore wetlands or create new ones to replace those lost to developmental pressures or erosion," says Ed Proffitt, chief of the Wetland Ecology Branch of the U.S. Geological Survey in Lafayette, Louisiana.

Northwestern University civil engineering professor Kimberly Gray is creating wetlands in the unlikely industrial setting of Chicago's South Side, but she cautions that recreated marshes aren't the same thing."

"It"s important for us to try to restore them, but I don't think we have in our power yet to go destroy one and recreate one that is comparable in substance and structure. When we create wetlands, they're usually not as diverse or robust," Gray says.

The struggle to meet the needs of people while recovering diminished wetlands has set up a curious dichotomy: Every day, permission to build new homes, businesses and farms in original wetlands continues to be granted by local or regional governments. Meanwhile, billions of tax dollars or private dollars are earmarked to restore other wetlands. Consider the ongoing restoration of Chesapeake Bay, where the fresh water of 48 rivers mixes with saltwater to produce the nation's largest estuary.

The splashing sound of fish breaking the watery surface and the harsh, noisy squawks of rails flying overhead make the Chesapeake's

wetlands among Michael Weinstein's favorite spots. Weinstein, director of the Sea Grant Program in New Jersey and an expert on wetlands and marsh habitats, is optimistic about the makeover: Fish immediately began using previously off-limits areas after a dike was intentionally breached. Yet, years of draining and damming destroyed nearly 60 percent of the wetlands in the three main bay states, sparking a goal of not just maintaining what's left, but adding even more wetlands.

More than 13 million people from six states live in the bay's watershed, and the next 625 years are expected to bring enough people to populate two more Baltimores and two Districts of Columbia, adding to area pollution. "Just one year of stormwater from the Washington, D.C. metropolitan area alone dumped between one million and five million gallons of oil, 400,000 pounds of zinc, 64,000 pounds of copper, and 22,100 pounds of lead into the bay," the EDF reports.

More than one in two Americans now lives on or near the coast, requiring an average of one-half acre of land apiece for new schools, post office and other public services, Weinstein notes. By 2050, 70 percent of Americans are expected to live on the coast. "So the pressures are ever increasing," Weinstein adds.

## Uneasy Neighbors

That people and wetlands make uneasy neighbors is nothing new to Burkett Neely. A woman called him to complain that an endangered wood stork had relieved itself in her backyard pool in tony Boca Raton, Florida. What could Neely say? At the time, Neely tended the northern Everglades as manager of the Loxahatchee National Wildlife Refuge, west of Boca Raton. He knew the stork was—and is—an endangered species. You can't kill it, or even bother it, he says. As urban sprawl marches closer to the marshy refuge, I think you're going to see all kinds of conflicts," adds Neely. Neighbors already pine for mosquito-spraying which is only marginally effective, since it isn't allowed in the refuge. "Living next to a swamp, you deal with swamp creatures," Neely replies.

The Everglades are close to the largest wetlands in the nation, despite being reduced to half their original size. Restoring the "River of Grass" is expected to become the largest freshwater wetlands restoration project in the world: It will take at least 20 years and an estimated $1 billion. It's also overseen by the U.S. Army Corps of Engineers—the same agency that did most of the swamp drainage a half-century ago.

But already, the Everglades may be losing some of their luster with politicians who favored the restoration Last year, Congress provided $76 million for buying land as a buffer between the Everglades and urban sprawl. This year, a Senate bill slashed that to $40 million for fiscal year 1999, and a House bill provided even less—$20 million. Buying land is widely recognized as crucial in restoring the Everglades, contends the National Audubon Society. Expect more homes and businesses to move in otherwise, the organization warns.

As South Florida adds a new resident every 12 minutes through the year 2020, geographers contend the population center of the region won't be the coastal cities of Miami or Fort Lauderdale—but farther west, near the wetlands of the Everglades. Four out of five new residents are expected to live in or fairly near suburban Sunrise, home to the new arena of the Florida Panthers professional hockey team.

"For the most part, we have come a long way from the old view that wetlands were mosquito-plagued, swamp wastelands full of snakes and alligators, and that their only worth was to be drained or filled for construction or agriculture," Proffitt says.

In its simplest form, the threats to wetlands seem to boil down to a curious circle. People need a place to live, work, shop. They look for affordable, attractive choices—which may be in former wetlands. Developers build homes where demand indicates people want to live. So more people move into new ranch houses in former wetlands. More builders build there. Soon, you have a suburb where herons once stood like statues, waiting silently for a meal to float by.

At any point, people could stop buying homes or doing business in former wetlands. That would encourage developers and businesses to stay in centralized cities. Or developers could stop building in wetlands—that would force homebuyers and businesses to look elsewhere. And government agencies could stop granting permits to develop them.

Maybe the cycle can be stopped by the folks in Washington, D.C. But don't bet on it. That town is itself the site of a former wetland.

CONTACT: Clean Water Network, 1200 New York Avenue NW, Suite 400, Washington, DC 20005/(202)289-2395; National Audubon Society, 700 Broadway, New York, NY 10003/(212)979-3000; Environmental Defense Fund, 257 Park Avenue South, New York, NY 10010/(800)684-3325

---

## Questions:

1. What is the rate at which American wetlands are being destroyed?
2. What are some of the points under the "quick permits" that are raising the concern of environmentalists?
3. How do Man-made wetlands compare to the originals?

Answers are at the back of the book.


## Activity:

Go to **http://multimedia.pnl.gov:2080/staff/leroberts/wetlands/status.html**
Click on the U.S. map picture.
What percent of wetlands have your state and surrounding states lost?
What are some of the results of wetland degradation listed on this webpage, and what are the main causes?

The Kissimmee River in southcentral Florida was channelized primarily for flood-control purposes over the past half century. This managed flood regime was successful in reestablishing wetland vegetation communities and some fish and avian communities. Now, these experimental measures are being studied for their successes and shortcomings to be a model (or potential restoration component) for other river-floodplain ecosystems. It has also helped to develop a comprehensive restoration program.

# Hydrologic Manipulations of the Channelized Kissimmee River

By Louis Toth, Stefani Melvin, and Joanne Chamberlain

Historically, much of southcentral Florida was dominated by a contiguous wetland system that extended from the headwater lakes of the Kissimmee River basin to Florida Bay. During the past half century, this wetland landscape has been compartmentalized with a network of canals, levees, and water-control structures. This network is used to manage hydrologic regimes of the regional hydrosystem, primarily for flood-control purposes (Light and Dineen 1994, Toth and Aumen 1994).

Modifications of the physical configuration and hydrology of the South Florida landscape have affected the Everglades, Lake Okeechobee, and the 7800 km² Kissimmee River basin, where an extensive flood-control project was constructed from 1962 to 1971. Lakes in the river's headwater basin were connected by canals and partitioned into floodstorage reservoirs; a 90 km long, 9 m deep, and 100 m wide drainage canal was excavated through the river-floodplain ecosystem; and the channelized river was transformed by levees and water-control structures into five pools with stabilized water levels.

The Kissimmee River flood-control project lowered both average and peak flood stages, reduced or eliminated water-level fluctuations, and drastically modified discharge regimes throughout the basin, but the hydrologic impacts were particularly severe within the channelized river-floodplain ecosystem. In contrast to the upper Mississippi River, in which maintenance of constant water levels upstream of navigation dams has increased open water and marsh habitats (Sparks 1995, Sparks *et al.* 1998), two-thirds (10,000 ha) of the Kissimmee's historic floodplain wetlands were drained by the lowering and stabilization of water levels (Toth *et al.* 1995). Even the most extreme post-channelization flood flows have been contained entirely within the banks of the constructed canal, and most of the floodplain wetlands that remain occur in the lower impounded portion of each pool and at the confluence of major tributary slough systems.

The modified hydrology and loss of wetland habitat have severely affected the structural and functional integrity of the floodplain (Toth 1993). Use of the river-floodplain system by wintering waterfowl has declined by 92% (Perrin *et al.* 1982). The naturalized cattle egret (*Bubulcus ibis*), a species that is primarily associated with cattle on pastures and other ruderal terrestrial habitats, has replaced the diverse complement of wading birds, including the endangered wood stork, that once used the floodplain wetlands (Toland 1990). A nationally renowned largemouth bass fishery and populations of other game fish species continue to decline, and the fish community is dominated by species tolerant of the altered habitat conditions of the channelized system (FGFWFC 1994). In addition to the loss of fish and wildlife habitat, drainage of floodplain wetlands has affected trophic resources that fueled the riverine food web (Toth 1990). Elimination of the nutrient-filtration function that was provided by the river's floodplain wetlands has exacerbated elevated nutrient loadings and transport to Lake Okeechobee, which is undergoing accelerated eutrophication due to intensive agricultural land uses (e.g., dairies) in contributing watersheds (Toth and Aumen 1994, Aumen 1995).

Shortly after the Kissimmee River was channelized, a sequence of hydrologic manipulations was initiated to reduce or alleviate some of the impacts to the system's ecology and natural resources. In this article, we discuss the successes and shortcomings of these experimental measures as potential restoration plan components for river-floodplain ecosystems.

## Historic Hydrology and Ecology

Historical flooding regimes along the Kissimmee River were unique among North American rivers (Toth et al. 1995). Water levels on the floodplain typically varied according to subtropical, rainfall-driven seasonal cycles. Except during rare droughts, significant portions of the floodplain often remained continuously inundated. As much as 77% of the floodplain had mean annual hydroperiods (length of inundation) of at least 265 days, with depths commonly exceeding 1.0 m on the inner portions of the floodplain, which flanked the river channel. Peripheral floodplain elevations had more variable hydroperiods but were generally inundated during at least a portion of the wet season (July–November).

Prolonged floodplain inundation regimes were facilitated by protracted basin inflows and geomorphic characteristics that led to slow drainage rates and sustained extensive hydrologic connectivity between the river channel and floodplain. Although discharges exceeded the capacity of the river channel during 35–50% of the historical period for

which hydrologic records are available (1934–1960), the flat topography and absence of a continuous natural levee along the river kept portions of the floodplain inundated at less than bankfull stages. The floodplain was also inundated by inflows from lateral tributary sloughs, in which water was delivered to the floodplain as overland sheet flow. Rates of drainage were moderated by the low gradient (0.07 m/km), meandering channel, dense wetland vegetation, and high water-retention capacity of the organic floodplain soils (Parker 1955). Prechannelization stages typically receded at rates of less than 0.3 m/month (Toth et al. 1993).

These unique hydrologic characteristics were the principal determinants of the ecological structure and function of the Kissimmee River floodplain. The historic floodplain supported a mosaic of habitats (Pierce et al. 1982), including backwater lakes and ponds and three dominant wetland plant communities, which were distributed according to lengths and depths of inundation (Toth et al. 1995). Floodplain elevations exposed to prolonged, deep hydroperiods were covered by coastal-plain willow (Salix caroliniana) and buttonbush (Cephalanthus occidentalis) shrub communities and by a predominantly herbaceous, broadleaf marsh composed of pickerelweed (Pontederia cordata), arrowhead (Sagittaria lancifolia), cutgrass (Leersia hexandra), and maidencane (Panicum hemitomon). Plant species diversity was greatest along peripheral floodplain elevations, where shorter and shallower hydroperiods selected for wet prairie communities composed of a mixture of forbs, grasses, and sedges. Other wetland habitats included cypress (Taxodium distichum) swamps, red maple (Acer rubrum) and popash (Fraxinus caroliniana) forests, and a sand cordgrass (Spartina bakeri) ecotone along the upland boundary of the floodplain. Although the distributions of these habitats were determined by prevailing hydroperiods, the persistence of this mosaic of floodplain habitats depended on the stochastic and widely varying inundation regimes. These regimes included extreme flood flows and, especially, rare but periodic droughts, which were the principal sources of disturbance (Pickett and White 1985) that regulated habitat dynamics of the historic floodplain landscape.

Hydrologic characteristics also affected use of the historic floodplain by fishes. By providing productive feeding areas, spawning sites, and refugia for young fishes, the availability of floodplain habitats can be a key factor influencing recruitment and population dynamics of many riverine fish species (Junk *et al.* 1989, Schlosser 1991, Gehrke 1992). In the prechannelized Kissimmee River, protracted floodplain inundation provided vast habitat for small-bodied fishes, including live bearers (e.g., Eastern mosquitofish, *Gambusia holbrooki*; least killifish, *Heterandria formosa*; and sailfin molly, *Poecilia latipinna*) and species that depend on vegetation for spawning (e.g., Florida flagfish, *Jordanella floridae*; Everglades pygmy sunfish, *Elassoma evergladei*; and swamp darter, *Etheostoma fusi forme*).

The densely vegetated floodplain habitats also provided refugia for larvae and juveniles of larger riverine species. In a 1957 survey, 90% of the fish found in a floodplain marsh were less than 100 mm long, and young game fish species comprised 41% of the sample (FGFWFC 1957). Use of the floodplain, particularly the broadleaf marshes, by large-bodied fish species was probably limited by vegetation density to the deeper marsh habitats that flanked the river channel. However, as in larger river-floodplain systems (Junk *et al.* 1989), accessibility increased during rising hydrographs, when large portions of the floodplain provided spawning habitat for pikes (*Esox americanus* and *Esox niger*) and gar (*Lepisosteus occeus* and *Lepisosteus platyrhincus*) and productive foraging areas for game-fish species such as bluegill (*Lepomis macrochirus*), largemouth bass (*Micropterus salmoides*), and black crappie (*Pomoxis nigromaculatus*; Trexler 1995).

The variety of floodplain habitats and range of water depths supported a diverse complement of avian species, including three endangered species: snail kite (*Rostrhamus sociabilis*), wood stork (*Mycteria americana*), and bald eagle (*Haliaeetus leucocephalus*; FGFWFC 1957, USFWS 1959, Howell and Heinzman 1967, Perrin *et al.* 1982). Deep, open-water habitats provided overwintering foraging areas for coots (*Fulica americana*) and migratory diving ducks (Chamberlain 1960, Bellrose 1968). Deep water may also have been a critical requirement for the reproductive ecology of several other bird groups. Many wading birds that lived on the historic floodplain form breeding colonies in low shrubs and trees surrounded by deep water, which protects nests from mammalian predation

(Frederick and Collopy 1989). The snail kite also prefers to nest over deep-water wetlands with long hydroperiods (Sykes *et al.* 1995), which provide a stable habitat for production of their principal food source, the apple snail (*Pomacea paludosa*).

In addition to providing nesting sites for wading birds, willow and buttonbush swamps provided habitat for neotropical passerines, including migratory warblers and flycatchers. The densely vegetated broadleaf marshes probably supported resident populations of limpkins, rails, and bitterns, although historic data on these cryptic species are limited. However, much of the avian utilization of the historic floodplain was concentrated in the shallower, outer elevations (particularly wet prairie habitats), where the dynamics of the advancing and retreating water's edge played a prominent role in the reproductive and foraging ecology of dabbling ducks and of a variety of short- and long-legged wading birds (Weller 1995). The formation of isolated floodplain pools during receding hydrographs provided concentrated fish and invertebrate prey for nesting wading birds and overwintering waterfowl. The temporal availability of these concentrated energy resources is critical to spring waterfowl migrations (Fredrickson 1991) and to wading bird nesting success and recruitment (Powell 1987, Frederick and Collopy 1989, Bancroft *et al.* 1990). Drying pools also provided prey for bald eagles.

## Experimental Hydrologic Manipulations

Although the historically dominant broadleaf marsh and willow communities were well adapted to persist in the stabilized-water level regimes of the channelized system, the more diverse wet-prairie communities that depended on seasonal wet-dry cycles along the periphery of the floodplain were largely eliminated. The first post-channelization hydrologic manipulations began in 1971, the year the flood-control project was completed; they were intended to explore the potential for mitigating this loss of wet prairie by attempting to encourage germination and reestablishment of annual plant species through seasonal (60–90 day) drawdowns of water levels in several of the channelized pools. Although a diverse assemblage of both annuals, particularly wild millet (*Echinochloa walteri*), and perennials colonized exposed river banks, shoals, and areas in which vegetation was burned, mechanically removed, or chemically treated, these drawdowns failed to change plant community composition in otherwise undisturbed wetlands, in which dense stands of established dominant plants, such as pickerelweed, Cuban bulrush (*Scirpus cubensis*), and water hyacinth (*Eichornia crassipes*), prevented reestablishment of other plant species (Goodrick and Milleson 1974). However, the rapid recovery potential of other floodplain resources was documented during the reflooding period as densities of small fish, crayfish, and grass shrimp (*Palaemonetes paludosus*) showed linear increases with water levels (Milleson 1976).

The ability to raise pool stages and thereby reinundate drained floodplain was limited by infrastructure constraints (e.g., the stability of the water-control structures), but water-management options were increased through the construction of diked floodplain impoundments. In one 672 ha impoundment, in which water levels were manipulated over a 1.5 m range (0.9 m higher than the stabilized pool stage), the reestablishment of seasonal inundation regimes led to ecologically significant changes in plant community structure on much of the previously drained floodplain. The portion of the community represented by annuals and other wetland plants, including important waterfowl food sources, increased, while coverage of the dominant pasture grasses that had become established on this section of floodplain decreased (Perrin *et al.* 1982). In another experimental impoundment (49 ha), which was created to evaluate the nutrient removal and retention capability of reflooded floodplain, pumped overland flow and an annual 0.9 m water level-fluctuation schedule led to reestablishment of a marsh that consistently retained 71% of the phosphorus inputs (Moustafa *et al.* 1996).

Experimental water-level manipulations (seasonal, pool-stage fluctuations) continued throughout the 1970s, but by the early 1980s more extensive restoration plans were being developed and evaluated (USACE 1985). Between 1984 and 1989, a demonstration project was constructed to evaluate more thoroughly the feasibility of restoring key biological resources and the relative utility of several recommended restoration approaches (Toth 1993). As in previous manipulations, a primary objective of the demonstration project was to reestablish floodplain inundation regimes. A 120 ha impoundment was created on a floodplain pasture, and pool stage fluctuations were used to seasonally inundate another 525 ha of drained floodplain. Three weirs were constructed across the flood-control canal to divert flow through adjacent remnant river channels. During high-discharge periods, water-surface profiles were elevated upstream of each weir, resulting in "backwater effects" that provided additional floodplain inundation.

The demonstration project's effects on vegetation communities were most evident on those portions of the floodplain that were subjected to prolonged inundation. The most striking change occurred in the impoundment, in which reestablishment of a 7–9-month annual hydroperiod led to rapid colonization by the broadleaf marsh species that had existed on this portion of the floodplain prior to channelization. The distribution of wetland plant species, particularly dotted smartweed (*Polygonum punctatum*), maidencane, and coastal-plain willow, also expanded in response to the increased floodplain inundation that resulted from pool stage fluctuation and the backwater effects of weirs (Toth 1993). As in the impoundment, a coastal-plain willow community reestablished on the same portion of floodplain where it had occurred historically, and coastal-plain willow replaced wax myrtle (*Myrica ceri fera*) as the dominant riparian species along the banks of river channels adjacent to the weirs. These results demonstrated the viability and recolonization potential of both the vestigial seed bank and remnant propagules of hygrophytic species (Wienhold and van der Valk 1989) and verified the feasibility of restoring historic wetland plant communities on the floodplain.

However, reestablished seasonal hydroperiods did not eliminate many of the upland plant species that had colonized the drained floodplain since channelization, including ragweed (*Ambrosia artemisiifolia*), dog fennel (*Eupatorium capillifolium*), wax myrtle, purple rattle-bush (*Sesbania punicea*), and caesar weed (*Urena lobata*). These invasive weed species are capable of persisting in a wide range of edaphic (e.g., soil moisture) conditions (Clewell *et al.* 1982, Marshall *et al.* 1985, Lowe 1986, Patton and Judd 1988), and they can provide formidable competition for reestablishment of wetland vegetation (Baird 1989), particularly in reflooded habitats with short, intermittent hydroperiods.

Fish and avian communities also showed mixed responses to the demonstration project. Higher pool stages led to increased densities of resident livebearing fish (e.g., mosquitofish) and some utilization of the floodplain by larger species, such as largemouth bass (FGFWFC 1994). However, fish species richness in all floodplain habitats affected by the demonstration project remained less than one-third of that in the historic floodplain. Fish colonization of the impoundment was impeded by the levee, which blocked hydrologic connectivity with canal and remnant river habitats. In other enhanced habitats, water levels did not get deep enough or did not remain sufficiently deep for long enough (Kushlan 1976) to accommodate extensive use of the floodplain, particularly the remaining densely vegetated broadleaf marshes, by the larger fish species that are found in the canal and remnant river channels.

Fish use of the floodplain may also have been limited by the chronically low dissolved oxygen levels that persisted within the demonstration project area. Before channelization, fish immigration onto the floodplain was probably tied to, and perhaps stimulated by, annual wet-season flooding, which flushed deoxygenated water from the floodplain, much as wet-season pulses of water rejuvenate the Sudd swamps of the Nile River (Howell et al. 1988). Thus, the demonstration project showed that a simple manipulation (rise) of water levels in the channelized pools does not reproduce the functionality of flood pulses over the floodplain landscape.

Within two years after reflooding, the demonstration project impoundment had the highest wading bird and waterfowl density in the channelized pools (Toland 1990). Although increased floodplain inundation caused by the weirs and by pool stage fluctuation also led to higher densities of wading birds, recent surveys indicate that the number of wading birds using the demonstration project area is not significantly different from numbers using other pools within the channelized system. However, the enhanced wetland habitat has led to a shift in avian community structure as the proportion of the community represented by cattle egrets has continued to decline, from 65% in 1978–1980

(Toland 1990) to 28% in 1996 (Stefani L. Melvin, unpublished data).

Most of the shortcomings of plant, fish, and avian responses to the demonstration project were due to the limited degree to which the project reproduced key hydrologic characteristics that determined the structure and function of the historic ecosystem. For example, pool stage fluctuations influenced hydroperiods on approximately 50% of the floodplain within the demonstration project area, but they did not reestablish historical depth regimes or, perhaps more important, the wide range of stage variability that maintained the mosaic of wetland habitats on the prechannelized floodplain. Backwater effects of the weirs increased the range of stage variability but only slightly increased the amount and duration of floodplain inundation. The influence of weirs on these inundation characteristics was limited by the flood-control canal, which rapidly drained the floodplain after peak discharges (i.e., during declining legs of discharge hydrographs). The resulting spiked hydrographs reduced the functionality of reflooded habitats by limiting the time available for development and production of integral components of the floodplain food web, such as invertebrate shredders (Toth 1993). Rapid stage recession rates also precluded the formation of drying pools with concentrated densities of fish and invertebrate prey, which are critical to the foraging ecology of wading birds and waterfowl. In addition, two major fish kills, in 1985 and 1988, resulted from anoxic conditions caused by rapid drainage of water off the floodplain (Toth et al. 1990).

Based on the results of the demonstration project, a new strategy for restoration of wetland communities was tested in another impoundment (228 ha), which was created on the channelized floodplain in 1990. This impoundment included an outlet water-control structure for managing water levels according to an operation plan that would reestablish historical inundation characteristics based on a model of pre-channelization relationships between antecedent rainfall and river stages. Before the model-based water-level manipulations were initiated, the impounded floodplain was subjected to 17 months of continuous inundation, which effectively eliminated the upland plant species that had become established on this portion of drained floodplain. Within two years after the manipulations, a broadleaf marsh developed over most of the impounded floodplain, and a diverse wet-prairie community composed of predominantly hygrophytic species became established on seasonally inundated peripheral elevations. As in the demonstration project, the impounded floodplain provides enhanced habitat for avian species; however, due to the lack of connectivity with the river, it continues to have a depauperate fish community.

## Implications of Managed Hydrologic Manipulations

More than 20 years of experimental studies have demonstrated the restoration potential and limitations of hydrologic manipulations in the channelized Kissimmee River. Managed flooding regimes have been successful at reestablishing wetland vegetation communities and enhancing some associated functional values, including secondary production and fish and avian utilization. However, restoration of the complex structure and dynamics of natural floodplain ecosystems requires the reestablishment of a broader range of spatial and temporal hydrologic regimes than is possible through manipulations of highly managed hydrosystems such as the Kissimmee River.

As in most altered systems, societal and infrastructure constraints have limited the scope of imposed flooding regimes to prescribed water level-fluctuation schedules based on average historic conditions. These manipulation schedules have not replicated the range of predisturbance hydrologic variability, particularly the frequency and duration of high stages and associated depths. Maintenance or restoration of the full range of hydrologic variability components, including the magnitude, frequency, duration, timing, predictability, and rate of change of flood regimes, are needed to sustain the mosaic of habitats and associated biodiversity and ecological functionality of riverine floodplain ecosystems (Poff *et al.* 1997). The construction of floodplain impoundments allowed more innovative manipulations but compromised geomorphic features (e.g., connectivity to the river channel) that are also critical to river ecosystem structure and function (Sparks 1995, Toth 1995).

## The Kissimmee River Restoration Plan

Hydrologic manipulations within the channelized Kissimmee River were supported, if not driven, by a politically active restoration initiative that began during the latter stages of channelization and was influential in securing both state and federal legislation in support of these and other restoration-related studies (Woody 1993, Toth and Aumen 1994). Although initially somewhat narrowly focused (e.g., on loss of floodplain wetlands and associated nutrient-filtration functions), the restoration vision expanded as evidence of the range of ecological impacts increased (Toth *et al.* 1997).

Results of the experimental hydrologic manipulations of the channelized system provided the scientific basis for the sociopolitical decisions that led to the development of a comprehensive restoration plan. These manipulations demonstrated the feasibility of restoring lost biological resources and led to the adoption of an ecosystem restoration goal of reestablishing the ecological integrity of the river-floodplain landscape (Toth 1995). The implicit scope of this goal, and the documented importance of reestablishing the physical form of the riverine system and the full complement of historic hydrologic characteristics, eliminated piecemeal measures such as weirs, impoundments, and pool stage fluctuation as potential restoration components. In 1990, the state of Florida endorsed a dechannelization plan for reconstructing over 100 km² of river-floodplain ecosystem by eliminating (backfilling) 35 continuous kilometers of the flood-control canal, removing two water-control structures, and reestablishing historical inflow regimes from the headwater lakes (Koebel 1995). The plan includes an extensive land acquisition program, which will reduce the need for flood protection in the basin (Toth and Aumen 1994). Authorization for a federal-state partnership for the implementation of this plan was provided by the 1992 Water Resources Development Act (PL 102-580).

The reconstruction of the Kissimmee River began in 1994, with a pilot dechannelization project that backfilled a 300 m long section of the flood-control canal and removed spoil (dredged sand and shell deposits from the canal excavation) from approximately 5 ha of adjacent floodplain (Toth 1996). Although the scale of this initial dechannelization was too small to restore a significant portion of the surrounding ecosystem, use of the reestablished floodplain habitat by both birds and fish has provided supporting evidence for the prospects of recovery of important riverine resources. Fish (primarily *Centrarchidae*) spawned on the newly created floodplain immediately after the spoil was removed, and the project area supports considerably higher avian and fish species richness than other existing floodplain wetland habitats.

Successful environmental restoration and management programs require adaptive, scientifically based planning (Walters 1986, Toth and Aumen 1994). The iterative planning, implementation, and monitoring of experimental hydrologic manipulations within the Kissimmee River system provided the necessary information for developing a technically and scientifically sound restoration plan. This adaptive research and evaluation paradigm will continue to be applied during the implementation of the Kissimmee River restoration project and will provide the scientific foundation for fine-tuning sequential phases of the reconstruction and for adaptively managing the recovering and restored ecosystem.

## Acknowledgments

We would like to thank Rebecca Chasan, Dale Gawlik, Karl Havens, Penelope Firth, Seth Reice, and an anonymous reviewer for constructive reviews and suggestions, and Cheri Craft for finalizing figures.

## References Cited

Aumen NG. 1995. The history of human impacts, lake management, and limnological research on Lake Okeechobee Florida (USA). Archiv fr *Hydrobiologie, Ergebnisse der Limnologie* 45: 1–16.

Baird K. 1989. High-quality restoration of riparian ecosystems. *Restoration and Management Notes* 7: 60–64.

Bancroft GT, Jewell SD, Strong, AM. 1990. Foraging and nesting ecology of herons in the lower Everglades relative to water conditions. West Palm Beach (FL): South Florida Water Management District. Final Report.

Bellrose FC. 1968. Waterfowl migration corridors east of the Rocky Mountains in the United States. Urbana (IL): Illinois Natural History Survey. Biological Notes no. 61.

Chamberlain EB. 1960. Florida waterfowl populations, habitats, and management. *Florida Game and Fresh Water Fish Commission Technical Bulletin* 7: 62.

Clewell AF, Goolsby JA, Shuey AG. 1982. Riverine forests of the South Prong Alafia River system, Florida. *Wetlands* 2: 21–72.

[FGFWFC] Florida Game and Fresh Water Fish Commission. 1957. Recommended program for the Kissimmee River basin. Tallahassee (FL): Florida Game and Fresh Water Fish Commission.

[FGFWFC] Florida Game and Fresh Water Fish Commission. 1994. Lake Okeechobee-Kissimmee River-Everglades resource evaluation project. Tallahassee (FL): Florida Game and Fresh Water Fish Commission. Federal Wallop-Breaux Completion Report no. F-52-5.

Frederick PC, Collopy MW. 1989. The role of predation in determining reproductive success of colonially nesting wading birds in the Florida Everglades. *The Condor* 91: 860–867.

Frederickson LH. 1991. Strategies for water level manipulations in moist-soil systems. Washington (DC): US Fish and Wildlife Service. Leaflet no. 13.4.6.

Gehrke PC. 1992. Enhancing recruitment of native fish in inland environments by accessing alienated floodplain habitats. Pages 205–209 in Hancock DA, ed. *Recruitment Processes. Australian Society for Fish Biology Workshop, Hobart, 21 August 1991.* Canberra (Australia): Bureau of Rural Resources. Proceedings no. 16. Australian Government Printing Service.

Goodrick RL, Milleson JF. 1974. Studies of floodplain vegetation and water level fluctuations in the Kissimmee River valley. West Palm Beach (FL): South Florida Water Management District. Publication no. 74-2.

Howell JC, Heinzman GM. 1967. Comparison of nesting sites of bald eagles in central Florida from 1930 to 1965. *The Auk* 84: 602–603.

Howell P, Lock M, Cobb S, eds. 1988. *The Jonglei Canal: Impact and Opportunity.* Cambridge (UK): Cambridge University Press.

Junk WJ, Bayley PB, Sparks RE. 1989. The flood pulse concept in river-floodplain systems. Pages 110–127 in Dodge DP, ed. *Proceedings of the International Large River Symposium.* Honey Harbor, Ontario (Canada): Canadian

Special Publication on Fisheries and Aquatic Sciences, no. 106.

Koebel JW Jr. 1995. An historical perspective on the Kissimmee River restoration project. *Restoration Ecology* 3: 149–159.

Kushlan JA. 1976. Environmental stability and fish community diversity. *Ecology* 57: 821–825.

Light SS, Dineen JW. 1994. Water control in the Everglades: A historical perspective. Pages 47–84 in Davis SM, Ogden JC, eds. *Everglades: The Ecosystem and Its Restoration.* Delray Beach (FL): St. Lucie Press.

Lowe EF. 1986. The relationship between hydrology and vegetation pattern within the floodplain marsh of a subtropical, Florida lake. *Florida Scientist* 49: 213–233.

Marshall DL, Levin DA, Fowler NL. 1985. Plasticity in yield components in response to fruit initiation in three species of *Sesbania* (*Leguminoseae.*) *Journal of Ecology* 73: 71–81.

Milleson JF. 1976. Environmental responses to marshland reflooding in the Kissimmee River basin. West Palm Beach (FL): South Florida Water Management District. Technical Publication no. 76-3.

Moustafa MZ, Chimney MJ, Fontaine TD, Shih G, Davis S. 1996. The response of a freshwater wetland to long-term "low level" nutrient loads-marsh efficiency. *Ecological Engineering* 7: 15–33.

Parker GG. 1955. Geomorphology. Pages 127–155 in Parker GG, Ferguson GE, Love SK, eds. *Water Resources of Southeastern Florida.* Washington (DC): US Government Printing Office. US Geological Survey Water Supply Paper no. 1255.

Patton JE, Judd WS. 1988. A phenological study of 20 vascular plant species occurring on the Paynes prairie basin, Alachua County, Florida. *Castanea* 53: 149–163.

Perrin LS, Allen MJ, Rowse LA, Montalbano F III, Foote KJ, Olinde MW. 1982. A report of fish and wildlife studies in the Kissimmee River basin and recommendations for restoration. Okeechobee (FL): Florida Game and Fresh Water Fish Commission, Office of Biological Services.

Pickett, STA, White PS, eds. 1985. *The Ecology of Natural Disturbance and Patch Dynamics.* Orlando (FL): Academic Press.

Pierce GJ, Amerson AB, Becker LR Jr. 1982. Pre-1960 floodplain vegetation of the lower Kissimmee River valley, Florida. Dallas (TX):

Environmental Consultants, Inc. Biological Services Report no. 82-3.

Poff NL, Allan JD, Bain MB, Karr JR, Prestegaard KL, Richter BD, Sparks RE, Stromberg JC. 1997. The natural flow regime. *BioScience* 47: 769–784.

Powell GVN, 1987. Habitat use by wading birds in a subtropical estuary: Implications of hydrography. *The Auk* 104: 740–749.

Reed PB Jr. 1988. National list of plant species that occur in wetlands: 1988 (Florida). St. Petersburg (FL): US Fish and Wildlife Service, NERC-88/18.09.

Schlosser IJ, 1991. Stream fish ecology: A landscape perspective. *BioScience* 41: 704–712.

Sparks RE. 1995. Need for ecosystem management of large rivers and their floodplains. *BioScience* 45: 168–182.

Sparks RE, Nelson JC, Yin Y. 1998. Naturalization of the flood regime in regulated rivers. *BioScience* 48: 706–720.

Sykes PW Jr., Rodgers JA Jr., Bennetts RE. 1995. Snail kite (*Rostrhamus sociabilis*). In Poole A, Gill F, eds. *The Birds of North America No. 171.* Philadelphia: Academy of Natural Sciences.

Toland BR. 1990. Effects of the Kissimmee River Pool B restoration demonstration project on Ciconformes and Anseriformes. Pages 83–91 in Loftin MK, Toth LA, Obeysekera JTB, eds. *Proceedings of the Kissimmee River Restoration Symposium, October 1988, Orlando Florida.* West Palm Beach (FL): South Florida Water Management District.

Toth LA. 1990. Impacts of channelization on the Kissimmee River ecosystem. Pages 47–56 in Loftin K, Toth LA, Obeysekera J. eds. *Proceedings of the Kissimmee River Restoration Symposium, October 1988, Orlando, Florida.* West Palm Beach (FL): South Florida Water Management District.

Toth LA. 1993. The ecological basis of the Kissimmee River restoration plan. *Florida Scientist* 56: 25–51.

Toth LA. 1995. Principles and guidelines for restoration of river/floodplain ecosystems-Kissimmee River, Florida. Pages 49–73 in Cairns J Jr, ed. *Rehabilitating Damaged Ecosystems. 2nd ed.* Boca Raton (FL): Lewis Publishers/CRC Press.

Toth LA. 1996. Restoring the hydrogeomorphology of the channelized Kissimmee River. Pages 369–383 in Brookes A, Shields FD Jr, eds. *River Channel Restoration:*

*Guiding Principles for Sustaining Projects.* Chichester (UK): John Wiley & Sons.

Toth LA, Aumen NG. 1994. Integration of multiple issues in environmental restoration and resource enhancement projects in south/central Florida. Pages 61–78 in Cairns J Jr, Crawford TV, Salwasser H, eds. *Implementing Integrated Environmental Management.* Blacksburg (VA): Virginia Polytechnic Institute and State University Center for Environmental and Hazardous Materials Studies.

Toth LA, Miller SJ, Loftin MK. 1990. September 1988 Kissimmee River fish kill. Pages 241–247 in Loftin MK, Toth LA, Obeysekera J, eds. *Proceedings of the Kissimmee River Restoration Symposium, October 1988, Orlando Florida.* West Palm Beach (FL): South Florida Water Management District.

Toth LA, Obeysekera JTB, Perkins WA, Loftin MK. 1993. Flow regulation and restoration of Florida's Kissimmee River. *Regulated Rivers* 8: 155–166.

Toth LA, Arrington DA, Brady MA, Muszick DA. 1995. Conceptual evaluation of factors potentially affecting restoration of habitat structure within the channelized Kissimmee River ecosystem. *Restoration Ecology* 3: 160–180.

Toth LA, Arrington DA, Begue G. 1997. Headwater restoration and reestablishment of natural flow regimes: Kissimmee River of Florida. Pages 425–442 in Williams JE, Wood CA, Dombeck MP, eds. *Watershed Restoration: Principles and Practices.* Bethesda (MD): American Fisheries Society.

Trexler JC. 1995. Restoration of the Kissimmee River: a conceptual model of past and present fish communities and its consequences for evaluating restoration success. *Restoration Ecology* 3: 195–210.

[USACE] US Army Corps of Engineers. 1985. Central and southern Florida, Kissimmee River, Florida. Jacksonville (FL): US Army Corps of Engineers. Final feasibility report and environmental impact statement.

[USFWS] US Fish and Wildlife Service. 1959. A detailed report of the fish and wildlife resources in relation to the Corps of Engineers' plan of development, Kissimmee River basin, Florida. Appendix A in Central and Southern Florida Project for Flood Control and Other Purposes. Part II. Kissimmee River Basin and Related Areas. Jacksonville (FL): US Army Corps of Engineers. Supplement no. 5.

Walters C. 1986. *Adaptive Management of Renewable Resources.* New York: Macmillan.

Weller MW. 1995. Use of two waterbird guilds as evaluation tools for the Kissimmee River restoration. *Restoration Ecology* 3: 211–224.

Wienhold CE, van der Valk AG. 1989. The impact of duration of drainage on the seed bank of northern prairie wetlands. *Canadian Journal of Botany* 67: 1878–1884.

Woody T. 1993. Grassroots in action: The Sierra Club's role in the campaign to restore the Kissimmee River. *Journal of the North American Benthological Society* 12: 201–205.

## Questions:

1. Within two years after reflooding, the demonstration project impoundment had the highest _____ and _____ density in the channelized pools.

2. In a 1957 survey, what percent of fish found in a floodplain marsh were less than 100 mm long?

3. Where is the Kissimmee River basin located?

## Activity:

Go to the Kissimmee River Restoration Project Website at:

**http://www.sfwmd.gov/org/erd/krr/index.html**

Find out what is currently being done on the upper and lower basin projects. See what research opportunities are available. List some of the frequently asked questions and their answers.

# 19

Sediments are the number one pollutant in our rivers and streams. Erosion of soils, due to agriculture and development, is occurring faster than the rate of formation. In addition, agriculture and development not only expose the soil to erosional processes, but also cause changes in soil properties and hydraulic characteristics. These changes make it increasingly harder for pedologists to understand the soil's natural processes. Moreover, it is some of the most unique soils, which could hold an abundance of geologic history, that are highly prized for agricultural uses.

# Do Soils Need Our Protection?

By Ronald Amundson

On clear days, the low mesa north of the city of Merced, Calif., is unremarkable—overwhelmed by the snow-capped peaks of the Sierra Nevada to the east and even by the Coast Ranges, barely visible to the west. Yet, when one leaves the road and crosses through this landscape at sunset on a spring day, the surrounding mountains are forgotten in the expanse of Mima mounds and vernal pools that extend as far as one can see. Technically, this landform is the China Hat member of the Laguna formation, a Pliocene-age fluvial deposit thought to be associated with erosion that began during the last major uplift of the Sierra Nevada. To the north and south, a series of imperfectly preserved inset river terraces record the waxing and waning of subsequent glacial events in the high Sierra. The ages of these river terraces, and the soils that form on them, encapsulate the diversity of soils in the San Joaquin Valley, a large structural basin filled with sediments similar in age and composition to the terraces of the rivers and streams draining the Sierra Nevada and Coast Ranges.

These terraces share other characteristics of soils in the San Joaquin Valley. From all directions, they are being encroached upon by agriculture and urban development. The San Joaquin Valley and its northern counterpart, the Sacramento Valley, are part of California's Great Central Valley, the most important agricultural region in the United States. The Great Central Valley has also recently become home to one of the most rapidly growing suburban populations in the nation. The environmental upheaval wrought by agricultural and suburban development has not gone unnoticed, particularly from a biotic perspective. But the fate of the undisturbed soils seems to have been forgotten.

Why should the loss of undisturbed landscapes be of concern to earth scientists? Because the soils are our imperfect and poorly understood record of Earth's Quaternary continental history. The world's large depositional basins, where this soil record is commonly the longest and most complete, are also the areas most suitable for agricultural and urban expansion. As a result, human society has greatly reduced, or even completely destroyed, unique soil types that may serve as scientific and educational resources for future generations.

## Geological View of Soils

Soils can be defined as historical objects that cover Earth's surface. The age and events that shape this mantle of soils differ greatly from one location to another, producing an almost infinite

From "Do Soils Need Our Protection?" by R. Amundson, Geotimes, March 1998, pp. 16-20. Reprinted with permission.

variety. Considered somewhat differently, any of Earth's soils may be the result of a singular set of events in earth history that may not be repeated exactly again in the future.

The study of soil properties and processes in situ on the landscape is the science of pedology. Pedology is only a century old, originating in Russia in the late 19th century. In its formative period, pedology was largely a historical science, concerned with observational studies that use effects (soil properties) to unravel causes (processes and mechanisms of soil formation). More recently, as the effect/cause relationships have become better understood, the properties of buried fossil soils, or paleosols, have been interpreted in light of the relationships gleaned from earlier observations.

The paradigm of pedology states that a change in any of the factors of soil formation (the conditions that dictate the rate and types of soil-forming processes) inevitably leads to a new or different soil. From this perspective, human activities such as cultivation, irrigation, and urbanization will have a profound effect on soil properties. Simple tillage of the soil surface reduces the carbon and nitrogen content of the soil and changes its hydraulic characteristics. Cultivation is but the most innocuous of the many types of agricultural and urban engineering activities to which soil is subjected. These activities greatly reduce or eliminate the hope of using a given soil in a scientific or educational framework to better understand natural processes.

The future of pedology as a branch of the natural sciences depends upon the availability of natural soils for study. Given the youthfulness of the science and the rapid advances in chemistry, physics, and biology that continue to be made available to the field, it is fair to speculate that we are only just beginning to understand the processes by which soils form, the manner in which they influence global water and atmospheric chemistry, and the way in which they record information about their history. In light of the great potential for scientific advancement, the attitudes toward soil use in the United States contrast sharply with the historical value of soil.

Rates of undisturbed soil conversion have been startlingly rapid. As a pedologist in California for the past 15 years, I have witnessed enormous change.

## Development in California

John Wesley Powell recognized that development of the West depended upon the availability of water and irrigable soils. California's relative abundance of both, along with a well-known equitable climate, has helped make it the most populous state in the nation. Agricultural development from the mid-1800s to the early 1900s quickly utilized much of the potential cropland of the state. While total cropland stabilized long ago, the population of the state continues to grow at an almost exponential rate. California's present population of approximately 30 million people is expected to grow by 60 percent in the next 30 years.

The high cost of housing and declining quality of life in major coastal urban centers have created a trend toward suburban expansion into the unincorporated areas, towns, and small cities of California's agricultural valleys. This expansion, and resultant loss of prime farmland has received deserved attention by agencies such as the American Farmland Trust, which issued a disturbing report in 1997. The report projected that by the year 2040, nearly a third of the Great Central Valley would be in urban or urban-margin land use.

While the loss of agricultural soil to urban use is a long-standing issue, I know of no study devoted to the rates of loss of undisturbed soil. Undisturbed soil is presently being lost throughout the state by urban expansion and by agricultural development, driven by agriculturists moving out of the path of urban sprawl or by improvements in agricultural technology.

## Loss of Soil Diversity

A drive through the agricultural heart of California's Great Valley gives little indication of the rich geological and pedological history underfoot. In the valley, soils range in age from a few hundred years along modern river flood plains to a few scattered river terraces of PlioPleistocene age. As the soils age, they acquire deep weathering profiles, brilliant red colors, high clay contents, silicacemented hardpans, and Mima mounds and vernal pools. The most common soil within this chronological sequence is the San Joaquin soil series, formed in an approximately 250,000-year-old glacial outwash deposit. Approximately 500,000 acres of the eastern Great Valley consist of San

Joaquin soils, making this soil the most areally extensive in the state.

Although it is California's most abundant soil, undisturbed areas of San Joaquin soil (or any other valley soil) are increasingly difficult to find, due to the intensity of land use. Various means of engineering these highly developed soils for agriculture have been implemented since the turn of the century. Horse-drawn and early tractor-drawn land levelers were first used to level the Mima mounds. The hardpan was initially fractured with dynamite. Today, the methods of choice are caterpillar-drawn chisel plows and laser-controlled land levelers. The net result is a level, homogenized landscape that bears only minor resemblance to the soils that once covered the region. Today, if one is interested in observing the Mima mounds of the San Joaquin sod, it must be done in landscape fragments preserved in the midst of a sea of agriculture.

## Earth Science and Soil Preservation

On paper, California may be the ideal location to conduct pedological studies. It possesses an enormous diversity of climate and bedrock and, having escaped continental glaciation, has surficial deposits that cover time spans matched by few places in North America. Yet actually conducting pedological studies is, at best, frustrating and, at worst, impossible.

Why do soil maps of California show large expanses of various soil types (such as the San Joaquin series) while it is paradoxically so difficult to actually observe undisturbed examples of them in the field? The reason lies in the manner in which the effect of agriculture and land use on soil properties is viewed. For pragmatic reasons, it was decided early in the establishment of the present U.S. Department of Agriculture soil classification system that cultivation should not change the classification of the soil. This decision now has unfortunate implications for soil-preservation activities.

Charles Darwin wrote that "rarity, as geology tells us, is the precursor to extinction"—an observation that is now accepted by the majority of our citizens and politicians. Yet, it is difficult for earth scientists

to develop arguments that certain soil types are becoming rare and endangered when our soil maps instead indicate an abundance.

As an earth scientist, I frankly find it surprising that I have found myself advocating this issue. "Soils? They're everywhere!" might have been my response a decade or so ago. However, what I see now is a world increasingly dominated by human-altered soils. Disturbed soils are, indeed, "soils," but the information they bear may not be what we wish to leave as a legacy to our scientific progeny.

Soils are geological bodies that take thousands to millions of years to develop. And, unlike living species, they do not reproduce nor can they be recreated. It is therefore imperative that discussions of soil diversity and preservation begin in our government and educational institutions. We will need to learn from the biodiversity battles waged by our biologist colleagues as we enter the political and economic arenas where preservation debates of all types are ultimately staged.

## Additional Reading

"Biological diversity, soils, and economics" by M. Huston. *Science*, v. 262, p. 1676-1680, 1993.

"Do we treat our soil like dirt?" by B. Gibbons. *National Geographic*, v. 166, p.351-389, 1984.

*Farming on the Edge* by A. A. Sorensen, R. P. Greene, and K. Russ. American Farmland Trust (Washington, D.C.), 1997.

"Late Cenozoic stratigraphic units in northeastern San Joaquin Valley, California" by D.E. Marchand and A. Allwardt. U.S. Geological Survey Bulletin 170, 1981.

"My friend the soil" by H. Jenny and K. Stuart. *Journal of Soil and Water Conservation*, May-June 1984, p. 158-161.

"Soils" (theme issue). *Geotimes*, June 1996.

"Using nighttime DMSP/OLS images of city lights to estimate the impact of urban landuse on soil resources in the U.S." by M. L. Imhoff, W.T. Lawrence, C.D. Elvidge, T. Paul, E. Levine, M.V. Privalsky, and V. Brown. *Remote Sensing of Environment Journal*, v. 59, p. 105-117, 1997.

**Questions:**

1. What is pedology?

2. Why do human activities present a problem for pedologists?

3. Why is it important to preserve soils?

Answers are at the back of the book.

**Activity:**

In your backyard, dig a 12-inch hole, preferably with straight clean edges, and observe the layers formed in the soil.

## 20

Human manure has been used for centuries by farmers. However, human feces contains many pathogens, such as *Cryptosporidium*, *Escherichia coli* 0157, *Salmonella*, and *Giardia*, all of which can kill. Unfortunately, no one can agree on how to make it safe. British water companies have been using raw sewage sludge, which has had plastic and detritus filtered out. They claimed this was safe provided it was not spread on fields where fruits and vegetables were already growing and if farmers waited three weeks before harvesting or letting animals graze on the sludged land.

The food industry disagreed, pointing out that *E. coli* survives in animal manure in fields for months, putting people at risk of infection. So, by 2002 the water companies must give the farmers treated sludge. Now, the problem is what treatment will make it the safest? Some possible techniques are pasteurization, mesophilic anaerobic digestion, thermophilic aerobic digestion and quick lime. But, still no agreement can be made as to which technique or combination of techniques is the most effective.

# Waste Not

By Debora MacKenzie

It's summer. You're by the sea or at the lake. Take a moment to appreciate the clear, sparkling waters. They are probably the cleanest they have been this century. Sewage outfalls that used to blight coastlines and sicken swimmers are disappearing in the industrialised world as we demand more and more purification before waste water is released.

But we are not producing any less of the prime ingredient in sewage. So where is all the shit going, if not into water? See that sandwich in your hand? That's right. In Europe and North America, between one and two-thirds of sludge—the biosolids from sewers—is now spread on farmland.

In principle this is fine. For centuries, human manure has been as highly valued by farmers as any other. Modern Western sensibilities are more delicate, but there's no avoiding the fact that if we don't want faeces in our rivers and coastal waters we must find somewhere else to put them.

And as sewage outflows get cleaner, there is more sludge to dispose of. By 2005, the amount produced in Europe will have increased by 50 per cent as European Union water directives take effect. Agricultural use will have to grow as other disposal routes close. In Britain, 55 per cent of a million tonnes of sludge produced this year will be spread on farmland. By 2002, 61 per cent of 1-5 million tonnes will end up there.

So is it safe? Human faeces contain microbes that cause diseases in humans, especially gut diseases. If these pathogens are getting into our food then it's not just the thought of excrement on food crops that could turn your stomach. In the US, France, Germany and elsewhere, all sludge is treated before being used as fertiliser. It undergoes various forms of heating, drying, fermentation or other treatments that kill germs—to varying degrees—and make sludge more chemically stable. Britain alone allows farmers to put raw sewage on their land, though pressure from food retailers is likely to put a stop to this by 2002. But the question remains as to how best to treat it.

About a quarter of all sewage sludge spread on British farmland this year was raw—little more than settled sewage with the plastic and detritus filtered out. The water companies that produce it say this is safe, as long as users follow voluntary guidelines based on European Union requirements. These prohibit the spreading of sludge on fields where fruit or vegetables are already growing, and say farmers must wait three weeks before harvesting fodder or letting animals graze on sludged land, and ten months before harvesting crops that touch the soil and are "normally eaten raw" —strawberries, for example. The EU rules are the same for raw and treated sludge, except raw sludge must be ploughed or injected into soil. Earlier this year, the British government told an inquiry by the House of Commons Select Committee on the Environment that this is safe.

The food industry disagrees. "Using raw sewage sludge to fertilise food crops is simply not acceptable any more," says Alec Kyriakides, chief microbiologist at the British supermarket chain Sainsbury's. For example, he says there is little proof that *Escherichia coli* O157, a bacterial strain that can cause fatal food poisoning, dies in soil during the waiting periods laid down in the guidelines. It is known to survive in animal manure in fields for months (This Week, 14 June 1997, p 12). The same Commons inquiry heard that dumping untreated human sewage and abattoir waste on farmland may put people at risk from *E. coli* O157 infection. And there is evidence that ploughing raw sewage into soil may actually help pathogens to survive by keeping them in dark, damp conditions.

If the food industry won't buy crops fertilised with raw sludge, farmers won't use sludge. And many water companies are counting on farmers as other disposal routes start to close down. Sea dumping has been illegal in North America since 1993 and will become illegal in the EU at the end of this year. Landfill sites are filling up, and rotting sludge emits methane, a potent greenhouse gas. It can be burned, but burning sludge can release air pollutants, and because it takes energy to dry the sludge before it can be burnt, there is no net energy generation. Incineration costs three to six times the current cost of spreading sludge. So the water companies have just reached an agreement with a consortium of British food companies, promising to give farmers only treated sludge by 2002.

## Safe Sludge

How exactly it will be treated remains to be decided. "We have nothing against recycling sewage sludge, as long as the companies can show it is safe," says Kyriakides. "So far, they haven't." To that end, the water companies have also agreed to fund new research. To satisfy the food industry, this will have to determine which treatments reduce the pathogens in sludge to safe levels although "safe" might be hard to define, as the water companies now say even untreated sludge is safe.

A report on sewage by Britain's Royal Commission on Environmental Pollution in 1996, and the Commons inquiry this year, both decided that safe means killing as many germs as possible. They called for all sludge used in agriculture to be pasteurised, which involves heating at 70 °C for an hour. The Water Services Association, the umbrella group for British water companies, rejects this, saying it is expensive and unnecessary.

The US requires pasteurisation for top-grade sludge—including what the National Parks Service puts on the White House lawn. "The Clintons are walking around on poo," says Alan Rubin, who is in charge of sludge at the US Environmental Protection Agency (EPA). "But it's very clean Poo." Germany effectively pasteurises all sludge used in agriculture. But the US, like France and other European countries, also allows sludge to be spread after less rigorous treatments.

The most common technique—and the one that most British water companies want to use— is called mesophilic anaerobic digestion. This means letting the sludge stew in a closed container at between 20 °C and 40 °C for four or five weeks, while bacteria break down reactive chemicals and produce heat and methane. The methane is used to generate power at the sewage works, and the digested sludge is chemically stable. Typically, says Rubin, this reduces bacterial levels a thousand-fold, but still leaves plenty alive. The Royal Commission concluded that the process is "not particularly effective" at destroying viruses or the eggs of intestinal worms.

Italian researchers found that sludge treated by mesophilic anaerobic digestion contained eggs from the intestinal worm *Ascaris*, *Salmonella* bacteria and enteroviruses. A similar study in Western Australia turned up *Streptococcus* bacteria, *Salmonella* and infective oocysts, of the parasite *Giardia*. All but the *Giardia* actually multiplied in sludge stored for a year, although storage is used by some British water companies to further treat digested sludge.

Most of these pathogens can be removed using a third treatment known as thermophilic digestion. The sludge is left to stew for several days at around 60 °C, either anaerobically or aerobically. In the latter, the fermenting sludge produces its own heat, and this technique is increasingly common in continental Europe. But British water companies still use mesophilic digestion almost exclusively, despite a recommendation made five years ago by their own research lab, the Water Research Centre. It said that any sludge containing oocysts of the parasite *Cryptosporidlum* should be treated by pasteurisation or thermophilic aerobic digestion before being spread on land. *Cryptosporidlum* is a protozoan that causes pain, fever and diarrhoea, can be fatal, and is carried by up to 45 per cent of people in industrialised countries. Hundreds fell ill with it in Britain this year. Animals are usually blamed, but humans excrete it as well.

In May last year, Anglian Water, a privatised water company in southeast England which spreads 40 per cent of its sludge raw, told customers to boil drinking water after finding *Cryptosporidium* oocysts. If these remain in treated water, there will be plenty in sludge. Anglian spokesman Graham Frankland says that by 2000, the company will treat all its sludge with mesophilic anaerobic digestion, plus quicklime (calcium oxide). This reacts with water producing heat and raising the pH of sludge. The Royal Commission said this "significantly reduces" bacteria, but "seems less effective" against viruses.

Frankland says that no process short of oven-drying and turning sludge into pellets kills all germs. These processes also destroy the soil-improving qualities of sludge, says Frederic Cartegny of SEDE, a French company that manages sludge application for farmers. Anglian plans to invest £150 million in sludge treatment over the next two years, says Frankland, but "only because of public perception," he adds. "Untreated sludge is safe." The proof, says the Water Services Association is that "there has

never been a single case of human disease linked to the use of sewage sludge".

The US National Research Council agrees. In 1996 it announced that "there have been no reported outbreaks of infectious disease associated with . . . adequately treated . . . sludge applied to agricultural land". And that includes sludge treated using mesophilic anaerobic digestion. The Environmental Protection Agency also finds no difference between the health of families on farms that use sludge, and farms that don't. But as Ellen Harrison at Cornell University in Ithaca, New York, points out, pathogens applied to land in sludge are most likely to reach humans via groundwater, possibly far from the sludged field. And the National Research Council admits it would be hard to see any cases of common intestinal diseases caused by sludge, such as gastroenteritis, against the high background incidence of infections spread by other routes.

## Contaminated Crops

Even uncommon diseases caused by sludge would be hard to trace. But we do know that people have been infected with cholera and typhoid by crops watered with sewage—where germs are less concentrated than in sludge. Last year in the US, imported crops that had been irrigated with sewage, including salad and berries, were implicated in gut upsets. And in 1985, the WHO revealed a link between sewage irrigation and the intestinal worm *Ascaris*, which causes pain and diarrhoea. In west Jerusalem, shifting boundaries and farming practices led to the levels of *Ascaris* infection increasing when local vegetable crops were watered with sewage, then decreasing when the sewage was turned off.

James Smith of the EPA says *Giardia*, hookworm, *Cryptosporidium*, and germs causing typhoid, cholera, hepatitis A, polio and amoebic dysentery have all been found in sludge in the US. In Britain, there has been a tendency to assume that serious gut pathogens are uncommon outside the tropics, says Mick Brown of the pressure group Water Watch. "Water companies have said they only need to treat sludge from an area with what they call a cosmopolitan population," he says. "They think rural English communities don't produce germs worth worrying about."

But Smith points out that hospital toilets flush into ordinary sewers. Healthy travellers can carry cholera and typhoid, and *Cryptosporidium* and *E. coli* O157 are solidly British. Earlier this year, people from two villages in Scotland claimed that sludge used to reclaim nearby mining land not only smelled "indescribable", but had also caused a rash of local illnesses, including scarlet fever and meningitis. Their claims remain just that—it is difficult to prove any association.

The water companies insist that even if pathogens persist in sludge, they die during the 10 months before crops can be harvested from sludged fields. The US, more cautiously, won't let the public walk on sludged land for a year. Smith says the eggs of some parasitic worms can persist in soil for seven years—and just one causes infection. Bacteria can persist for a year, he says, viruses for six months.

But those may be underestimates. John de Louvois of the Public Health Laboratory Service told the Commons inquiry that most survival studies were carried out before microbiologists could reliably detect pathogens in soil. The few studies performed with sensitive, recent techniques show that some pathogens last longer than had been thought. Enteroviruses, he says, can survive "many hundreds of days" in soil. "All the major bacterial pathogens associated with sewage," he says, are now known to lurk in soil at levels below the detection capabilities of the studies used to establish current safety guidelines—but they can still multiply and cause infection.

David Kay, professor of environmental science at the University of Leeds, says the scientific literature contains "almost no good information to tell us how rapidly viruses die" in sludged soil, and little data even for well-studied coliform and streptococcus bacteria. Caution, and two British inquiries, suggest we should pasteurise the muck. Over the next three years, the water companies will try hard to convince us with their research that sewage—even if laden with germs—is safe on fields.

---

## Questions:

1. When was sea dumping (dumping sewage into our waterways) outlawed in North America?
2. In the U.S. which form of treatment is required for the use of sludge?
3. How much more would it cost to incinerate sludge rather than spread it on the land?

Answers are at the back of the book.

## Activity:

Critical thinking: Design a model community in which human waste can be utilized as fertilizer for local food production and farming. Keep in mind that you need to maintain low cost so this method can be implemented in presently occupied communities.

# 21

Plastics are very important for consumers and scientists alike, but are not biodegradable and end up in our landfills. Green composites may be the answer. Green composites consist of fibers from different plants, such as pineapple leaves, banana stems, kenaf stems and sugar cane. Fibers from these various plants are embedded in a resin matrix making it feel like, look like, and possess the strength of current plastics, except it will be biodegradable and go into our compost piles rather than landfills.

# The Greening of Plastics

By Jeanne Mackin

By early next century, if Anil Netravali has his way, we may be putting broken or outmoded pieces of the family car—the cracked tape deck holder, the broken window handle—out on the compost pile, along with last night's onion peels and apple cores. Netravali, an associate professor in the Department of Textiles and Apparel, is investigating the properties of several natural fibers that, when imbedded in a moldable, degradable matrix, may become the reinforced plastic of the future—one that's biodegradable, or a "green" composite.

Remember the scene from *The Graduate* when a winking uncle advised the graduate that plastics were the future? To a large extent that fictional uncle was right. Plastics did become a large part of the history of the second half of the twentieth century, both for consumers and scientists. Without plastics and other new lightweight composites, the space program would have been all but impossible. In fact, much of the research leading to the new materials and fibers of the twentieth century focused on flight and space exploration and eventually filtered down to more mundane consumer items.

The problem is that those now ubiquitous plastics tend to hang around a very, very long time, even after the product is defunct. Landfills are clogged with the detritus of unusable or just unwanted plastic consumer goods. Recycling is one way of reducing this stream of solid waste, though recycling plastic brings its own problems and environmental issues.

Why not, asked Netravali, make composites that decompose? Plastics that, in fact, could end up in the compost heap instead of the landfill? Enter green composites, as Netravali terms this new and needed material for the manufacture of consumer goods. While small groups of scientists in Europe are also investigating more environmentally friendly plastics, Netravali probably is the only researcher whose ultimate goal is a totally biodegradable plastic.

"Green composites could eventually become inexpensive composites for mass-produced items," Netravali says. "And they would be environmentally friendly because they would be biodegradable."

Netravali came to Cornell, "for one year," he explains, as a postdoctoral associate in 1984, and he has been here since then as a faculty member in the college's Fiber Science Program. The green composite program began officially in 1996.

The substance that nonscientists call plastic is part of a larger group of manufactured substances technically known as polymers. Polyurethane foam, nylon, polyester, and spandex are examples. Reinforced polymer

**From "The Greening of Plastics," by J. Mackin, 1997, pp. 9-11. This article appeared in the Human Ecology Forum, Cornell University. Reprinted with permission.**

composites are versatile and provide a high strength-to-weight ratio, making them invaluable in, for instance, the space program and aviation. *Voyager*, the first aircraft to circle the globe without refueling, was made of a high-strength, lightweight composite. Green composites, made of biologically and chemically active rather than inert materials would, of course, be weaker and less durable. But there are many other applications for such products that are noncritical, where a biodegradable composite could be as feasible as a tougher, nonbiodegradable alternative.

"You wouldn't use a green composite in an application where, if it broke, it would be a significant loss," Netravali says. "You don't want an airplane wing to break, for instance. But there are many other instances of use when if a piece of plastic breaks, it is just inconvenient, not critical. Those uses would be suitable for green composites."

Current research in the fiber science laboratory at Cornell focuses on finding the most advantageous combination of fiber and resin—one that would provide a finished product that is as strong as possible, durable, affordable, and biodegradable—and the most efficient method of combining the two substances. Netravali's research includes fibers obtained from various parts of plants that have already been used by some populations, mostly in Southeast Asia, to make textiles: sugar cane and kenaf steam, pineapple leaves, and banana stems.

"Exotic, but nice," Netravali describes them. "They are strong and degradable." Silk is not used because it is too expensive and involves too much processing. Cotton fibers are short and weaker than those of the pineapple and banana plants, which produce longer fibers.

"We need to incorporate long fibers in composites to obtain maximum advantage of their strength. The ends of the fibers are not load bearing, so shorter fibers with more ends weaken the composite," Netravali explains. Each fiber end, each break, represents a weak link in a chain, and the fewer the breaks the better.

The fibers will be imbedded in a matrix made of resin, just as straw was mixed into mud to make the first bricks. The fibers provide strength, the resin provides form. The resins Netravali uses in his research are polyvinyl alcohol and biopol, a commercial resin made from microorganisms. Because the fibers and resins are both degradable and contain no oil or hydrocarbons, once in the compost pile they'll break down as easily as the morning's orange peel or last night's potato peels. That's the theory. The trick is to find the perfect combination. Getting there is a series of painstaking steps that must determine, among other things, how much load a single fiber can bear after it has been treated with resin.

To demonstrate, Netravali and his graduate research assistant, Shuiyuan Luo, have set up the lab to determine the fiber interface and composite strength. Interfacial shear strength is a critical factor in obtaining better composite properties such as toughness and transverse strength. The long, whitish-blond pineapple fiber is already being used commercially as a geotextile. But how does pineapple fiber respond to resin? What is the interface strength between resin and fiber?

"In a composite, the fiber provides the strength," Netravali explains. "It, not the resin, takes the load."

First, the thread of pineapple fiber must be measured. Natural fibers vary greatly in thickness, unlike synthetic ones such as acrylic, which are made to a uniform diameter. The measurements are taken with a vibroscope, an instrument that vibrates the fiber and measures the frequency at which it resonates to determine diameter. After measuring diameter, the thread is stretched taut in an Instron until it breaks, which gives a measure of tensile strength. Next, a bead of resin is applied to the fiber and pulled out. The test is slow, measuring electronically in microscopic increments to what extent the fiber can be stretched and how much load it can bear before the bond between resin and fiber breaks down.

"We repeat these tests, twenty, thirty times," Netravali says, "then take an average from the different test results." By summer, the testing will have advanced beyond individual fiber and resin bonds to laminate composites. Ultimately, the fibers must be completely imbedded in resin, and five to ten layers laminated on top of each other to produce a laminate green composite.

"It will feel like, look like, and have the strength and other mechanical properties of other plastics, but it will be biodegradable," Netravali says.

The green composite Netravali envisions as the end result of his research will not just be a replacement for current plastics, which use increasingly scarce supplies of petroleum and end up clogging landfills. Green composites could also be a needed substitute for wood in many products such as packaging materials and crates.

"Wood takes twenty-five years to generate," Netravali says. "A pineapple stem takes a year. Green composites would help preserve forests."

Perhaps before the end of this century, Netravali speculates, we'll have reached an understanding of the properties of products made from green composites, and even have some products in the marketplace. "This is probably too optimistic. The main thing is how much effort we put into this program. Time is the only challenge," he says.

Well, perhaps not the only challenge. A major problem Netravali anticipates is convincing industry and consumers of the product's usefulness and the practicality of goods made from them. To some people, biodegradable and plastic are a contradiction of terms.

"We'll have to prove that green composites are worth trying, worth consideration. We'll have to overcome that perception of use," Netravali says.

And initially, products made of green composites may be more expensive than nonbiodegradable plastics. But as green composites gain acceptance and their production increases, the marketplace rule that as production increases costs decrease will help eventually to make the green composites less expensive. Graphite fibers, for example, one of the most common fibers now used for reinforced composites, cost $180 a pound when they were first developed; now the price is down to about $10 a pound.

Netravali has already opened communications—and stirred interest in his research—with several industries, including one of the world's largest chemical companies, Monsanto. "They are being very supportive," he says.

Even if demand and consumer interest are initially slow, Netravali is convinced that green composites will be a large part of the future of plastics simply because we need them. We need to cut back on use of petroleum in industry, reduce plastics in the solid waste stream, and find other natural substitutes for wood, because, as Netravali says, quoting Carl Sagan, "The world is not disposable."

"So plastics should be," he concludes.

---

## Questions:

1. What is meant by 'green' composite?
2. In what way could green composites be a substitute for wood and how would that be beneficial?
3. What problem does Netravali anticipate?

Answers are at the back of the book.

## Activity:

Take a piece of a plastic bottle, a plastic sandwich bag and a paper cup and bury them in your yard (be sure to mark the spot). In 4 weeks, dig up the items buried and compare their rates of decay.

**22**

Plastics are everywhere, from Tupperware to cars. Americans enjoy the convenience of plastic immensely, and consume 14 billion pounds of plastic every year. A significant amount of these plastics go into the production of disposable products, and, therefore, into landfills. However, plastics are not biodegradable. Recycling of plastics is another method of disposal, but plastics can only be recycled into new products a limited number of times. Burning of plastics is a third option that may provide an excellent source of energy with cleaner emissions due to advances in technology to control air pollution.

# Is Combustion of Plastics Desirable?

By Bruce Piasecki, David Rainey, and Kevin Fletcher

What should be done with the plastics in our garbage? This question mirrors in miniature the complex choices facing policymakers about what to do with solid wastes in general. For plastics, the issue is clear: Some amount of plastic waste, primarily resins that cannot be economically recycled, must either be buried in the ground or burned in a waste-to-energy plant. And considering that a good deal of energy may be derived from plastics, one must ask whether it might be wiser for industrial consumer societies to burn more of it and bury much less.

In 1993, plastics of various sorts accounted for approximately 9 percent (by weight) of all garbage, and the United States Environmental Protection Agency (EPA) estimates that this number will rise to over 10 percent by the year 2000. Right now, many people see recycling plastics as the best option. But empirical evidence shows that not more than 10 percent of the plastics in a town's solid waste actually gets recycled, and average municipal recycling rates are closer to 18 percent across America. People do not always participate in recycling programs, and market forces have not yet made plastics recycling attractive enough for the practice to become more widespread than it is. The same trends are seen in Europe, Asia and the nations of the former Soviet Union. Some of the rare

exceptions have been observed in towns in Germany and Japan.

Of the three waste-management options, burying is the least desirable. Plastics are now buried in landfills designed to minimize the kind of biological activity that would normally degrade wastes, so waste—not just plastics—dumped in landfills remains there, frequently for decades. Faced with the options, burning starts to look better, especially when one considers that plastics can be a relatively clean, reliable and economical energy source. Burning waste starts to seem particularly attractive as the world begins to grapple with the question of greenhouse gases generated by the combustion of fossil fuels.

A plastic bottle in a landfill just takes up space. The materials and energy used to produce it will never be recovered—even partially. If the same plastic bottle is recycled, the materials used in its manufacture will be partially recovered. But if a bottle is processed in a waste-to-energy plant, a portion of the initial energy invested in making the bottle will be recovered in the form of electricity, steam or heat. This becomes more significant in an age of climate change.

Of course, burning plastic carries with it certain controllable environmental risks. Burning plastic releases compounds such as

From "Is Combustion of Plastics Desirable?" by B. Piasecki, D. Rainey and K. Fletcher, American Scientist, July/August 1998, pp. 364-373. Reprinted with permission.

polyvinyl chloride, acid gases, dioxins and carbon dioxide, which are potentially harmful to people and produce unwanted consequences to the earth's atmosphere and weather. In addition to organic toxins, incinerated plastics may release heavy metals, such as lead and cadmium, into the atmosphere.

As policymakers consider the question of how to handle plastic wastes in the future, they will grapple again and again with how poorly scientists understand the significant trade-offs in solid-waste management choices. Carol Browner, administrator of the EPA, cautions that Americans are in an age of "environmental adolescence." Still, given the potential environmental, resource-recovery and economic benefits of burning plastics, it is instructive to examine each of the leading technical concerns regarding this procedure.

## Power from Plastics

Municipal solid waste has an energy content that is recoverable, making it suitable and valuable for combustion. In fact, when garbage is burned in a waste-to-energy facility, there is rarely a need to add supplemental fuels to maintain combustion. Of all types of garbage, plastics release the most energy per unit of weight. Plastic accounts for only about 8 percent of the municipal solid waste burned, but it already provides about 34 percent of the energy liberated from combustion. Surprisingly, wood and paper account for relatively little of the liberated energy

Compared with combusting most carbon-based fuels, such as oil or coal, waste is a clean power source. A modern waste-to-energy facility generates less sulfur and nitrogen oxides—both precursors to acid rain—than do most existing coal- and oil-fired power plants. Even when compared with natural gas, energy from waste looks good, emitting fewer nitrogen oxides and only slightly more sulfur oxides. That is why people must balance these environmental benefits into their management choices.

Currently municipal-waste combustors contribute less than 1 percent to the total carbon dioxide emissions in the United States. Even if all of the country's solid waste were burned in waste-to-energy plants, as opposed to the roughly 16 percent now burned, the conversion of waste to energy would contribute only about 4 percent of total carbon-dioxide emissions. This is a small figure compared with the volume of carbon dioxide emissions produced during the production of petroleum, steel, cement and chemicals.

However, burning plastics is not emission free, and the two main foci of legitimate concern have been the release of chlorine and heavy metals into the environment. Plastics made of polyvinyl chloride, or PVC, contain 40 percent chlorine. Chlorine is a component of both hydrogen chloride, an acidic gas, and of polychlorinated dibenzo(p)dioxins and furans (PCDD and PCDF, respectively), compounds known to cause or suspected of causing cancer and other adverse health effects in laboratory animals, wildlife and people. A particularly intense aspect of this debate centers on the concern that chlorine and chlorine compounds in the environment might combine with other chemically active agents in the environment to make toxic compounds, including dioxins and furans.

In addition to chlorine, heavy metals are also released when plastics are burned. These metals are derived from lead- and cadmium-based pigments and stabilizers used in the plastics. Heavy-metal concentrations appear to be about 10 parts per million (ppm) of cadmium in PVC, about 200 ppm of lead in PVC and about 100 ppm of lead in polyethylene. These concentrations are, of course, lower than concentrations of heavy metals found in the metal components of municipal solid waste. However, metals tend to be removed from solid wastes prior to combustion, which in effect increases the relative contribution of heavy metals from plastics.

Still, when deciding on the desirability of burning plastics, it is crucial to determine the actual quantities of these substances released during burning. Recent studies reveal some answers that will come as a surprise to many.

In a 1994 study conducted for the Association of Plastics Manufacturers in Europe, Frederick E. Mark reported last year on the results of the controlled combustion of solid waste in a commercial waste-to-energy facility in Würzburg, Germany. In his study, Mark looked at the emission profile of waste in which the plastic component ranged from a low of 10 percent of the total weight, to the medium case of 17.5 percent and, finally, to the high-composition level of about 25 percent.

The study was designed to test the effects of burning plastics within a real operating environment. For all tests, the incinerator was operated at the full thermal-load capacity of the boiler. Among other things, this means that as plastic was added above the base amount of 10 percent, the total amount of solid waste fed into the incinerator was reduced. The plastic materials tested contained a mixture of common commercial plastic polymers, including polyethylene, polystyrene, polyethylene terephthalate and polyvinyl chloride. Mark monitored the gases emitted from the incinerator for a variety of pollutants.

The level of hydrochloric acid emission was essentially identical for all three cases. The same is true for emissions of dioxins and furans. In all cases, after treating the flue gases with pollution-controlling lime and activated carbon, stack-gas concentrations of dioxins were well below the rigorous German emission limit of 0.1 per cubic meter.

Mark notes that the dioxin levels he measured in his study fall into the lower part of the legally acceptable range of the European solid-waste combustion industry. He goes on to conclude that the findings indicate that the furnace is well run and well designed. The results from the Mark study underscore the fact that the controlled combustion in and pollution-control equipment of a modern waste-to-energy facility easily reduce hydrochloric acid and prevent additional dioxin formation even when plastics constitute a relatively high proportion of the total waste composition.

With the exception of glass and metal, all principal components of municipal solid waste contain chlorine. The vast majority of materials, whether natural or manufactured, contain chloride or chlorine. Even vegetable matter contains chloride at levels that range from 200 to 10,000 ppm, so burning yard waste will release some chlorine into the environment. Although this is far less than the 400,000 ppm of chlorine emitted by polyvinyl chlorides, it is important to recognize the variety of sources for this ubiquitous chemical. Removing chlorinated plastics from municipal solid waste bound for burning will indeed reduce the chloride concentrations in the raw gas of an incinerator. However, such reductions will likely be insufficient to allow a waste combustor to operate without acid-gas emission controls, or to result in any material difference in dioxin formation and emission. This significant physical evidence must inform policy choices regarding the viability of plastics combustion in waste-to-energy plants.

The general management point is this: The amount of dioxin emitted from a waste-to-energy incinerator is influenced by many factors. In spite of the emphasis on chlorine in current public debates of waste-to-energy proposals, chlorine in the waste stream should be one of the least important of these. The important factors parse into two major classes—combustion control and post-combustion control. The former includes designs and practices intended to optimize combustion efficiency. The latter includes air-pollution-control devices and additives intended to minimize the release of acid gases and particulates into the environment. It is this latter arena that presents the greatest opportunities for improvements.

Returning then to the evaluation of commercial waste-to-energy combustors, it is important to note that Mark's results compare quite well with other studies. The New York State Energy and Research Development Authority reported on a series of tests performed at the Vicon incinerator in Pittsfield, Massachusetts. Dioxin formation was found to correlate roughly with temperature and oxygen levels, but it did not correlate with the amount of polyvinyl chloride added to the waste stream. Tests of commercial incinerators in France, Belgium and Italy similarly found that polyvinyl chloride concentrations had no measurable effect on emissions of dioxins and furans. In 1993, the U.S. Department of Energy reviewed the available literature and concluded, "Regardless whether or not [polyvinyl chloride] is present in [municipal solid waste], when [municipal solid waste] is incinerated, available control measures can limit dioxin emissions to levels that are below current regulatory concern. The presence or absence of [polyvinyl chloride] in the [municipal solid waste] stream will not reduce the need to employ control measures." This official Department of Energy position restates the logic, public safety and managerial appeal of burning plastics in controlled settings.

In addition, after reviewing several studies, along with information from tests performed at the waste-to-energy combustors in Westchester County, New York, and Marion County, Oregon, the U.S. Office of Technology Assessment concluded that "plastics do not appear to play a major role in the formation of dioxins and furans within the combustion chamber." In spite of these significant findings, the presumption that burning plastic releases toxic levels of dioxins and chlorides continues to inform many policy debates. One sees this popular misrepresentation of the issues in a host of public-interest campaigns to stop incineration as a policy option, as well as more focused, yet unscientific efforts to rid the world of chlorine.

## Generating Energy

The Mark study, and others like it, are starting to demonstrate that emissions from plastics can be much less toxic than is popularly perceived. Yet, the essential ingredient in these findings is the requirement for a modern combustion facility with adequate pollution controls.

Each waste incinerator is different in its specific design; however, they all share some general characteristics. Waste destined to go into a mass-burn incinerator is moved by crane from a refuse pit to a storage hopper, which feeds a charging mechanism. The charging mechanism controls the rate of flow of the hopper and drops the waste by gravity onto combustion grates on the bottom of the boiler. In the combustion zone, the solid waste is burned, which liberates heat and reduces the waste volume by an average of 85 percent.

Normal operation requires that adequate amounts of air be present in the combustion zone in order to ensure complete combustion. Combustion air typically enters through a grate and at ports above the grate and mixes with the burning waste. The combustion process generates products of combustion at high temperatures, ranging between 1,800-2,000 degrees Fahrenheit (982-1,093 degrees Celsius). Because municipal waste contains so many different components—wood, paper, cloth and plastics, to name a few—it is an extremely heterogeneous fuel and must travel along the grate for a considerable time before it is combusted completely

One result of combustion is that the waste is now divided into two states: gaseous and solid. The gaseous products of combustion, called the flue gases, leave the lower grate section and flow upward to the furnace section. The residual solid mass on the grate drops into a container that collects what is called bottom ash.

The gaseous component of the waste is very hot, and this heat is extracted and used to produce energy. The gases pass by steel tubes with water inside. The heat is transferred through the metal tube to the water inside, which absorbs the heat and transfers it to a steam-producing system. It is important that the temperature be adequately controlled and that the flue gases be retained in the combustion zone for the required amount of time to ensure proper combustion and to reduce the toxins in the emissions. Generally flue gases should be retained in the combustion zone for 2 seconds, and the combustion temperature should be at least 1,800 degrees Fahrenheit. In the upper sections of the furnace there are water-cooled surfaces to pre-heat the boiler water and additional sections to superheat the steam. This superheated steam is generally used to turn steam-powered turbine-generators, which produce electricity

The flue gases include toxic gaseous acids and organic molecules. Large waste-to-energy plants incorporate numerous air-pollution control devices, such as carbon filters and lime slurries to reduce the toxic load in these gases. For example, sulfur, a precursor to sulfur dioxide, as well as sulfur dioxide itself, can be removed by either wet or dry scrubbers. Wet scrubbers use an alkaline reagent, such as limestone, to remove approximately 85 percent of the sulfur dioxide. Dry scrubbers use lime to remove approximately 75 percent of the sulfur dioxide. Scrubbers can also remove hydrochloric acid, one of the waste products evolved from plastic combustion specifically. There is always the possibility that incomplete combustion will produce carbon monoxide. The key to eliminating this gas from the emissions is to ensure the proper functioning —combustion at the correct temperature and with the correct degree of oxygenation—to guarantee complete combustion of the waste products.

## Ashes and Dust

The flue gases that leave the furnace section contain particulate matter, including ash and trace metals. (The larger particles, too heavy to remain airborne in the gases, generally leave the grate section as bottom ash.) Smaller, lighter particles are carried in the flue gases, as is mercury gas. These are the particles that are most likely to pose a risk to human health. Heavy metals pose a particular concern, since they have the potential to leach into groundwater. However, it is important to note that this is more likely to happen to heavy metals disposed of in a landfill. In contrast, metals that go through a waste-to-energy facility can be recovered or controlled and are less likely to become an environmental hazard. Waste-to-energy facilities often use simple technology to separate out and recover recyclable metals, such as iron.

Currently there exist two means of removing these particles. Electrostatic precipitators use electrical charges to energize the particulate matter that then can be attracted to oppositely charged plates, where the particles are collected. Using this method it is possible to collect about 99.7 percent of the particulates in the flue gases. The second method employs cylindrical bags, called baghouses, that mechanically filter out particulates from the gas.

Of particular concern in waste combustion are trace amounts of heavy metals that either enter the bottom ash or vaporize into the air emissions. Scrubbers and baghouses are both effective in reducing the metal compounds in the flue gases. The great majority of waste-to-energy facilities in operation in the United States today make use of these advanced pollution controls.

Unlike dioxins and furans, metals are neither destroyed nor formed during combustion. The release of metals into the air from an incinerator bears a more direct relationship to the metals put into the furnace minus, of course, those metals removed between the furnace and the exiting gas. As they leave the furnace, the flue gases cool, and essentially all of the potentially toxic metals become part of the ash released by the combustor. The efficiency with which heavy metals are removed from the ash is determined by the efficacy with which particulates in general are removed from the emissions. Mass-balance studies show that approximately 99 percent of particulate-bound heavy metals can be removed through existing affordable means.

Plastics contribute primarily cadmium and lead to the total heavy-metal load of incinerator ash, but, unfortunately, there are still few reliable data on the exact percentage of heavy metals in ash derived from combusted plastics. The Würzburg study previously discussed did test the ash derived from combustion of the medium- and high-polymer wastes. In this study, Mark reported that compared with the total concentration of regulated heavy metals in the combustor feed stream, heavy-metal contributions from polymers were insignificant.

Furthermore, he found that increasing the amounts of plastics in the feed did not increase heavy-metal concentrations found in the ash. Finally, he concluded that, overall, heavy-metal concentrations matched the typical ranges common for Western European municipal waste combustor operations. Since the amount of particulate matter is not yet measured directly, measuring levels of heavy metals in the emissions is the best current indicator of how much particulate matter in general is being produced by waste-to-energy combustors.

## New Directions

Like manufacturers of batteries, electronics and other products, plastics manufacturers are increasingly seeking to reduce the amount of heavy metals in their products. They are looking for substitute pigments and heat stabilizers that do not contain heavy metals. Few companies still sell lead-based pigments, but cadmium-based pigments are more difficult to replace, since the pigments containing cadmium are more resistant to fading.

Researchers have had more success in identifying replacements for cadmium- and lead-based heat stabilizers. But lead has been especially difficult to replace. A 1992 EPA report notes that "the replacement of lead-based heat stabilizers in electrical cable insulation and jacketing ... has been difficult due to the critical properties of weathering, humidity resistance and thinness of the jacket that lead imparts."

In response to public attention and the threat of further regulation, new technologies and market advantages have driven the plastics industry to actively seek replacements for heavy metal additives. In addition to preparing for potential U.S. regulation, companies are also watching international regulatory developments. For example, Sweden's government is beginning to restrict plastics additives. In 1993, Bo Wahlstrom and Beryl Lundqvist, in a text developed for the Swedish Environment Institute, reported that cadmium for surface treatment in plastics and in dyes is banned in Sweden and has been for many years. They anticipated similar prohibitions to be introduced broadly throughout the European Community. They noted at the time that the EPA was expected to consider a proposal to restrict plastic additives in the U.S. and to monitor the transition to lead-free pigments and drying agents in paint production. The follow-up actions by the EPA are still pending. As such, when the new regulations are introduced, U.S. firms wishing to remain competitive in the European market will be required to further reduce the amounts of heavy-metal additives in plastics. This will spur further innovation across the industry

In addition to seeing changes in the composition of the plastic product, the public can anticipate seeing some changes in the waste-to-energy industry. The industry has learned from experience that it must demonstrate the integrity of its operations to an increasingly sophisticated public in order to remain a player in the vital arena of solid-waste management

Continued improvement is evident in technologies to control air pollution and in ash management—the two most ardent public concerns. That the public is justified in its concerns is borne out by comparisons between older and newer waste-combustion facilities. Recent enhancements to the combustion process, for example, have enabled waste-to-energy operators to significantly reduce dioxin formation. A 1992 study conducted by Herman Vogg for the International Symposium on Chlorinated Dioxins and Related Compounds showed that the overall pollution factor for an older plant was 25 nanograms per cubic meter of raw emission gas, versus one-tenth that value, or 2.5 nanograms per cubic meter, in a newer facility. Vogg's data also showed that modern combustion techniques and equipment reduce dioxin levels in raw gas by as much as 90 percent.

One new feature of many recently completed facilities, and of those still in the planning stages, controls mercury in emissions. The EPA emissions standards for metals, which were released in 1995, require mercury controls on facilities, since mercury emissions cannot be effectively reduced by traditional control methods.

Advances in ash management are making it more likely that beneficial uses for the by-products of solid-waste combustion will increase in the future. However, in the case of *Chicago v. the Environmental Defense Fund*, the U.S. Supreme Court ruled in 1994 that operators must now test ash for compliance with federal standards before disposing of it in a nonhazardous-waste landfill.

The Court ruled that ash from waste-to-energy plants that use solid waste as their fuel may be hazardous, even though the products from which it is derived are not. This is because contaminants in the ash residue are more concentrated and more likely to leach into the ground water than are the preburned components of waste. The Court ruled that the Resource Conservation and Recovery Act, which is the dominant federal legislation that defines solid waste and makes distinctions between manageable and hazardous wastes, does not exempt ash derived from municipal-solid waste from the regulation of hazardous waste. Prior to that ruling, the EPA classified ash generated from nonhazardous waste as nonhazardous. The Court's ruling was based on an argument made by the Environmental Defense Fund that the Northwest Waste-to-Energy plant in Chicago violated the Resource Conservation and Recovery Act by dumping ash that exceeds the federal standards for lead and cadmium in landfills not designed for hazardous waste. The Supreme Court's decision directed the EPA to develop regulations concerning the type and frequency of testing necessary to determine the toxicity of ash.

Reporting on the Court's decision on May 3, 1994, Linda Greenhouse wrote in the New York Times that over the short run, ash-disposal costs may increase from three to ten times. Engineers estimate that over the long run, however, the economic impact of the Court's decision should be minimized if ash-stabilization and reuse technologies continue to advance as expected. At some point, the ash might be put to some practical uses, as it is now in Europe. In the Netherlands, combustor ash is an ingredient in the blocks used to construct dikes. And elsewhere in Europe, ash is used for building roadbeds. Combustion will remain an important component of environmental management in the next century, and waste feeds will increasingly include plastics.

## Trade-offs

Managing waste will always entail some tradeoffs. All of the three options—recycling, landfilling and combustion—have some disadvantages. Even landfilling, which produces no emissions, fails to take advantage of the energy value inherent in plastic. Waste combustion, on the other hand, recovers the energy in plastic materials and reduces the volume of disposed solid waste by up to 90 percent of its initial preburn volumes. However, this management option generates emissions and produces an ash residue that must be managed.

As demonstrated by recent test burns, improvements in combustion and air-pollution-control technology have dramatically reduced the health risks from emissions and ash. Recent studies have shown that plastics—in quantities even higher than those normally found in municipal-solid waste—do not adversely affect levels of emissions or the quality of ash from waste-to-energy facilities.

In addition, waste-to-energy facilities may be a relatively economical source of fuel, and may be a more economic solution to waste management than the other available options. A waste-to-energy plant generally produces electricity that is sold to the electric utilities for approximately six cents per kilowatt-hour, a rate that is competitive with those offered by nuclear power plants and power plants that generate energy by burning fossil fuels.

Waste-to-energy facilities also have an advantage over landfills, since they generate revenue that partly offsets other costs related to running the plant. Assuming that municipal solid waste contains 5,000 Btus per pound, a waste-to-energy plant produces about 500 kilowatts per ton of municipal solid waste. Thus a plant burning 50 tons each hour produces 25,000 kilowatts per hour, generating a revenue stream of $1,500, or $30 per ton of municipal solid waste. In addition to the energy recovered, burning waste has the advantage over landfills, as it avoids a tipping fee, the price that garbage collectors pay to the landfill to deposit waste there. Tipping fees generally run between $60 and $90 per ton. Overall, burning waste generally costs between $90 and $120 per ton, which includes the capital cost, the operating cost and the cost of ash disposal.

The EPA estimates the U.S. will generate approximately 220 million tons of municipal solid waste by the year 2000. Even their optimistic estimate that 35 percent of plastics will be recycled requires that the rest be burned in waste-to-energy facilities or buried in landfills. Although landfills become more expensive to construct, the cost of building a waste-to-energy facility is offset by the revenue generated by producing a usable product.

Of course, the public should continue to ask pressing questions about the appropriateness and safety of waste-disposal options for plastics. Citizens should continue to hold waste-to-energy

and landfill operators to a high standard of operating integrity. Productive relations between citizens and industry can, and do, positively affect the performance of waste-management facilities. The need for sustained public scrutiny and review is clear. Just as performance standards have been aggressively improved in the combustion industry, plastics manufacturers and others are continuing to respond to public concerns by increasing their use of nontoxic additives.

In the final analysis, communities and consumers will often need to make decisions in this age of environmentalism based on evidence that may seem incomplete—or even contradictory. But the evidence regarding plastics combustion in modern waste-to-energy plants is clear. Waste-to-energy should remain an acceptable, even desirable, option for managing plastic wastes.

## Acknowledgment

The authors would like to thank Laura Green, president of Cambridge Environmental, for her help with the early drafts of this manuscript.

## Bibliography

DeFre, R. 1986. Dioxin levels in the emissions for Belgian municipal incinerators. *Chemosphere.* 15:1255.

Franklin Associates. 1994. *The Role of Recycling In Solid Waste Management to the Year 2000.* Stamford, Conn. Keep America Beautiful, Inc., p. 4.

Greenhouse, L. 1994. Justices decide incinerator ash is toxic waste. The *New York Times*, May 3, p. Al

International Ash Working Group. 1994. *An International Perspective on Characterization and Management of Residues from Municipal Solid Waste Incineration.* New York: The International Ash Working Group.

Karasek, F., A. Viau and M. Gonnord. 1983. Gas chromatographic-mass spectrometric study on the formation of polychlorinated dibenzo-p-dioxins and polychlorobenzenes from polyvinyl chloride in a municipal incinerator. *Journal of Chromatography* 270:277.

Kiser, J., M. Charles, J. Bridges, G. Carr and C. Velzy. 1994 WTE: Citizens respond to plants in their neighborhood. *Solid Waste Technologies*: May/June.

Kiser, J. 1993. *The ISWA Municipal Waste Combustion Directory: 1993 Update of U.S.*

*Plants*. Washington, D.C.: Integrated Solid Waste Association.

Mark, F. E. 1994. *Energy Recovery through Co-Combustion of Mixed Plastics Waste and MSW.* Denmark: Association of Plastics Manufacturers of Europe.

Meikle, J. 1995. *American Plastic: A Cultural History.* Newark, New Jersey: Rutgers University Press.

Office of Technology Assessment. 1989. *Facing America's Trash: What's Next for Municipal Solid Waste.* OTA-0-424. Washington D.C.: Government Printing Office.

Roffman, H., and R. Cambotti. 1991. The Woodburn MWC-ash study, fourth year results as of September, 1991. *Proceedings of the 4th International Conference on Municipal Solid Waste Combustor Ash Utilization.* Arlington, Virginia: University of New Hampshire, Environmental Research Group, pp. 117-142.

Tellus Institute and JSC Center for Environmental Health Studies. 1994. *Incineration: Decisions for the 1990s.* Boston: Tellus Institute and JSC Center for Environmental Health Studies.

U.S. Department of Energy. 1993. *Polyvinyl Chloride in Incinerated MSW: Impact upon Dioxin Emissions.* April: Report No. NREL/TP 430 5518, NREL, Golden, Colo.

U.S. Environmental Protection Agency. 1990. *Characterization of Municipal Solid Waste Combustion Ash, Ash Extracts, and Leachates.* EPA-530-SW-90-029A. Washington, D.C.: EPA.

U.S. Environmental Protection Agency: 1990. *Methods to Manage and Control Plastic Wastes.* Washington D.C.: EPA.

U.S. Environmental Protection Agency. 1992. *Characterization of Municipal Solid Waste in the United States.* Washington, D.C.: EPA.

U.S. Environmental Protection Agency 1995. *Municipal Waste Combustion: Background Information Document for Promulgated Standards and Guidelines; Public Comments and Responses.* Research Triangle Park, N.C.: Office of Air Quality Planning and Standards.

Valenti, M. 1993. Today's trash, Tomorrow's fuel. *Mechanical Engineering.*

Vogg, H. 1992. *Arguments in Favor of Waste Incineration.* 1992 Annual European Toxicology Forum. Copenhagen, Denmark.

Wahlstrom, B., and B. Lindqvist. 1993. Risk reduction and chemical controls. In*: Clean Production Strategies: Developing Preventive Environmental Management in the Industrial Economy,* ed. T. Jackson. Chelsea, Mass.:Lewis Publisher.

---

## Questions:

1. How much of the plastic in a town's solid waste gets recycled?
2. How is combusting waste better than combusting coal, oil, and natural gas?
3. Which emissions from the burning of plastic are people most concerned about?

Answers are at the back of the book.

## Activity:

Contact your local Waste Management Company and/or Beautification Board, and find out the percentages of plastics that are landfilled, recycled, and burned.

# 23

The acid rain problem in the United States is finally on the mend, after 20 years of control. Sulfur dioxide emissions are now down by half. Economists are trying to figure out why the control is so cheap. The answer could be that the United States has employed a flexible, freemarket approach to curb acid rain and is now trying to spread these lessons worldwide. Could this approach work on greenhouse gases?

# Acid Rain Control: Success on the Cheap

By Richard Kerr

Back in the 1970s, sulfuric acid seemed to be consuming the environment. Spewed from power plant smokestacks, it rained or drifted down on lakes, streams, forests, buildings, and people in ever-increasing volumes, killing fish and trees, disfiguring stone buildings, and corroding the lungs of people.

But today, after 20 years of control, acid rain is a problem on the mend. In the United States, emissions of sulfur dioxide—the chief precursor of acid rain—are down by half. The nation is on track for another round of reductions beginning in 2000, and, with some significant exceptions, lakes and forests are on the road to recovery. Perhaps even more surprisingly, U.S. acid rain control has been a bargain: The latest cost estimates are about $1 billion per year— dramatically lower than earlier forecasts of $10 billion or more, and about half as much as even the lowest estimates.

As negotiators gather this week in Buenos Aires to try to figure out how to cut greenhouse gas emissions, the story of U.S. acid rain control offers a case study in the successful regulation of a wideranging pollutant. Economists are still trying to understand just why control is proving so cheap, but they agree that at least partial credit must go to the unusually flexible U.S. regulations and their use of the free market. In the 1990 Clean Air Act Amendments, Congress

told power plant operators how much to cut emissions but not how to do it, and established an emissions trading system in which power plants could buy and sell rights to pollute.

It was "a radically different way to go about environmental regulation," says economist A. Denny Ellerman of the Massachusetts Institute of Technology (MIT). "The lessons learned are pretty impressive." The United States is now trying to spread those lessons worldwide. Indeed, in Europe, where acid rain reductions appear to be more expensive than in the United States, regulators are taking a close look at the U.S. model. A flexible system of emissions trading also serves as the crux of U.S. proposals for reining in greenhouse warming—although no one is sure whether such a system can be scaled up to work across many different countries. "We proved the concept," says Joseph Kruger of the Environmental Protection Agency (EPA) in Washington, D.C. "If the acid rain program hadn't been such a success, we wouldn't be talking about trading greenhouse emissions."

## A New Flexibility

The prospects for economical acid rain reductions by any means looked bleak in the 1980s, says Joseph Goffman of the Environmental Defense Fund in Washington, D.C. In the late '80s, when it was thought that sulfur dioxide emissions—then totaling 25 million tons a year—would have to be reduced by 10 million tons a year, he recalls, estimates of the cost were running from many hundreds to $1000 for every ton shaved off the total, or a cool $10 billion a year. Those high prices were based on complying with the standard type of "command and control" emissions regulations, in which regulators made all the decisions. In the 1977 Clean Air Act, for example, regulators decided on a control technology—a "scrubber" that strips the sulfur dioxide from the spent combustion gases before they go up the stack—and they also decided which plants needed scrubbers.

Under a command-and-control scheme, "you've fixed the technology in place," says Goffman. "You've eliminated innovation. We did this in the '70s and '80s because that was all we knew how to do. For a while it worked well," until the easy, cheap reductions had been made. By the late 1980s, regulators had started to look for cheaper options.

When Congress contemplated the next round of emissions cuts, the $10 billion price tag triggered sticker shock. Instead of instituting ever more draconian and expensive command-and-control regulations, Congress took a new tack in the 1990 Clean Air Act Amendments: It commanded reductions but let power plant operators figure out the cheapest way to control emissions. The reductions were to come in two steps. Starting in 1995, 110 mostly coal-burning plants out of thousands in the country—then emitting about 4 pounds of sulfur dioxide per million British thermal units (mBtu) of heat—would be cut back to only 2.5 pounds/mBtu. In Phase II, starting in 2000, more plants are to fall under the plan and emissions will be tightened to 1.2 pounds/mBtu. The total release expected in 2010 is 8.95 million tons per year, a reduction of 10 million tons per year from the amount projected to be released without controls.

Congress made the rules even more flexible by authorizing a limited number of emission allowances, "right-to-pollute" coupons that could be bought, sold, or saved. Such trading with a cap on total releases means emitters are "strictly accountable for the end result," says Kruger, "but they have flexibility in the way they get there."

## Cost and Effect

But as the final Clean Air Act Amendments neared passage in 1990, just how much money the new rules would cost was a matter of sharp debate. At the high end, some lobbyists, columnists, and industry advertisements were touting vaguely documented figures of "$3 billion to $7 billion per year, with the price tag rising to $7 billion to $25 billion by the year 2000," according to environmental policy analyst Don Munton of the University of British Columbia. The lower end of these estimates compares with the estimated cost of simply putting scrubbers on the 50 dirtiest plants. That was thought to cost $7.9 billion per year, according to a 1983 Office of Technology Assessment study, or $11.5 billion per year, according to an industry study (figures in 1995 dollars).

More rigorous cost projections came in lower. These generally fell within the range of a 1990 study for the EPA made by ICF Inc. of Fairfax, Virginia, that found annual costs (in 1995 dollars) could be as low as $1.9 billion per year through to the 2010 goal or as high as $5.5 billion per year. But the lower figures were not widely believed at the time. When EPA testified to Congress just before passage that the annual cost in 2010 could be roughly $4 billion, notes Kruger, "we were roundly criticized for being overly optimistic."

It turns out that those figures weren't optimistic enough. Two groups of economists—Dallas Burtraw and colleagues at the Washington think tank Resources for the Future (RFF) and Anne Smith of Charles River Associates in Washington, D.C., Jeremy Platt of the Electric Power Research Institute (EPRI) in Palo Alto, California, and Ellerman—have recently compared those early analyses with actual costs. In 1996, after the first 2 years of the Phase I limits, emissions from participating power plants dropped to 5.4 million tons, 35% below the legal limit for those plants of 8.3 million tons. And it was done at a cost of about $0.8 billion per year, according to two independent estimates by Ellerman and by Curtis Carlson and colleagues at RFF.

Phase I was expected to be cheaper than later reductions, but estimates of the longterm costs through 2010 have also been dropping. By 1995, ICF's estimate for the EPA had dropped to $2.5 billion per year. EPRI's 1997 estimate was down to $1.6 billion to $1.8 billion per year, and Carlson and colleagues' 1998 estimate is $1.0 billion—a far cry from many earlier estimates and below EPA's early projections.

Why is acid rain reduction so economical, at least so far? Economists are still exploring the answer, but they agree that the biggest advantage was the overall flexibility of the program, which allowed power plants to exploit unexpected opportunities. The emissions trading system has been just one factor in this flexibility, these analysts conclude, but its impact is likely to grow in the years ahead as reductions become increasingly harder to achieve.

The chief benefit of the trading system is that it puts free market forces to work, economists explain. "It's very much like [the way] a bank operates," says Ellerman. Emitters have a checking account system, and the EPA limits the amount of "currency" in the system. Everyone is free to find the best buy in emissions reduction as long as they don't "overspend" their allowances. "You no longer have a bureaucratic nightmare" like that of command and control, he says.

The allowance system broadens a power plant operator's options. An operator might install a scrubber—the cheapest available, as there are no regulations on types of scrubbers—or perhaps switch from a coal supply high in sulfur to a low-sulfur one, whatever option is cheaper per ton of emission reduction. Because allowances can be bought and sold, emissions can be cut wherever it's cheapest to do so—even at another company's plant in a different state. Each emitter just needs enough allowances to give to the EPA at the end of the year to cover the tons released. If a plant emits fewer tons than allowed, it can save leftover allowances for later.

Trading has saved money, reducing costs by perhaps 30%, according to Burtraw and others, but it's by no means an ideal system. "The trading program has worked well, but I wouldn't say it has worked perfectly," says Burtraw. Although increasing, trading has been light and largely limited to swaps between plants within the same company, perhaps because state regulatory commissions new to the system didn't steer utilities to the lowest cost option allowed by outside trading.

For the most part, economists suspect that the trading system hasn't come up to speed because operators have had a choice of other unexpectedly inexpensive options. For one

thing, "scrubbers turned out to be a lot cheaper than people thought," says Ellerman. New instrumentation and controls reduced staffing requirements, and units fitted with scrubbers are in use more often than expected, reports economist Richard Schmalensee and his MIT colleagues. And although relatively few trades actually occur, the trading system reduces overall scrubber costs by doing away with the need for backup scrubbers: If a scrubber goes down, plant operators can buy allowances to cover the added emissions. Overall, Schmalensee found, the cost of scrubbing in phase I has been 40% lower than estimated in 1990.

More unexpected savings came from fuel switching. Much of the Appalachian and Midwestern coal that fed the plants of the Ohio Valley—the biggest source of sulfur dioxide in the country—had a sulfur content of several percent. By switching to coal containing 1% sulfur or mixing low-and high sulfur coal, plant operators could avoid scrubbers. In 1990, most observers believed that fuel switching would be limited. They expected that burning fuel with 1% sulfur would damage hardware—a prediction not borne out by experience—and that the price of low-sulfur coal would rise once the Clean Air Act upped demand.

That hasn't happened yet thanks to developments that, at least initially, were unrelated to acid rain control. Low-cost, low sulfur fuel had been available in the West for some time, notably in the Powder River Basin of Wyoming; the expense in the 1970s was in getting the coal to the East, where the big markets were. Then the Staggers Act of 1980 largely deregulated railroads. Coal transportation costs have fallen 35% since the 1980s, notes Burtraw. By 1990, the amount of low-sulfur coal burned had doubled, but the implications for acid rain control were underappreciated by most policy-makers and power plant operators, says Munton. In fact, Smith and her colleagues say, because plant operators shied away from the unknowns of the fuel-switching option in favor of more familiar scrubbing, phase I reductions cost significantly more than they had to.

Although these external factors rather than trading have apparently dominated savings in phase I of acid rain control, most observers credit the innovative flexibility of the Clean Air Act Amendments with letting this mix of solutions develop. Once Congress gave plant operators complete freedom to cut emissions, "all the compliance vendors—low-sulfur fuel suppliers, scrubber manufacturers, and natural gas producers, for example—had to compete very hard to win," says Goffman. "The more choice you give to more people, the better the outcome." Burtraw agrees that "a big thing about the trading program is the flexibility that allows firms to take advantage of changes in prices and technology."

How that flexibility will work out in phase II remains to be seen, as there are new uncertainties in the offing. EPA is considering restrictions on fine atmospheric pollutant particles, some of which form from sulfur dioxide. Reduction of greenhouse emissions could also shrink sulfur emissions, if the United States adopts energy-conservation or fuel-switching measures. But the prospect of unknown steps has the power industry worried, says Platt.

Meanwhile, Europeans have also been successful in reducing their sulfur dioxide emissions, halving them between 1980 and 1993, says EPA international liaison Rhona Birnbaum. But they have not fully embraced trading. Instead European countries have adopted diverse approaches ranging from limited trading to pure command-and-control regulations. No one has calculated the costs of this mixed approach to date, but estimates for Europe's ambitious 2010 goals—to cut sulfur emissions damaging sensitive ecosystems by 60%—are quite high: about $1100 per ton of sulfur dioxide,

according to Mary Saether of the European Union in Brussels. As a result, Europeans are showing increasing interest in American-style allowance trading. "People come to the United States and want to know how this works and how it is generalizable," says Burtraw.

He notes that the United States succeeded in making the concepts of trading and flexibility hallmarks of the Kyoto agreement to reduce greenhouse gas emissions. And some of the solutions might be similar to those used in the acid rain case: As producers switched to low-sulfur coal, so they might switch to natural gas, which produces less warming per unit of energy produced. Technology and efficiency improvements, particularly in developing nations, might be a relatively cost-effective way to reduce greenhouse gas emissions.

But the parallels are not perfect, Ellerman cautions. For starters, it's not clear that a trading system will work with a half dozen greenhouse gases, where trades among different industries and across the world would be required. And a key factor in the greenhouse case is the stringency of the emission cap—the final figure of allowable emissions. If it's too low, flexibility is reduced along with the price competition it encourages. As Ellerman and his colleagues have written, emissions trading "is not a panacea that inevitably makes costs of emissions control simply disappear into thin air." But for reining in pollution without choking industry, it looks like a good place to start.

---

## Questions:

1. How much sulfur dioxide (million tons per year) is expected to be released in 2010?
2. What is the chief benefit of this U.S. emissions trading system? Explain.
3. What is the latest cost estimate for U.S. acid rain control?

Answers are at the back of the book.


## Activity:

Go to: **http://www.epa.gov/docs/acidrain/ardhome.html**
Learn about the environmental effects and $SO_2$ emissions trading.

Find your state or a neighboring state and see what phase I allowances are given for some of their plants.

**24**

Acid rain has received much attention in the past decade. It forms from $SO_x$ and $NO_x$ molecules emitted into the atmosphere by cars and industry which combine with water droplets in the air or on the ground to form an acidic substance. Acid rain has been known to damage plant life and aquatic systems, but it also takes its toll on urban stone. Though acid rain contributes to urban stone decay, there are many other factors that come into play including surface water, groundwater, and salt.

# The Complexity of Urban Stone Decay

By Erhard Winkler

Stone decay has been a part of urban life throughout history. More than 2,000 years ago, Roman architects and engineers wrote about the adverse effects of the elements on their buildings, statuary, and networks of paved roads.

Stone, hewn from rock forged deep within Earth, naturally breaks down at Earth's surface. Regardless of location, dry, semiarid, and humid climates all take their toll on building stone. However, predictable geologic weathering has had major competition over the past 200 years. Since the beginning of the Industrial Revolution, the rate of most types of stone decay has increased dramatically in our cities. Expanding municipalities, climbing populations, and burgeoning factories have had a direct impact on city environments and surrounding areas. Factories—once powered by water or wind—turned to wood-burning, then coal-burning, furnaces to push production and smoke stacks to new heights. The soot and mixture of gases from these factories blackened and corroded the foundations of our cities. We are all familiar with the photographs and paintings of the darkened buildings of 19th-century European and American cities.

Transportation has also had a major impact on the urban landscape. Increased numbers of trucks and cars have spewed vast quantities of noxious gases into city air. Carbon monoxide,

ozone, nitrous and sulphuric oxides—all products of automobile and truck exhaust—react to corrode building surfaces when they mix with atmospheric moisture (rain and humidity).

Fortunately, we have recognized the problem and are attempting to remedy the situation. In the United States and much of Europe, strict rules have been implemented to reduce the major problems caused by transportation and industrial pollution. These polluting agents include increased levels of ozone and acid rain that damage the facing stones of buildings, even buildings that are far from a sulfur source. We are still plagued with smaller, yet ubiquitous, problems like the unsightly damage caused to the stone bases of buildings by de-icing salts. Understanding the physical and chemical nature of the problem will help us develop means to maintain the beauty of facing stone.

## Physical Properties of Stone

Architects and engineers are interested in the physical properties of building stone. They learn that each major group has a set of characteristic properties that allow them to determine the best stone for a particular project. When choosing a stone type for a project, the most significant properties are rock pores and porosity, water sorption, bulk specific gravity, rock hardness,

From "The Complexity of Urban Stone Decay," by E. Winkler, Geotimes, 1998, pp.25-29.
Reprinted with permission.

**138**

compressive strength, tensile strength, the moduli of rupture and elasticity, and thermal properties of the rock and constituent minerals. The dry-to-wet strength ratio indicates the expected durability; the ultrasound travel velocity readily correlates with density and strength. Light transmission is occasionally desired for special effects.

Once a decision is made to use a particular stone, architects and engineers must be on guard for the natural deformities of the rock. Brittle rock fracture, such as jointing and faulting, can render an ordinarily durable rock useless. The presence of plumose markings or discoloration along joint surfaces may indicate that a rock has been subjected to internal stresses and groundwater penetration. These stresses can weaken the stone, thus making it a poor choice for construction. No one wants to see signs of deterioration in stone after it has been set into the side of a multistory building.

It is important to know the source of the stone. A quick glance around a quarry will reveal much about the potential behavior of a particular stone. For example, the "tenting" of granite slabs caused by stress-releasing sheeting parallel to the surface can give direct evidence of the future appearance of the granite cladding.

## Urban Stone Decay

Today we have evidence that stone decay occurs about twice as rapidly in cities as in rural areas. Stone-damaging moisture is generally derived from two sources: acid rain and humidity in the atmosphere, and corrosive groundwater. It is well known that acid rain with varying pH levels can cause accelerated dissolution of carbonates and iron minerals. In addition, rising groundwater, rain-water seepage, and leaky plumbing fixtures are common sources of damage-causing moisture. Even high relative humidity can push moisture into the narrow capillari of facing stone. Nor does water work alone. Dissolved salts from obvious sources, such as de-icing salt on sidewalks and roads, as well from more subtle origins like ocean spray or groundwater seepage, can wreak havoc on stone surfaces by producing a variety of unsightly discoloring, including efflorescence, spalls, and flakes.

### Acid Rain

Rainwater has always been mildly acidic, with an average pH near 5.6. As rain droplets fall through the atmosphere they absorb many of their constituent gases. This process actively cleanses the air of noxious components. It also changes the chemistry of the rain water.

Domestic heating, as well as industrial and automotive combustion, have produced a large amount of other gases, such as sulfates, nitrates, ozone, and carbon dioxide. The interaction of these gases changes the composition of the atmosphere. The presence of ozone ($O_3$) tends to oxidize sulfur dioxide ($SO_2$) to sulfite ($SO_3$). $SO_3$ forms sulfuric acid droplets, one of the most corrosive and reactive ingredients in the atmosphere.

A detailed study of the pH distribution of rainfalls in the city of Montreal, Canada, in 1990 recorded ranges from 3.56 to 4.6 with an average pH of 4.26. Two isolated rainfalls reached 6.9. Minimum values are usually not given but play an important role because such low values accelerate the dissolution of carbonates and ferric minerals like hematite, goethite and particularly the amorphous hydroxide, limonite (natural rust)—which increases exponentially with acidity. Dust, rich in calcite and transported by winds from the arid and semiarid Southwest, may have a neutralizing effect on acidity as far away as Montreal. The few extreme incidences of acid rain, however, may be sufficient to dissolve calcite and ferric iron in spite of the presence of calcite dust. The broad range recorded for Montreal can be considered typical of most urban areas. It is thus the minimum pH (maximum acidity) that determines the dissolution and transport of carbonates and ferric oxides. While acid rain affects many of the carbonate building stones, a more alkaline rain damages silica-rich stones such as granite.

### Relative Humidity

Moisture in the atmosphere is the only source of entry for water molecules into stone capillaries smaller than 0.1 micron. The $H^+$ sides of the molecule will attach to the negatively charged sides of the capillaries. Additional water molecules will attach to the negative side of the wall molecules, forming an "ordered" icelike structure. Such ordered water is considered responsible for the high pressures of moisture expansion. Fine capillaries are characteristic of granite rocks with microcracked quartz, caused by 4.5 percent contraction during cooling from the originally liquid magma. Porosity, moisture content, and rock strength of quartz-rich igneous rocks are closely related to one another Expansion of liquid water trapped in capillaries larger than 0.1 micron can also be quite destructive.

Moisture in humid atmospheres can cause problems wherever there is an abrupt drop in temperature. Anyone who wears glasses knows the frustrating experience of having them fog up as soon as you step from an air-conditioned building into the moist heat. The same process occurs with stone buildings when warm moist air hits cooler stone. The basilicas of San Marco in Venice and in Pisa are just a few of many examples where granite columns flake near the church entrance because of the condensation of warmer, humid air.

## Surface and Groundwater

Surface water from flooding or rising groundwater can bring corrosive substances into contact with building stone. Liquid water can act as a solvent of the water-soluble grain cements of carbonates, limestone, siderite, marble, and dolomite. Capillary action draws water up from below. A capillary only one micron in diameter can transport water to a height of 10 meters and a length of about 15 meters. In this manner, salts in solution may be transported and will effloresce and corrode the stone surface of buildings far from their sources.

Dissolution, redeposition, and frost action are all potential problems. As the drying powers of sun and wind cause surface moisture to evaporate, moisture in the stone capillary travels to the stone's surface. The stone's inner moisture, in turn, evaporates, leaving the dissolved constituents behind in the form of surface induration and efflorescence. The dissolved material is often derived from the dissolution of grain cement or microscopic salt crusts left behind from previous soakings. The presence of these crusts means that degradation is occurring beneath the stone's surface.

## Salt Damage

Water is the prime catalyst for stone decay, yet damage can increase exponentially if the water contains dissolved salts. The presence of salts in stone is not always evident, but is more common than we realize. Salts are usually hidden within the stone substance, but often effloresce as coatings that are powdery white, grainy, and often irregular. Damage can be observed as

stones crumble near the ground surface; the decay may become so deep that entire stone blocks have to be replaced. Such salts in urban areas can be of different solubility and sources.

## Insoluble Crusts

Calcite may be leached from blocks of limestone in mortar and concrete, or by acid rain or condensation water. Later, it is redeposited as secondary external stalagmitic crusts. These deposits are difficult to remove from any stone surface. Limewater dissolved from carbonate stone blocks can seep into a granite base course and deposit calcite in the capillaries of the granite, causing flaking.

Wind-blown mineral dust, with mostly calcite, clay minerals, and silt-size quartz, helps to neutralize acid rain. Urban industrial dust sources often include considerable quantities of iron, which settles on stone surfaces as hematite or limonite or as black iron sulfide after reacting with atmospheric sulfate. Acid rain can "solution-attack" such crusts, carrying iron into mostly pervious sandstones where it readily oxidizes to rust. It is impossible to clean such a stone surface. The Cotta'er sandstone of Dresden, Germany, is encountering this problem.

## Airborne Sprays

Even as far away as 200 miles from ocean shores, salts may infiltrate stone as airborne ocean spray. Oceans and inland lakes supply mostly halite, which settles on top of buildings and visibly effloresces on upper stone and brick masonry. The surfaces of many buildings in New York City, for example, are crowned with irregular white salt rims, carried by spray from the Atlantic. Dust blown from desert floors is another source of salt. Rainwater, however, tends to wash much of the soluble salts downward into streams and groundwater.

## Excess Fertilization

Surface and underground salts may originate from excess fertilization of lawns and landscape plants, appearing as mostly residual sodium chloride rejected by vegetation. Capillary action can push these salts to astonishing heights within masonry. The least soluble sulfates precipitate near the ground while the less common nitrates settle above. The solubility of these salts determines how high they will rise and precipitate. The chlorides of maximum solubility rise highest.

The concentration of salts can be considerable along irrigation canals in semiarid and arid floodplains and deltas. The Nile floodplain and delta, for example, have accumulated salts by the recycling of irrigation water and excessive evaporation. Such salts have entered obelisks, the Sphinx, and many other ancient monuments in the Egyptian desert, resulting in decay that is often beyond repair.

Today, de-icing salts are most plentiful in urban areas. They are often spread in excessive quantities along roadways and sidewalks. These salts are either crushed halite (NaCl) or more reactive and highly soluble ($CaCl_2$). Calcium chloride does not crystallize easily in humid climates but remains in solution and attracts moisture from the atmosphere, forming permanently corrosive wet areas instead of efflorescence. The low solubility of halite readily crystallizes and effloresces, causing mechanical dissociation of granites and sandstones near the upper fringe of the salts.

Rock will continue to be a favorite building medium. Its varied textures, colors, and inherent durability make it the choice of architects, civil engineers, and artists. Knowing more about its properties and the effect of natural and human activities on those properties will help us preserve our urban edifices for millennia.

## Questions:

1. How much faster does stone decay occur in urban areas versus rural areas?
2. What is the average pH of rainfall? Is this slightly acidic or basic?
3. How can surface water and groundwater be damaging to stone structures?

Answers are at the back of the book.

## Activity:

Walk around your campus or town. Take note of all the structures made of stone. Can you see any evidence of stone decay? Look into the average pH of rainfall within your region. Does this surprise you?

# PART 5
# Global Climate Change—Past, Present, and Future

**25**

Americans are known to love the independence they acquire through their cars. It is this cultural trademark which helps the U.S. consume almost one fourth of all oil consumed daily worldwide. Is the end of oil in sight, and does science have feasible alternatives to replace the legacy of oil? Oil is a non-renewable resource, and the majority of its reservoirs have been found. One thing is for sure: oil conservation needs to be taking place now.

# Spending Our Great Inheritance—Then What?

By Walter Youngquist

During more than 500 million years, geological processes accumulated a rich bank account for us—oil. The "account" actually was set up as numerous accounts—some large, some small—in various parts of the world. In 1859, Col. E.L. Drake initiated the modern search with his now-famous well near Titusville, Pa. Soon the hunt spread across the United States and then around the world. With increasingly sophisticated equipment to read the clues about where this inheritance was hidden, we have been increasingly successful in finding it.

Just how successful have we been? How much of this oil inheritance have we found and how much is left to find? In their article, "The End of Cheap Oil," published in the March 1998 issue of Scientific American, exploration geologists Colin J. Campbell and Jean H. Laherrère of Petroconsultants in Geneva noted that the world has consumed more than 800 billion barrels of oil and has discovered or has in reserve another 850 billion barrels. They estimate that only about 150 billion barrels remain to be discovered. Apparently, we have been very successful in our search, having already consumed, by their estimate, nearly half of the world's ultimate resource of about 1,850 billion barrels of oil.

Now that we're close to having consumed half the world's oil, how soon will we reach peak production? This question has been the subject of discussion for many years, with various forecasts of the peak of world or regional oil production offered. Many have already proved wrong. One estimate, however, was correct. In 1956, as Campbell and Laherrère point out, Shell Oil geologist M. King Hubbert predicted that the United States would peak in oil production around 1970. His forecast was widely ignored or scoffed at by the general public, and by many geologists, but Hubbert was right on the mark.

When the future of oil is discussed, the common question asked is "How long will oil last?" This is the wrong question. Insignificant amounts of oil will probably be produced in the year 2100 and perhaps beyond. The critical date is when the peak of oil production is reached and the world's demands can no longer be supplied. From then on, there will be less and less oil to divide, in contrast to the current happy situation where we have more and more to divide. It is probable that the decline of world oil production will affect more people in more ways than any other event in human history.

Because various estimates of the date of world oil peak production have been wrong, it is sometimes assumed that forecasts such as Hubbert's will be wrong. This may be true, but

From "Spending Our Great Inheritance--Then What?" by W. Youngquist, Geotimes, 1998, 24-27. Reprinted with permission.

the question is "How wrong?" With many more production-based data points now available than in the past, production curves are becoming well established.

The peak of world oil discoveries passed in the 1960s, so the downward trend of that curve has already been established. Simply continue the classic bell curve and you'll find a representation of the total amount of available oil.

The theoretical graph of the production life of a finite resource indicates approximately 30 years from peak of discovery to peak production. Applying these curves to oil, with new technology such as horizontal drilling, 3-D seismic, and improved secondary recovery methods, we can predict that peak production (after the world oil discovery peak in the mid-1960s) will occur in about 40 to 45 years. But Campbell and Laherrère state: "Barring a global recession, it seems most likely that world production of conventional oil will peak during the first decade of the 21st century." Their estimate agrees with what many others, myself included, have said. In his article, "Crude Oil and Alternative Energy Production Forecasts for the Twenty-first Century: The End of the Petroleum Era," J.C. Edwards sets the peak at 2020—a more optimistic forecast than others, but still clearly within sight.

As an interesting sidelight to the time of peak, Chevron Corporation in 1997 announced the discovery of an oil field offshore of Angola. They stated it could hold as much as a billion barrels, and appeared to be the largest find the company had made in the last 10 years. A billion-barrel oil field is indeed a prize. But in "An Analysis of U.S. and World Oil Production Patterns Using Hubbert Curves," a paper recently submitted for publication, Albert A. Bartlett calculates that adding a billion barrels to the world oil supply would move the peak of world oil production back just 5.5 days! His assessment indicates the magnitude of the world's current oil appetite and how difficult it is becoming to feed it. Important regions that have seen their maximum time of production include the United States (1970), North America (1984), and the former Soviet Union (1987).

Individual countries (other than the United States) that have already peaked in oil production include Libya (1969), Iran (1973), Romania (1976), Trinidad (1977), Brunei (1979), Peru (1981), and Egypt (1993). The list of producers on permanent decline is growing and will eventually include the Persian Gulf nations, which now hold the bulk of the world's remaining oil. The difference in oil-well production between the United States and Saudi Arabia is striking: Average daily production per well in Saudi Arabia is about 5,600 barrels. Average U.S. daily production per well is 11.3 barrels.

## Out of the Oil Business

Whatever forecast of the world oil production peak is accepted, there are two overriding facts: The world is now consuming about 26 billion barrels of oil a year, but in new field discoveries, we are finding less than 6 billion barrels a year. The date of the peak of world oil production is important, but also important is the sobering fact that it will occur within the lifetimes of most people living today—and much sooner than is generally expected. There is little time left to begin to adjust lifestyles and economies to the coming post-petroleum era.

The United States can no longer write those big checks against its oil bank account. Oil reserves have declined from a maximum of 39 billion barrels in 1970 to the present 22 billion barrels, and total daily production has dropped in that same period from more than 9 million barrels a day to 6.4 million barrels. We now import more oil than we produce. So we have increasingly been writing our oil-supply checks against the accounts of others—chiefly the Persian Gulf countries, Nigeria, Mexico, and Venezuela. But when world oil production peaks, the oil checks that all of us can write must become smaller and smaller. Eventually, those checks will be insignificant, relative to world needs. We will have spent our oil inheritance. Then what?

## Alternative Energy Sources

If the public briefly thinks of oil as a finite resource, the popular placebo is: "The scientists will think of something." Just what have we thought of up to now? The chart below answers that question.

## ALTERNATIVE ENERGY SOURCES

| NON-RENEWABLE | RENEWABLE |
|---|---|
| Oil sands/heavy oil | Wood/other biomass |
| Gas hydrates | Hydropower[1] |
| Shale oil | Solar |
| Coal | Wind |
| Nuclear fission, fusion [2] | Tidal |
| Geothermal [3] | Ocean thermal energy conversion (OTEC) |

1 Renewable only to life of reservoir.
2 If ever accomplished, may be regarded as renewable, since fuel supply is huge.
3 So far, all electric quality reservoirs are in declining production.

This is essentially the complete alternative energy spectrum. There are no indications in the foreseeable future of other significant energy sources.

The question is how well can these sources individually or collectively replace oil? The topic is large, but some salient facts can be noted. The world uses about 72 million barrels of petroleum a day. Just replacing that volume with an equivalent energy source becomes a huge task. Petroleum equivalents can be made from coal but doing so on any significant scale would involve the largest mining project the world has ever seen.

There are 2 trillion barrels of kerogen (not oil) in the Colorado Plateau oil shales. But trying to modify kerogen into oil has cost oil companies billions of dollars in experimental projects. All have been abandoned, leading to the expression: "Shale oil—fuel of the future and always will be." The Athabasca oil sands of Canada contain 2 trillion barrels of oil (real oil). Today some 500,000 barrels a day are produced. Scale this up 10 times and you have 5 million barrels a day. The problems to achieve that scale are enormous, and 5 million barrels a day must be measured against the 19 million barrels a day used by the United States and the 72 million barrels a day used worldwide. Oil sands will help—a little, for a time.

## Renewable Resources

Ethanol is a net energy loss—it takes 70 percent more energy to produce than is obtained from the product itself. Other biomass resources show, at best, very low net energy recovery. In their comprehensive study, "Feasibility of Large-Scale Biofuel Production," Mario Giampietro, Sergio Ulgiati, and David Pimentel write: "Large-scale biofuel production is not an alternative to the current use of oil and is not even an advisable option to cover a significant fraction of it."

The two most popularly suggested energy alternatives, wind and solar, suffer because they're undependable, intermittent sources of energy, and the end product is electricity. We have no way to store large amounts of electricity for use when wind and sunshine are not with us. Geothermal and tidal energy are insignificant energy sources in total but can be locally important. Nuclear energy can be a large power source if the safety aspects can be guaranteed (and this may be possible)—but again, the end product is electricity. There is no battery pack even remotely in sight that would supply the energy needed to effectively power bulldozers, heavy agricultural equipment such as tractors and combines, or 18-wheelers hauling freight cross-country.

Can electricity be used to obtain hydrogen as a fuel from water? It can, but hydrogen is difficult to store and dangerous to handle. And there is no energy system now visualized to replace kerosene jet fuel, which propels a Boeing 747 about 600 miles an hour nonstop on the 14-hour trip from New York to Capetown (currently the longest plane flight). We continue to seek the holy grail of energy — fusion — but containing the heat of the sun at 10 million degrees Centigrade is still only a far-off hope.

## A Gap

Which brings us back to the peak date of oil production. Even if we assume that alternative sources could somehow fill the gap left by the departure of oil, the time frame needed to put these into sufficient production to replace oil as it declines clearly indicates a large gap at best. The British scientist and statesman Sir Crispin Tickell has defined our situation well: "We have done remarkably little to reduce our dependence on a fuel [oil], which is a limited resource and for which there is no comprehensive substitute in prospect." All alternative energy sources must be drawn upon, but oil will be sorely missed.

We are consuming what is, in many ways, an irreplaceable resource. We have all seen the bumpersticker on huge recreational vehicles: "We are spending our children's inheritance." That RV, and the more than 600 million gasoline- and diesel-powered vehicles now in the world, are doing just that—as they guzzle oil.

We are most fortunate to be living in a brief, bright interval of human history made possible by an inheritance from half-a-billion years of oil-forming Earth processes. We rarely give thought to the greatly depleted balance in the oil account we are leaving to future generations. When checks can no longer be written against that inheritance, world economies and lifestyles will undergo great changes. Life will go on, but it will be quite different from the present. Most people living today will see the beginning of those times.

Fortunately, as Campbell and Laherrère state, oil production will not decline abruptly. We are simply about to run out of the cheap oil we have enjoyed. This gives us time to develop as many alternatives as possible and to think about changing consumption patterns and lifestyles (such as increased use of mass transit), to arrange for a "soft landing" in the post-petroleum era. However, with the peak of world oil production now clearly in sight, the time to begin to make adjustments is now.

## Additional Reading

"An Analysis of U.S. and World Oil Production Patterns Using Hubbert Curves" by A.A. Bartlett (in press).

"Crude Oil and Alternative Energy Production Forecasts for the Twenty-first Century
—The End of the Petroleum Era" by J.C. Edwards. *AAPG Bulletin*. v. 81, n. 8, 1997, p. 1292-1305.

"Feasibility of Large-Scale Biofuel Production" by Mario Giampietro, Sergio Ulgiati, and David Pimentel. *BioScience*. v. 47, n. 9, October 1997, p. 587-600.

*GeoDestinies: The Inevitable Control of Earth Resources over Nations and Individuals* by Walter Youngquist. National Book Company (Portland, Ore.), 1997.

"Nuclear Energy and Fossil Fuels" by M.K. Hubbert. *Drilling and Practice*, American Petroleum Institute, (Washington, D.C.), 1956.

*The Coming Oil Crisis* by C. J. Campbell. Multi-Sciences Publishing Co. & Petroconsultants S. A., (Essex, England), 1997.

"The End of Cheap Oil" by C. J. Campbell and J. H. Laherrère. *Scientific American*, March 1998, P. 78-83.

"The Future and Its Consequences," by Sir Crispin Tickell. The British Association Lectures, The Geological Society, London, 1993, p. 20-24.

## Questions:

1. When did the U.S. reach peak production of oil?

2. How many barrels of oil does the U.S. consume daily? How many barrels of oil is consumed worldwide?

3. What are the problems with solar and wind power that prevent them from being significant energy sources?

Answers are at the back of the book.

## Activity:

Calculate your car's gas mileage. When filling up your tank, record your car's mileage (A). The next time you fill your tank record the mileage (B), again, and the gallons used to fill it. Subtract A from B and then divide by the gallons to get miles per gallon. Compare the gas mileage to other types of cars and inner city driving and interstate driving.

# 26

The Pleistocene holds interest for many due to not only the evolution of hominids, but also the ice ages. The ice ages, which were not a continuously cold environment, fluctuated between colder and warmer periods, or glacials and interglacials. There is sufficient evidence to support the existence of at least four glacial periods. Currently, there is much debate over whether or not the earth is out of the ice ages or still within an interglacial. Debate stems from not knowing the cause of these climatic fluctuations. Theories include the Milankovitch cycles of the earth, cosmic dust, and ocean currents. Surprisingly, recent evidence supports an ice age cycle driven by the Southern Hemisphere instead of the Northern Hemisphere ice sheets.

# Deep Freeze

By Gideon Henderson

Ice skating in Central Park would have been a breeze 20,000 years ago. Or anywhere else in Manhattan, for that matter. The last ice age was at its peak, and giant sheets of ice stretched as far south as London and New York. Around the world, average temperatures were about 6° colder than today, the sea lapped coastlines more than 100 metres lower and ocean currents ran more slowly. The atmosphere was dry and dusty and the whole climate was unstable, prone to sudden swings between mild and freezing spells. Yet a hundred thousand years before that, things were very similar to today. And 20,000 years before that, the Earth was in the grip of another big freeze.

This lurching from cold to warm and back to cold again has been going on for the past million years. But despite decades of research, no one agrees why they happen. And yet, if we can understand these dramatic shifts in the natural climate system, we might also learn more about how the Earth will respond to our own climate-changing efforts.

Until recently, the main contender was an explanation combining the effects of changing sunlight patterns and the action of currents in the North Atlantic. Then, in 1996, two scientists from California suggested that these climate changes could all be down to dust pouring onto the Earth from space. Though this idea appears to have fallen by the wayside, the challenges it posed to the traditional model triggered a frenzy of research activity. In its wake a new explanation is rapidly gaining ground.

The conventional explanation of the ice ages dates back to the 1920s, when a Yugoslavian astronomer called Mulitin Milankovitch suggested that small wobbles in the Earth's orbit might be to blame. Milankovitch noticed that the Earth's orbit around the sun is distorted by the gravitational pull of the Moon and the other planets. This changes three things about the way the Earth orbits: the shape of the orbit—more oval or round—how much the Earth's spin axis tilts away from vertical, and the time of year when the Earth is closest to the sun.

Though the distortions make only a small difference to the total amount of sunlight reaching the Earth, they have a big effect on the amount of sunlight arriving at different parts of the globe, and at different times of the year. The upshot is that regular cycles in the sunlight pattern have periods of 100,000 years in the ovalness, 41,000 years in the tilt, and 23,000

From "Deep Freeze," by G. Henderson, New Scientist, 1998, pp. 28-32. Reprinted with permission.

**149**

years in the timing of closeness to the Sun. Milankovitch thought that the Earth's climate is controlled by these changes in the pattern of sunlight.

Sure enough, scientists looking at how climate changed in the past have found signals with all three of these frequencies. For the 23,000 and 41,000-year cycles, the pattern and relative sizes of climate changes around the Earth match well with the pattern of sunlight changes. But with the crucial 100,000-year cycle—the timescale on which ice ages occur—Milankovitch's theory hits a problem. The climate changes going into and out of ice ages are ten times the size of the ones on the other two cycles. Changes because of the ovalness have the right timescale to explain the ice ages, but they are far too small.

## Chilling Out

To get round this, Milankovitch suggested that something else is amplifying the effect of changes in sunlight patterns. His favourite candidates were the large ice sheets that form on the continents of the northern hemisphere during the ice ages. When there is relatively little sunlight falling on these areas during the summer, ice sheets don't get a chance to melt, so they keep growing year after year. A big, growing ice sheet cools the surrounding region both by directly chilling the air that blows over it, and by reflecting more of the Sun's warmth back into space. So a small initial change in temperature could trigger growth of the ice sheets, which would, in turn, cause more cooling.

But how could changes in this one region affect the rest of the planet? In the 1980s, Arnold Gordon and Wally Broecker, both at the Lamont-Doherty Earth Observatory of Columbia University in New York, discovered that the North Atlantic comes equipped with its own system for transmitting local climate changes around the world: it is the starting point for the deep-ocean currents that drive a global "conveyor belt" of ocean water, carrying heat along with it. Water sinks in the North Atlantic, flows south-ward, around Antarctica, and then northward into the Pacific before returning to the surface. Changing the local climate in the North Atlantic could easily change the rate at which the water sinks and hence change the whole circulation system.

So a general consensus began to emerge. Changes to sunlight patterns in the northern hemisphere changed ice sheets around the North Atlantic, which influenced the ocean conveyor belt, which cools the planet by changing the way heat is transported around the Earth.

But there have always been some nagging worries. For instance, although the average effect of changing the ovalness of the Earth's orbit is a sunlight cycle of 100,000 years, a more careful look shows that some of these cycles are 95,000 years long while others are 125,000. Climate records, however, show only 100,000-year cycles. It may be that the records just aren't detailed enough to pick out both cycles, or it might be that the ovalness of the orbit is not what causes the 100 000-year cycle at all.

In 1966, Richard Muller of the University of California at Berkeley and Gordon MacDonald of the University of California at San Diego, suggested a controversial new approach. They pointed out that the plane of the Earth's orbit wobbles relative to the average plane in which the whole Solar System rotates. This wobble gives rise to just one cycle exactly 100,000 years long. This new "inclination" model no longer suffers from the ovalness problem of two different cycles. But as the wobble makes no difference to the sunlight reaching the Earth, how could it affect climate?

Muller and Macdonald suggested that climate might be affected by cosmic dust entering the atmosphere. Perhaps, they say, when the Earth orbits in the same plane as the rest of the Solar System, more dust enters the atmosphere, blocking sunlight and triggering the ice ages. Then, when it tilts out of the plane, less dust arrives and the planet warms up again.

But was there enough cosmic dust to have an effect? One way to work out how much has arrived over the years is to look at ocean sediments. Franco Marcantonio of Tulane University of Louisiana in New Orleans and Ken Farley of the California Institute of Technology in Pasadena have each been working on this question (New Scientist, Science, 9 December 1995, p 22). The jury is still out, but it looks as if there are no significant changes in the cosmic dust flux during past climate cycles—bad news for the inclination hypothesis.

## Strange Changes

The lack of any mechanism to link inclination changes with climate has meant that most researchers have now discounted Muller and Macdonald's model. But the controversy it raised prompted many scientists to look back at the existing data, to collect new information and to question the traditional explanations for ice ages. Just how well does the theory about the northern hemisphere's ice sheets hold up to scrutiny?

One indication that all might not be well comes from research done by Chris Charles from Scripps Institution of Oceanography in California. In 1996, Charles and three of his colleagues made a puzzling discovery: changes in the North Atlantic and its conveyor belt seem to lag behind changes elsewhere. The researchers studied sediment that had been laid down in the South Atlantic over the past 80,000 years. They extracted the shells of tiny sea creatures called foraminifera from different depths in the sediment. As the foraminifera grow, their shells act as a record of the local chemical conditions—conditions that depend both on the temperature of the surface of the ocean where the creatures live, and on the amount of water passing through from the North Atlantic on the great ocean conveyor belt.

The shells revealed that whenever the temperature in the South Atlantic ocean changed, the amount of water passing through also changed. But the surprise was that the circulation change occurred slightly after the temperature change. In other words, the South Atlantic was changing before the North Atlantic. How could this be, if temperatures in the south were controlled by changing conditions in the north?

Charles suggested that the North Atlantic was itself responding to changes elsewhere. It was lagging behind, he believed, because the ice on the northern continents slowed down the currents' response—a bit like a storage cooler. As the sediments that Charles analysed only went back 80,000 years, he was not studying the full ice-age cycles but shorter climate oscillations occurring within the last ice age. But they demonstrate that the North Atlantic is not the sole driver of global climate.

## Core of the Problem

At Christmas, Charles joined around thirty scientists sailing for two months aboard the Ocean Drilling Program ship, the JOIDES Resolution, to the same part of the South Atlantic. This time, the researchers plan to drill longer cores into the sediment. When analysed, the hope is that they will back Charles's original

results and show whether the same effect is seen over more than just the past 80,000 years.

There is other evidence that suggests Charles is on the right track. In the past few years, Mike Bender and Todd Sowers of the University of Rhode Island compared climate records in ice cores taken from the north and south poles to try to see which changed first. When you are dealing with two cores, rather than one, it is much more difficult to be sure about the relative timing of climate changes. Bender and Sowers solved this problem by measuring oxygen that they found in small air bubbles trapped in the ice cores.

Oxygen is well mixed in the atmosphere, so any changes should show up simultaneously all over the world. By lining up the changes in oxygen, the researchers managed to compare the timing of the temperature records.

Once again, they were surprised by what they saw. The last time the planet warmed suddenly, at the end of the last ice age, the warming started in Antarctica and only later showed up in Greenland.

This approach shows that the southern hemisphere seems to warm first. But it still only gives relative timing. If you could compare the temperature records directly with the past sunlight pattern for different latitudes, it might be possible to pinpoint the exact location of the iceage amplifier. The problem is that it is extremely difficult to obtain reliable dates for changes in ocean sediments or in ice cores. So Isaac Winograd and Ken Ludwig and colleagues at the United States Geological Survey in Denver and Virginia turned to an unusual cave in Nevada where stalactites and stalagmites have been growing continuously for the past few hundred thousand years. Just like the foraminifera shells, these speleothems created a temperature record as they grew. But as well as temperature they also absorb natural radioactive isotopes which can pin down their age very precisely

In 1992, Winograd and Ludwig produced a record, carefully dated, of the timing of climate changes in these Nevada caves. They discovered that, when the Earth's climate makes the sudden shift out of an ice age, warming occurs before changes in the northern hemisphere sunlight patterns but at about the same time as changes in the southern hemisphere. Sceptics of this research have suggested that the Nevada record reflects the local climate rather than the global one. However, even if this is true, the record is certainly one more argument against an ice age cycle driven by the traditional northern hemisphere ice sheet model.

It seems that the evidence is pointing towards the southern hemisphere as the culprit. But is there a mechanism for amplifying the weak change in sunlight into a strong enough signal to cause the ice ages? Because there is less land in the southern hemisphere, ice sheets in the south don't change their size in the drastic way that those in the north do so they are not likely to be the cause. But the ocean surrounding Antarctica does contribute water to the conveyor belt as it passes by. So some scientists believe that ocean circulation changes in the south, rather than the north Atlantic, could be the key.

Another way in which the south could act as an amplifier is by changing the amount of carbon dioxide in the atmosphere, and therefore changing the amount of warming caused by the natural greenhouse effect. One mechanism for controlling carbon dioxide in the south was first suggested during the 1980s by the late John Martin who worked at Moss Landing Laboratories in California. He pointed out that some areas of the oceans, especially the ocean around Antarctica, have very low concentrations of iron—a key nutrient for the growth of marine plant life. If these areas had more iron dissolved in the water during the ice ages than they do now, there could be more biological activity in the water. This in turn would consume more carbon dioxide from the atmosphere, as the plants photosynthesised.

This idea was supported in an experiment in 1996 when a large team of scientists lead by Kenneth Coale, one of Martin's colleges from Moss Landing, added bucket-loads of iron into one such area of the ocean (This Week, 12 October 1996, p 4). This produced a big increase in the amount of biological activity, just as Martin had predicted, and a big increase in the carbon dioxide taken up. During the ice ages, the climate is drier and the winds stronger, so more dust blows into the ocean. So perhaps a small change in sunlight caused drier, dustier conditions and this increased the amount of dust—which contains plenty of iron—blowing onto the Southern Ocean, thereby sucking carbon dioxide out of the atmosphere to cause the ice age.

Proponents of the northern-hemisphere model are not ready to go quietly, however. Their model does explain a lot of what is known about the Earth's past. For instance, sea levels hit their highest values in the past just when the northern-hemisphere model predicts the most melting of the ice sheets. So now the race is on to reconcile the new data with the old model or to improve the Southern Ocean models to make them as convincing as the northern hemisphere model used to look.

To be really sure why London and New York were under ice 20,000 years ago, a theory would need to unpick the intricate combinations of temperature, carbon dioxide and ocean circulation—how the whole complex climate system fits together. When we succeed, we may even learn more about how the climate system will respond to the carbon dioxide released by mankind's activities. Not just where we've been, but where we're going too.

## Questions

1. What are the Milankovitch cycles?
2. Why do the Milankovitch cycles run into a problem with explaining the ice ages?
3. In what ways could the Southern Hemisphere help trigger an ice age?

Answers are at the back of the book.

## Activity:

Go on a virtual field trip sponsored by Hartwick College in central New York. Explore current landforms in New York with evidence of past glacial activity.

Go to **http://www.hartwick.edu/geology/work/VFT-so-far/VFT.html** Pay particular attention to ice age links and stop #1. Write a short summary of the virtual field trip.

# 27

The main question under debate is whether the Earth is warming due to natural climatic processes or due to an increase of greenhouse gas concentrations in the atmosphere because of human activities. If changing atmospheric chemistry is causing global warming, what are the future consequences and what can we do to remediate the situation?

Scientists are now testing subsurface temperatures in order to obtain a better idea of what the future has in store. As the Earth's atmosphere changes temperature, the uppermost part of the crust absorbs these temperature changes and propagates them downward. So, within the upper 500 meters of the Earth's crust we can trace back 500 years of surface temperature history.

The records can be obtained almost everywhere on the continents, enabling both local and global assessments of climate change and providing us information regarding human impact on the atmosphere (first observed around 1750).

# Underground Temperatures Reveal Changing Climate

By Henry N. Pollack and Shaopeng Huang

Probably no other scientific topic dominated the 1997 news so completely as did Earth's climate. Along with the recurring waves of El Niño-related storms, the topic of long-term global climate change was a favorite subject, particularly at year's end when the Kyoto conference on greenhouse-gas emissions took place. When the discussion of climate change engages not only scientists, but also industrialists, economists, and politicians, you know something of significance is on the table. The issue, of course, is whether the warming of Earth that has been occurring since the late 1800s is simply a manifestation of natural climate processes or a result of increasing concentrations of greenhouse gases due to human activities. A few voices still question whether Earth has warmed and if greenhouse-gas concentrations in the atmosphere are increasing. But for the most part, a consensus that both phenomena are real has emerged. The debate is now centered on whether the changing atmospheric chemistry is the cause

of the global warming, and if so, what will be the consequences in the future and what can or should be done to remediate the situation. Earth scientists continue to play an important role in setting the scientific foundations for the debate, in part by illuminating the characteristics of past climate changes so as to better understand the fluctuations in the climate system that occurred before there was any possibility of anthropogenic influences. As all geologists know, Earth is always changing—on time scales that vary from annual to decadal to millennial and into the millions and billions of years. If we are to identify and assess a possible human impact associated with the utilization of fossil fuels, we must focus attention on an interval of time that encompasses both the industrial and pre-industrial eras—approximately the past several hundred years of Earth history—and on the natural climatic processes that may fluctuate significantly in that interval. Conveniently, humans have been keen archivists over the most

From "Underground Temperatures Reveal Changing Climate," by H. N. Pollack and S. Huang, Geotimes, 1998, pp. 16-19. Reprinted with permission.

recent century, patiently taking Earth's temperature and measuring precipitation and other climatic variables at weather stations all over the globe. But for prior centuries, we must reconstruct climatic factors from natural rather than human archives. In this endeavor, climate proxies such as those found in tree rings, coral growth, ice cores, lake sediments, and loess deposits provide insights into the characteristics of past climate fluctuations, whether of natural or anthropogenic origin.

## Geothermal Archive

Important geophysical data relevant to reconstructing the surface temperature variations of the recent past reside in a rather unlikely place—in the rocks beneath our feet. How does the subsurface contribute to climate reconstruction? In a nutshell, it is the temperature at relatively shallow depths beneath Earth's surface. The underlying principle is simple: If Earth's atmosphere is warming, the rocks in the uppermost part of the crust will also warm and temperature changes that occur at Earth's surface will propagate slowly downward into the rocks beneath. Thus, present-day rock temperatures at shallow depths provide evidence of temperature changes that have occurred at the surface in the recent past.

The pace of heat conduction in rocks is such that the past 500 years of surface temperature history is imprinted on and contained within the upper 500 meters of Earth's crust. This record exists almost everywhere on the continents, where the atmosphere and the solid surface are in direct contact. The archive can be accessed by drilling a borehole and lowering a thermometer to obtain a profile of rock temperature vs. depth. This observed profile can then be interpreted to reconstruct the temperature history at Earth's surface that produced the present day rock temperature profile.

The idea that subsurface temperatures contain information about past climate change is not new. In the early decades of the 20th century, temperature measurements in the mines around Lake Superior were analyzed to see if the effects of the Pleistocene glaciation could be identified. Later, attempts were made to assess the magnitude of the climatic disturbance on measurements of the heat flowing outward from Earth's interior. But only in the last decade or so have geothermal researchers been making serious efforts to use borehole temperature profiles from all of the continents to reconstruct the history of Earth's surface temperature, particularly for the past five-to ten centuries. Geothermal studies have clearly moved into the arena of modern global climate change investigations.

A number of characteristics of this geothermal archive make it an attractive source of information about recent climate change. Perhaps first and foremost, the data comprise direct measurements of temperature to reconstruct a temperature history. Unlike many proxy data, such as those derived from tree rings or coral growth patterns, which require an empirical calibration relating ring thickness or isotopic ratios to temperature, the subsurface measurements need no empirical conversion. The primary observations are already direct measurements of temperature with a thermometer.

From the perspective of temporal resolution, however, the geothermal data are complementary to other proxies. Tree rings, for example, have excellent annual resolution but less ability to identify long-term trends. Geothermal data, however, are best used to illuminate longer-term trends of temperature, but only in rare circumstances can they offer information about annual variations. Another important aspect of the geothermal data is that there are thousands of boreholes worldwide in which temperature measurements have been made, providing a large and well distributed archive that enables both local and global assessments of climate change. The fortuitous availability of so many borehole temperature profiles stems from the focused international effort, roughly from 1960 to 1990, to measure and map the flow of heat outward from Earth's interior. Many of these sites were mineral exploration holes in remote areas, far from disturbances to the subsurface temperature regime brought about by other human activities such as urbanization and agriculture.

Temperatures in the shallow subsurface are governed principally by two processes—the flow of heat from the deeper interior and changes of temperature at the surface. In the absence of any climatic or other surficial perturbations, the temperature in the subsurface reflects only the deep crustal heat flow and is characterized (in a homogeneous medium) by a linear increase of temperature with depth, commonly known as the geothermal gradient. If temperature changes take place at the surface, they propagate downward, disturbing the temperatures represented by the geothermal gradient. But the disturbance gets smaller as it goes deeper and at some depth, becomes imperceptible. The shape of the disturbed temperature profile and the depth to which the disturbance can be observed are the essential characteristics that are analyzed to reconstruct the history of changing temperature at the surface.

## Other Disturbances

But one must be cautious in reconstructing climate from subsurface temperatures, because many other processes and factors can affect temperatures in the shallow subsurface. These processes include changes to the local microclimate caused by activities such as deforestation, agricultural expansion, and wetland destruction. Aspects of the local topography, hydrology, and vegetative patterns can also leave a subsurface temperature signature, which can be misinterpreted as changing climatic conditions. Nonuniform rock properties and subsurface geologic structures also distort the subsurface temperature field. Many of these nonclimatological disturbances can be quantitatively modeled and their significance assessed, if sufficient information about a site is available. Alternatively, one can average the individual results obtained from a number of boreholes spread across a region. Under the assumption that it is highly unlikely that all of the boreholes would have identical topography, vegetative patterns, geological structure, or hydrological disturbances, one can more safely ascribe a common signal seen in the ensemble of borehole temperatures to climate change.

## What Boreholes Reveal

What do borehole temperatures reveal about Earth's changing climate? A very special set of high-precision observations from two 3-kilometer-deep holes in the ice of central Greenland gives a glimpse of some 50,000 years of climate change. They have revealed that temperatures during the last glacial maximum were about 20-25 degrees Celsius colder than the present-day temperatures at that location. As Earth emerged from the last ice age, temperatures in the early Holocene exceeded the present day temperatures by 2-3°C. Gradual cooling followed. During the Little Ice Age

(1400-1850), temperatures were cooler than they are at present by about 0.8-1.0°C. Of course, what happens in Greenland is not necessarily indicative of what happens all over the globe, but at least these borehole temperature reconstructions provide a long-term framework with which to compare results from other locations at temperate and tropical latitudes.

The great majority of boreholes are not in ice and are typically much shallower, in the range of 300-600 meters depth. The time interval for which their temperature profiles provide the most information is roughly the past five centuries, an interval represented in the upper few hundred meters of the Greenland ice cores as well. The International Heat Flow Commission, an organization of geothermal scientists, has had a special working group assembling and analyzing borehole temperature profiles from sites around the world. The data collection now includes observations from several hundred sites, which have been searched for a five-century climate change signature. The preliminary results from eastern North America, central Europe, southern Africa, and Australia show, not surprisingly, a fair amount of regional variability. Some boreholes indicate some modest cooling, others some modest warming, and yet others more substantial warming. But with fully 75 percent of the boreholes registering a warming of some magnitude, the overall pattern of change emerges clearly. As a global ensemble, these boreholes document a temperature increase over the past five centuries averaging a little less than 1°C, a result very similar to what the Greenland boreholes show for their local setting.

When looked at in more detail, this ensemble of boreholes shows that of the total temperature change since 1500, most has taken place since about 1750, when the Industrial Revolution began. And about half (0.45°C) of the five-hundred-year change has occurred in this century alone. This estimate of 20th-century warming is fully consistent with the instrumental record of surface warming determined from meteorological observatories and weather station records, as documented by the Intergovernmental Panel on Climate Change (IPCC). The rate of warming in the 20th century is four times greater than the average rate of change over the previous four hundred years, and has earned for the 20th century the honor—dubious or otherwise—of being the warmest century of any since the year 1500. These geothermal interpretations thus provide a historical perspective that indicates the 20th century has not been just another century in terms of temperature change, but has been unusual.

What do these observed temperature changes suggest for the future? The magnitude of the change since 1500, globally a little less than 1°C, may provide some information about climate sensitivity—that is, the way temperature responds to changes in factors that affect it. Human impact on the atmosphere can first be observed around 1750 when greenhouse gases began to increase from pre-industrial levels. If the temperature changes since 1750 are assumed to be due principally to increasing concentrations of greenhouse gases and aerosols from the combustion of fossil fuels, then this observed warming provides a rough calibration of the way in which the global mean temperature responds to the anthropogenic changes in the atmosphere. If that sensitivity continues into the near future, the anthropogenic changes in the atmosphere anticipated over the next half-century will likely be accompanied by another 1.0-1.2°C of warming, an estimate rather close to the IPCC's "best estimate" of mid-21st century temperatures.

## Questions:

1. How many years of surface temperature history is imprinted on and contained within the upper 500 meters of the Earth's crust?

2. A historical temperature record can be accessed by drilling a borehole and lowering a thermometer to obtain what type of profile?

3. How many years of climate change were obtained through boreholes in the ice of central Greenland?

Answers are at the back of the book.

## Activity:

What steps could we, as a nation and the entire world, take to decrease greenhouse gas and its concentration in our atmosphere?

28

An increasing population, more advanced technology, and rising pollution levels are all contributing factors to changes in the Earth's climate. However, climatic change can also be attributed to natural phenomena, such as volcanic eruptions and the sun. Within the last century, the Earth has warmed by approximately one degree Fahrenheit. As a concerted effort to resolve the debate about the cause of the change, two teams of scientists developed computer models. These models were programmed to take into account that we are able to warm and to cool large portions of the Earth simultaneously. Both teams agreed that man-made global warming is most likely occurring. Even though the study has some weaknesses, it is probably accurate to assume that human influences are affecting climatic change.

# Verdict (Almost) In

By Carl Zimmer

Police detectives aren't the only people who look for fingerprints. Climatologists do, too: they've been looking for the collective fingerprint of humanity on Earth's climate. Most of them suspect that the 6 billion tons of carbon we pump into the atmosphere each year, in the form of carbon dioxide, could warm the planet through the greenhouse effect. In the coming century the warming could be dramatic; but is it detectable already? This past year two teams of climate modelers said yes: man-made global warming is happening—almost certainly, anyway, and it's getting more certain every year.

Certainty would be easier if it were just a matter of looking at the thermometer. "We know that Earth has warmed by roughly a degree Fahrenheit in the past century," says Benjamin Santer, an atmospheric scientist at Lawrence Livermore National Laboratory in California, "but you could have many different combinations of factors—volcanoes, the sun, carbon dioxide—that give you the identical temperature change." To exclude the natural suspects, researchers have been looking not just at the average global temperature but at the geographic pattern. The idea is that if we are

warming the planet by polluting it, we'd produce a different temperature pattern than the sun would.

The two teams that said in 1995 that they'd found our geographic fingerprints (Santer's at Livermore and a group at the British Meteorological Office) both used computer models. No other method is possible: you can't put the planet in a laboratory and run experiments on it. And though they used different methods, they were successful for the same basic reason: they took into account that we are able not only to warm the planet but also to cool large regions of it.

That's because each year we release not just 6 billion tons of carbon but 23 million tons of sulfur, mostly from fossil fuels and mostly in the form of sulfur dioxide. This gas turns into sulfate aerosols that reflect sunlight back into space even as carbon dioxide is trapping heat near Earth. The cooling effect of the sulfates is more regional—they tend to stay close to their sources, mainly in the Northern Hemisphere, while $CO_2$ spreads around the globe—but in just the past few years it has become clear that they have a big impact on the geographic temperature pattern. The Livermore and British teams were the first to include the effect in supercomputer climate models.

The Livermore researchers first simulated the atmosphere with preindustrial levels of $CO_2$ and measured the natural variability it might experience over the course of a few centuries. Then they added today's levels of $CO_2$ and sulfur. Overall, the combined gases did warm the planet, although not as much as $CO_2$ would alone. But the more striking result came when the researchers compared the geographic temperature pattern predicted by their model for today's polluted world with year-by-year records of the real world's climate over the past 50 years. They found that with each passing year the real world pattern grew more like the model—which makes sense, because the real-world levels of sulfates and greenhouse gases were climbing toward today's levels. Santer and his colleague calculate that the chances of this trend's being a coincidence caused purely by natural climate variability—and unrelated to air pollution—are slim at best.

The British team reached essentially the same conclusion by a different approach; you have to simplify something to model climate even on a supercomputer, and the two teams chose different things. The British used a more realistic ocean than the Livermore group did, one that could transport heat to and from it depths, but a less realistic atmosphere: rather than re-creating the complicated chemistry of sulfate aerosols, they simply estimated how much sunlight Earth would reflect for a given level of sulfur emissions. Then they put in the actual measured increases of atmospheric carbon dioxide year by year since 1860 and tracked the response of the model. After 1950, the real world temperature pattern conformed increasingly to the one predicted by the model—suggesting, just as the Livermore study did, that $CO_2$ and sulfates were taking increasing control of climate. The chance of natural variability's producing the pattern was less than 10 percent.

The British team let their model run into the future. As sulfur emissions rise and $CO_2$ rises faster, they found, global temperature should rise 2.3 degrees Fahrenheit by the year 2050. Without sulfates, it would rise 3.3 degrees. (North America would get a bigger break—it would warm only 3 degrees instead of 4.5.)

Other climatologists have been praising these studies, but the two teams themselves are quick to point out weaknesses. The Livermore correlations between model temperature pattern and reality are much looser in winter and spring than in summer and fall. The British correlations work on the global scale, but when the researchers analyze a region like Europe or North America in detail, the correlations fall apart. And though both models now include sulfates, neither includes soot or other haze-producing hydrocarbons, which can either cool the planet or warm it. No one understands these effects well enough yet to put them in a computer model.

Yet the fact that both teams found the same strengthening pattern may nevertheless hint that our influence on climate is making itself felt. "We've found emerging evidence that we're beginning to see a fingerprint, but we're not quite there yet" is the cautious conclusion of John Mitchell, who led the British team. "As far as understanding climate change goes, this is the end of the beginning, not the beginning of the end."

**Questions:**

1. Other than average global temperature, what can researchers look at as evidence of man-made global warming?

2. We release what chemicals, and in what quantities, into the atmosphere each year?

3. The cooling effects of sulfates is more regional—but how far does carbon dioxide spread?

Answers are at the back of the book.

**Activity:**

Keep a journal for a day of all the things you do or used that consume or have consumed fossil fuels. Don't forget less obvious items, such as plastic containers.

What can you do to consume less fossil fuel? What is your reaction to our dependence on fossil fuels?

29

The debate over global warming still continues. However, many scientists do believe a human-induced rise in global temperature is occurring. One of the leading greenhouse gases, carbon dioxide, is pumped into our atmosphere through the burning of fossil fuels and deforestation. Though all of the effects of global warming are not known, possible effects are a rise in sea level, an increase in weather severity, a northerly migration of climatic zones, and the extinction of species throughout the food chain. With atmospheric carbon dioxide levels on the rise and ten of the world's warmest temperatures on record since 1980, the effects of global climate change are being witnessed in our national parks.

# The Heat is On

By Lily Whiteman

In 1850, more than 150 glaciers could be found in what is now Glacier National Park in Montana. Today, that number is closer to 50. At this rate, within the next four decades, all 50 remaining glaciers will vanish from the park. By then, as Vice President Al Gore predicts, this harbor of ancient icefields will become known as "The Park Formerly Known as Glacier."

No longer just grist for disconcerting prophecies, global climate change is upon us. Indeed, impending temperature increases are expected to cause such damage that in 1997 Interior Secretary Bruce Babbitt billed them as "the largest, most pervasive and ominous threat" to confront civilization.

Scientists believe that the average temperature of the Earth's surface has jumped one degree Fahrenheit since the 1800s. And what a difference a degree makes! A deceptively small change in mean global temperature can lead to extremes in weather, such as the droughts in Florida, searing heat in Texas, and rainfall and flooding in Wisconsin and elsewhere witnessed this past year. Such processes could ultimately turn the Earth topsy-turvy with the planet's poles tossing off their ice caps and some temperate regions experiencing freezes. It can also have the greatest initial effect on the smallest links in the food chain: plankton and invertebrates.

National Park System units are already showing that one degree in varied ways. Scientists, for example, believe that global warming helped completely melt Glacier Bay, a 65-mile-long inlet in Glacier Bay National Park and Preserve in Alaska, which was frozen shut only 200 years ago. Glaciers are particularly good barometers of global warming because they are too stable to manifest ephemeral, annual, or even decade-long variations; they reflect only long-term changes. And these glaciated parks, which include North Cascades, Olympic, and Mount Rainier national parks, will lose more than their famous icefields.

As glaciers disappear, their flow of frigid meltwater into streams ceases, and the temperature of the stream water consequently rises. These temperature changes, in turn, affect the plants and invertebrates that support amphibians, fish, and waterfowl. Some high alpine invertebrates, such as Glacier National Park's caddisfly, could be lost if water temperatures rise too high or reduced water availability results in intermittent flow of streams.

From "The Heat is On," by L. Whiteman, National Parks, January/February 1999, pp. 33-37.
Reprinted with permission.

The melting of glaciers and polar ece, together with the expansion of warming waters, has already raised sea level by four to eight inches since the mid-1800s. Scientists expect another increase of up to three feet by 2100. Such a sea change could inundate a number of national park units, including Dry Tortugas and Everglades national parks in Florida and Padre Island, Cape Hatteras, Assateague Island, Fire Island, and Cape Cod national seashores, which line the gulf and east coasts from Texas to Massachusetts.

Sea level rise threatens more than the lighthouses and historic buildings found at these seashore parks. The increased water level would eliminate crucial habitat for endangered birds such as the piping plover and least tern. It could also inundate marshes, which are important marine nurseries and migratory bird stopovers, such as Nauset Marsh at Cape Cod National Seashore.

Some scientists, in addition, suspect that the unusual strength of several recent storms, such as Hurricane Andrew, reflected the influence of global warming. Although scientists generally refrain from attributing short-term weather to long-term warming that varies geographically and temporally, computer models predict that rising temperatures do increase the frequency of such catastrophes.

Also under surveillance as a possible result of a changing climate are the unusually hot summers of the past two decades. According to worldwide temperature readings from varied locations, including many national park units, the ten warmest years on record all occurred since 1980, and four of them since 1990. Once analyzed, data for 1998 will likely reflect another record-breaker.

With this ongoing heat wave, summers in the United States have become particularly hot and dry, and regional fire hazards have been aggravated. If this potential symptom of global warming continues for just another 50 years, most whitebark pines in Yellowstone National Park could wither in water shortages. The disappearance of these trees could affect grizzly bears, which feed on the pines' nuts.

Some food chains have already proved capable of magnifying even small temperature changes with particular speed and penetration. A case in point: During the 1950s, surface layers of the waters off Point Reyes National Seashore began warming by several degrees. In response, some plankton populations declined by as much as 80 percent. Marine creatures of all sizes, from squid and rockfish to whales, feed on plankton and are affected by its loss.

Consequently, several seabird species, for which Channel Islands and Point Reyes national parks are renowned, appear to be declining. For instance, the sooty shearwater, a gulllike seabird that annually migrates from Antarctica to the Northern Pacific, has declined by an alarming 90 percent.

On the Atlantic side, computer models predict that additional, though somewhat still undefined, warming could dull forces propelling the Gulf Stream, a current which has a tremendous effect on the weather in the northern latitudes. Evidence is mounting, for example, that spring in the Northern Hemisphere is arriving a week or more earlier than it used to, birds are migrating sooner, and trees are leafing earlier. What this shift may mean for the hundreds of parks along the eastern seaboard, a majority of which are historic or cultural, remains unclear.

"As temperature zones creep northward, some landscapes will die off, and an increase in sea level will have a significant effect on the migratory bird population as marshes and other havens disappear," says Eileen Woodford, NPCA's Northeast regional director. "What will happen to cultural sites is less clear. We know what effect acid rain has on cultural sites. But we do not know what effect a sustained temperarure increase will have on humidity levels, for instance, and what those humidity levels will do to historic buildings and artifacts."

What is driving global climate change? While not ignoring the plausibility of natural contributions, the Intergovernmental Panel on Climate Change (IPCC) points its authoritative finger toward human sources such as deforestation and the burning of fossil fuels, which includes coal and petroleum products that release so-called greenhouse gases like carbon dioxide. Emissions from automobiles and power plants—such as the Mohave Generating Station near Grand Canyon National Park—are among the biggest offenders. Just as the glass of a greenhouse traps incoming heat within the structure, so do these gases trap heat within the Earth's atmosphere. At current rates, a doubling of carbon dioxide levels will hike global temperatures another 3.5 degrees by 2100, according to IPCC.

**161**

Not only do human activities promote global warming, many of our activities cripple natural mechanisms for neutralizing its damage. Land use changes and pollution, for example, limited the ability of many animals to evolve with changing conditions, even in the relatively rare cases when the warming onslaught would allow for such gradual accommodations.

Global warming could compel some native species to abandon stricken park units altogether, but such migrants would not necessarily find sanctuary elsewhere. Many would likely be stumped by such obstacles as sprawling agricultural and urban areas separating relatively small protected areas. "Few animals or plants would be able to cross Los Angeles on the way to a new habitat," says former World Wildlife Fund biologist Robert Peters, giving just one example. Populations that can neither adapt nor outrun changing conditions—as have survivors of earlier, slower climatic shifts—remain primed for extinction. And species that already have been reduced in great number, such as the endangered Florida panther and piping plover, would be even less likely to adapt.

Everyone agrees, more information is needed. Researchers from the Biological Resources Division (BRD) of the US. Geological Survey are thus inventorying ecosystem characteristics, such as stream temperatures, glacial movements, and fire vulnerability, in about 80 parks. These study areas were selected for their ecological health, which reduces confusion from unrelated damage; the availability of historical records about them, which supports reconstruction of past conditions; and their inclusion of varied ecosystems from tropical to arctic.

By coupling field monitoring with computer analyses, BRD will define warming effects, as reflected in focused comparisons among past, present, and future conditions within many ecosystem types. Ongoing progress is publicized through news media, several Internet sites, pamphlets, and lectures.

After more precisely pinning down warming effects, BRD will recommend park-specific strategies to the National Park Service (NPS) for coping with the impacts of climate change. For example, some parks might be advised to acquire amphibious vehicles to accommodate encroaching ocean; others might be advised to bolster fire suppression to discourage migrations from burned habitats.

Nevertheless, simply designing coping strategies for warming, let alone following them, will burn a hole in the budgets of NPS and USGS's Biological Resources Division. Why? Because additional resources for these important activities have not been allocated to either agency. Warming-related projects just "get in line" for funding behind truckloads of other urgent park needs. Worse still, many BRD researchers complain that long-range issues, such as global warming, are frequently shortchanged in favor of creature comforts for visitors. "We can't compete with asphalt and toilets," says BFD ecologist Tom Stolgrem.

The Park Service has signalled that favoring creature comforts over resource protection may be changing. Although Congress has yet to appropriate necessary funds, Park Service Director Robert Stanton has proposed a draft plan for an ambitious five-year Natural Resources Initiative, a comprehensive strategy to shift the priorities within the Park Service toward resource protection.

Meanwhile, glaciers continue to back away from Glacier Bay without being monitored The clock is ticking.

## Questions

1. Why are glaciers good indicators of global warming?

2. At our current rate of emitting carbon dioxide, what will a doubling of carbon dioxide levels do?

3. What are some of the ecosystem characteristics being studied by the BRD, and why were those 80 parks chosen?

Answers are at the back of the book.

## Activity:

Go to **http://www.ncdc.noaa.gov/ol/climate/globalwarming.html**
Read over the list of frequently asked questions about global warming, and write down three things that you did not know.

# 30

Typically, volcanic eruptions cause global cooling by blocking sunlight. A group of scientists believe a volcanic eruption triggered a rapid increase in a global warming trend 55 million years ago. In a very fascinating time in historical geology that saw the first appearance of modern mammals, what could a spike in temperature do to terrestrial and marine organisms? Also, in an age when global warming and loss of biodiversity are of increasing concern, can this global warming and mass extinction event provide answers to how we are altering our climate and its effects?

# The Day the Sea Stood Still
## Ancient Eruptions, Global Warming, and Mass Death

By Tom Yulsman

It may have been the blast that changed the world.

One day in the Caribbean Sea, at the end of the Paleocene Epoch 55 million years ago, a volcano blew, spewing a climate-altering parasol of tiny particles high into the atmosphere. Such events are hardly rare. But this was no ordinary volcano: It was huge. And this was no ordinary time in the planet's history: The environment already was on the threshold of profound change.

According to a new theory proposed by marine geologist Timothy Bralower of the University of North Carolina at Chapel Hill, climatic fallout from the Caribbean eruption pushed Earth beyond that threshold, triggering what has been identified in the last 10 years as one of the most remarkable worldwide transformations known. In the dry argot of earth scientists, this event is called the Late Paleocene Thermal Maximum, or LPTM. In the language the rest of us speak, it is simply unbelievable. Its prelude, innocuously enough, was a long-term global warming trend that began about 60 million years ago and weakened the circulatory system of the world's oceans. Five million years of that warming seems to have left Earth's environment vulnerable to catastrophic change. Then, at the very end of the Paleocene, the

circulation system experienced the oceanic equivalent of a heart attack. The global network of ocean currents stopped doing its vital job of delivering cold, oxygen-rich water to the deep ocean. As a result, the abyss warmed and stagnated. This shock caused a mass extinction of marine organisms, including as many as half of all species of deep-sea foraminifera This family of ubiquitous one-celled sea animals forms one of the primary links in the oceanic food chain. The mass killings of forams and other species, Bralower says, represent "the biggest extinction event in the deep sea in the last 90 million years. Nothing else even comes close."

When the oceanic heart attack struck, Earth was considerably warmer than today, thanks to the five-million-year global warming trend. Global average air temperature was several degrees higher than now, chiefly because the poles were far warmer. Antarctica was glacier-free, possibly draped by forests, and surrounded by sea water with surface temperatures about 35 degrees Fahrenheit higher than now. Immediately after the heart attack, it became still warmer as a spike of very high global temperatures was superimposed on the already toasty Earth.

According to Gerald Dickens, a geochemist at James Cook University in Australia, a gargantuan gasp of methane, loosed from the sea floor as the deep ocean warmed, may be the best explanation for why Earth seems to have spiked the high fever. Oceanic and climate changes of the LPTM may have peaked in 10,000 years, a mere heartbeat on the geologic time scale. Methane ($CH_4$), like carbon dioxide, is a rather potent greenhouse gas. Per molecule, it can absorb 10 to 20 times more heat radiation than $CO_2$, trapping that warmth in the air.

## Mammals Suddenly Appear

Little was untouched by the higher temperatures. When the heat peaked on land, for example, the group of modern mammals that now dominate Earth, including primates that later would give rise to our own species, suddenly made their first appearances on at least two continents. "I and others in the earth-science community are beginning to feel that this is one of the most—if not the most—fascinating time intervals in the history of the Earth," Dickens says.

Research into events of this period is part of a larger effort to reconstruct climate changes that occurred millions of years ago. This initiative is driven partly by the need to know how we may be altering the climate through emissions of greenhouse gases. Conducting controlled experiments on the global environment obviously is out of the question. But nature has conducted plenty of her own, and the LPTM is one of the most remarkable.

"We really need geologic records of climate to understand the long-term causes and effects of climate change," Bralower says. "The Late Paleocene Thermal Maximum is relevant because it is the most abrupt warming event ever documented." That record, preserved in seafloor sediments, provides a warning that Earth's environmental system may not be as stable as we might like to believe, says James Kennett, an oceanographer at the University of California at Santa Barbara. "It's clear that Earth at times develops an environmental system that's extremely sensitive to change and can flip from one state to another, creating bedlam."

Hints that something strange had happened began turning up a decade ago. But convincing evidence came in 1991, in the form of hardened mud from the Antarctic sea floor. Kennett and geochemist Lowell Stott of the University of Southern California found chemical fingerprints in those sediments indicating that the ocean had warmed and changed its circulation pattern dramatically.

Kennett and Stott analyzed the difference in abundance of two different forms of oxygen, called isotopes, that are preserved within fossil foraminifera skeletons. As the tiny marine animals drift on currents they absorb from their surroundings more of the lightest isotope $^{16}O$, with eight protons and eight neutrons for an atomic weight of 16) and less of the heavier $^{18}O$ (which has two extra neutrons) when the water warms. Evidence of this change is preserved within the foram skeletons. Forams preference for $^{16}O$ in warm water occurs because atoms of this isotope vibrate faster than those of $^{18}O$. Physics dictates that, as water warms, a foram more easily absorbs the faster-vibrating atom. When forams die, their skeletons rain into the abyss and accumulate as seafloor mud. Over time, these sediments, become deeply buried and harden under pressure, locking away an oxygen-isotope record of past water temperature. Scientists can access that record by drilling into the sea floor for a core of sediment.

Kennett's and Stott's analysis of 55-million-year-old forams from Antarctic waters showed that, just before the Paleocene closed, the bottom waters were at 50 degrees, considerably warmer than today's near-freezing temperatures but still quite chilly. Then something forced the temperatures of those waters to rise nearly 20 degrees, possibly in less than 10,000 years. Meanwhile, surface waters also warmed, although somewhat less, from 57 to 70 degrees. At this point, bottom and surface waters were almost uniformly warm. This was shocking. With only a few exceptions, ocean waters almost always are layered, with warm water atop cold bottom water.

If you've ever gone swimming in a lake in the summer, you may have noticed that the surface water was tolerably warm while deeper water was numbingly cold. In fall, lake waters "turn over" as warm and cold layers mix. But this isn't supposed to happen in the ocean. For example, in the Caribbean today, surfaces average about 81 degrees whereas water half a mile below may be only 40-45 degrees. That difference keeps the layers stable.

Kennett's and Stott's evidence showed that the two layers did mix as they came to the same temperature at the end of the Paleocene. The ocean, Kennett says, "turned over just like a lake." The warming of deep waters that made the turnover possible occurred more than 6,000 feet below the surface. Kennett and Stott proposed that this deep heating must have been caused by a profound change in the global system of ocean circulation or perhaps a virtual halt.

---

## Questions:

1. Why are foraminifera good fossil indicators of warming ocean water in the past?
2. When the earth's climate warms, what happens to the circulation in the ocean?
3. How did a volcanic eruption trigger a spike in global warming?

Answers are at the back of the book.

## Activity:

Go to **http://eospso.gsfc.nasa.gov/NASA_FACTS/volcanoes/volcano.html**
Read the information. Explain how volcanoes cause global cooling. Drawing a picture may help.

# 31

El Niño is a large scale disturbance in the production and consumption cycle in the Pacific ocean. The North Atlantic Oscillation (NAO), El Niño's cousin, affects the circulation of seas at the North Atlantic's margins. It affects the average temperature all over the world. Fluctuations in temperature caused by NAO may account for some greenhouse warming, but it may also be controlled by the greenhouse effect.

Global warming and ozone depletion cause changes in the stratosphere. These changes, such as in wind and structure, could reflect changes in the NAO's behavior.

It appears that the NAO has the ability to recall what it does from winter to winter, so when each winter comes around it goes back to the state it was in the previous year. Climatologists are interested in possibly using the NAO's slower rhythms to predict climate rather than weather.

# The Storm in the Machine
## It defeated Hitler's assault on Russia and has brought violent weather to Europe

By Oliver Morton

In the vast open wastes of the South Pacific, El Niño is now reaching its climax. This is the biggest, most dramatic recurring climatic event there is—its effects are felt around the world. It is reasonably easy to grasp how El Niño works, at least in broad outline: a big glob of warm water, normally kept in the west by prevailing winds, breaks free as the winds collapse, and heads towards South America. The El Niño currently under way is the biggest on record, and is already causing havoc: crop failure in southern Africa, storms in Santa Monica, failed monsoons across Asia.

Climate scientists and meteorologists who specialise in the weather of the Atlantic Ocean might well feel a twinge of envy when they see such an impressive suite of effects. What do they have that can compare? Over the past few years it has become increasingly clear that the Atlantic has a dramatic event all of its own. And in some areas its effects could be just as important.

## World of Weather

El Niño's cousin is called the North Atlantic Oscillation. Changes in the NAO correlate with all sorts of variables around the ocean, from rainfall in Bordeaux to the amount of Saharan dust that ends up in the Bahamas and the richness of the fisheries off Iceland. The NAO affects the circulation of seas at the North Atlantic's margins. It leaves smudgy cyclonic fingerprints all over the northern hemisphere's climate. And it affects the average temperature of the whole world.

An analysis carried out by Jim Hurrell at the National Center for Atmospheric Research in Boulder, Colorado, which was published in 1996, shows that mild winters across Europe and Asia linked to the NAO account for a good chunk of the warming trend in global temperatures seen in the past few decades—with cumulative global effects on the same scale as those of El Niños. This news has been enough to make NAO researchers flavour of the month with the energy lobby in the US, which likes nothing better than natural explanations for global warming. But it may well find that the NAO is not quite the boon it had hoped for.

One of the greatest unanswered questions in climatology is the range of the climate's natural variations. How does it vary when left to itself? How long are the variations? How extreme? These natural fluctuations may account for some of what is seen as greenhouse warming, but at the same time they might mask it. They could even be under the control of the greenhouse effect, channelling its impact. In a greenhouse world they could become still more important.

The NAO is one of the most intriguing of those fluctuations. Its long-term swings have become more pronounced over the past century. It had a significant warming effect on European winters in the early 1990s, and it may now be on course to cool things down for a decade or so. To find out what the NAO will do next means trying to understand all sorts of subtleties in the ways the North Atlantic's winds drive and are driven by its currents. Which is why more and more climate researchers are getting into boats and putting to sea whatever the weather.

## When the Wind Blows

In terms of measurement, the NAO is a fairly simple thing—it is an index created by comparing the pressures in the Azores and in Iceland. In the first half of the 20th century meteorologists created a number of similar oscillating indexes. The Southern Oscillation, the difference between the pressure in Darwin and in Tahiti, measures the atmospheric component of the El Niño effect, which is why climate modellers tend to call the whole ocean and-atmosphere ensemble ENSO (El Niño/Southern Oscillation).

For the NAO, a high index signifies low pressure around Iceland and the reverse off Portugal, a situation that gives rise to strong westerly winds. In winter—which is when the NAO's signal is strongest—these westerlies bring heat from the ocean's surface into the European continent, along with storms like those that this winter saw in the New Year in Britain. So a high index tends to mean relative warmth in northern Europe, while a low index means weak westerly winds and a continental climate dominated by cold air from the north and east.

The difference is marked. David Parker of the Hadley Centre for Climate Prediction and Research in Bracknell, part of Britain's Meteorological Office, says that oscillations in winter between high values and low values of the index account for about 50 per cent of the variability in monthly average temperatures in central England. And it's not just the wind and the warmth, it's the rest of the weather too. Mild winters with high NAOs dry out the south of Europe, with a spate of them through the early 1990s explaining everything from poor olive harvests in Valencia to poor skiing in Val-d'Isère.

If you could predict how the NAO is going to rise and fall on a scale of weeks and months, then you would be doing all Europe a favour. Unfortunately, you can't. According to Tim Palmer Of the European Centre for Medium Range Weather Forecasts in Reading, no weather forecast could predict the exact evolution of the NAO over more than a few weeks. The problem is that the atmosphere over the North Atlantic is chaotic. Forecasts based on almost identical initial conditions give similar results in the short run, but in the long run they diverge hopelessly.

Computer models make the point neatly—and show up the difference between the NAO and its better-known cousin half a world away. If you run a computer model with the temperatures of the sea surface fixed, says Palmer, you will see something rather like the NAO flip-flopping around happily enough. The Southern Oscillation, though, will shut down completely if sea surface temperature remains constant. For ENSO, the ocean calls the shots, and nothing happens without its active participation. So at first sight the NAO appears to be an atmospheric phenomenon, inheriting all the unpredictability that air is heir to.

But the story isn't that simple. The NAO doesn't just jiggle about on timescales of weeks and months, it also has slower rhythms whose beats are measured in years, decades or more. These are the trends that interest climatologists, and raise the hope of predicting climate—as opposed to weather. The chart shows over a century of these long-term ups and downs: a dip into the negatives in the 1940s, which saw some of the coldest European winters of the century, including the ones that delayed Hitler's invasion of France and defeated his assault on Moscow; a protracted dip in the 1960s, the decade with the consistently coldest winters in Britain since the

1880s; and the odd, long period of very high average NAO states in the late 1980s and early 1990s which corresponded to particularly mild winter weather across Europe.

## Complex Rules

These long-term changes suggest that something more than atmospheric chaos is going on. Atmospheric effects tend not to persist all that long unless there is something pushing them. Though it is possible to get decade-long ups and downs in the NAO in computer models that mimic the atmosphere alone, many scientists in the field are pretty sure that that's not the whole story. They think the long-term memories that shape the NAO are held beneath the waves, in the cat's cradle of currents inside the North Atlantic. The NAO may not be the highly marshalled double act of atmosphere and ocean seen in El Niños, but its long-term shifts may well be the product of the same players following more complex rules.

Hurrell and Mike McCartney, who studies the NAO from the Woods Hole Oceanographic Institution in Massachusetts, have produced data suggesting that the ocean seems to be responsible for the NAO's ability to remember what it has been doing from one winter to the next. They took a set of years with similar NAOs and looked at what the NAO was like in the preceding autumn and the following spring, summer and autumn, finding that there is no fixed pattern at all. In the warmer parts of the year the NAO appears to wander around aimlessly. But the clue to the role of the ocean comes when you compare one winter with the next: when winter comes, the NAO tends to go back to the state it was in the previous year no matter what it has being doing in the meantime.

McCartney and various other oceanographers are convinced that the ocean reminds the atmosphere what to do from year to year. And since there are long-term patterns in the NAO, it makes sense to look to the ocean to see what might be responsible. There are some intriguing candidates. Donald Hansen of the University of Miami and Hugo Bezdek, who runs the Atlantic Oceanographic and Meteorological Laboratory at the US National Oceanic and Atmospheric Administration, showed in 1996 that big patches of ocean with anomalous surface temperatures—a fraction of a degree hotter or colder than expected for the time of year—slowly drift up across and around the North Atlantic. They follow a path from west to east, rather like that of the Gulf Stream and the North Atlantic Current, though they move much more slowly than the currents themselves. Then, at the same stately pace, they turn in a great circle around the sub-polar gyre of the North Atlantic, ending up somewhere west of the tip of Greenland.

McCartney and his colleague Ruth Curry soon showed that these anomalies are not restricted to the surface—they have deep roots (which get deeper as they make their slow and cooling progress towards the pole). They contain a lot of heat, and represent significant changes in the overall flow of heat from the subtropical Atlantic to the north. McCartney thinks that they may be responsible for the NAO's decade-long swings—they are in the right place carrying the right sort of energy and taking the right sort of time to pass by. As yet, however, there is no clear mechanism to explain quite how such circulation might force the NAO one way or another. Without understanding this in detail, longterm guesses about what the NAO will do next remain impossible.

Even though the effect that the ocean has on the air above it is still obscure, the effect that the atmosphere has on the waters below is becoming clearer. Bits of the story have been worked out in studies all around the ocean, including long-term observations of the Labrador Sea coordinated by the US, and pulled together in Britain by Robert Dickson of the Centre for Environment, Fisheries and Aquaculture Science in Lowestoft.

Over the past few years, Dickson has studied the changes in the ocean that went along with the shift from very low NAO indexes in the 1960s to consistently high indexes in the 1990s. He has found that these changes in average atmospheric pattern fitted into changes in the Sargasso, Labrador and Greenland Seas. Each of these areas is capable of creating large volumes of homogeneous water, a process which can have effects on the ocean elsewhere, and their propensity to do so seems to be linked—to the NAO. In the 1960s, when the NAO was low, the Sargasso Sea was producing the water it is famous for, which oceanographers call "18° C water", and there was deep convection going on in the Greenland Sea, producing cool salty water at depth. But nothing much was happening in the Labrador Sea. When the NAO was at its reverse extreme in the early 1990s, there was deep convection in the Labrador Sea, but little interesting activity anywhere else.

This deep convection in the Labrador Sea makes sense. Water sinks when it is cold and dense. Strong winds passing over the surface will always cool water, and when the NAO is high the westerlies rip across the face of the Labrador Sea. The same pattern of pressures doesn't provide such winds for the Greenland Sea, up above Iceland, and so when the NAO is high the Greenland Sea gets warm; it also gets less salty, which makes it less dense, and that also lessens convection. When the NAO changes to its negative state, though, the tables are turned—high pressure over Greenland brings strong, cold, dry winds from the North Pole down over the Greenland Sea, cooling it enough for convection. The same winds drive relatively fresh water and sea ice down the east coast of Greenland and round its tip into the Labrador Sea, where the freshness and lack of cooling winds make deep convection less and less likely.

During the early 1990s, the high NAO drove convection in the Labrador Sea at a cracking pace, producing large amounts of cool water at depth. But in the winter of 1995/96, everything changed. Straight after the highest winter index of the century came the lowest. The winter after that was ambiguous, and the current one is only halfway through, though it has certainly had its fair share of westerlies recently.

What is going on? Is the NAO undergoing a long-term shift towards the negative—suggesting a cooling of northern European winters for years to come—or is it simply in the middle of a blip?

A long-term shift in the climate towards low NAOs and cold European winters should, if Dickson is right, shift the production of cool, deep water from the Labrador Sea to the Greenland Sea. A European Union programme called ESOP-II has been coordinating a wide range of studies of the Greenland Sea over the past three years. Last year, as part of ESOP-II, a team led by Andrew Watson of the University of East Anglia released an inert marker chemical, $SF_6$, below the circling waters of the Greenland Sea. The way this has spread out vertically from its original depth suggests that moderate convection has indeed started again, though it has not reached the depths. And observations made early this winter by McCartney show that there may be twice as much water coming through the Denmark Strait between Iceland and

Greenland at depth than there was a couple of years ago, which might also indicate increased convection. There may be a big shift going on, but it's too early to say for sure—the models just aren't good enough.

Such shifts in circulation may not just be a result of the NAO's moods, they may be contributing to them. McCartney says that a large mass of cold water that sank in the Labrador Sea at the end of the 1980s and the beginning of the 1990s is now heading south along the eastern edge of America's continental shelf. Its effects have already been detected south of the Gulf Stream at Bermuda, and east of Miami.

According to numerical models created by McCartney's colleague Michael Spall, this cold water at depth could make its effects felt in the Gulf Stream passing over it. This could in turn create colder surface temperatures that would then generate anomalies like those Hansen and Bezdek found wandering across the Atlantic. If so, changes in the NAO's output would be leading directly to changes in the factors that force its development. The snake may be biting its tale, producing a closed cycle where the factors which encourage one NAO state lead, through oceanic and atmospheric routes, to changes that encourage the opposite state.

But if there is such a closed cycle, it is unlikely to be closed tight. The different processes—the creation of sea surface temperature anomalies, the flip-flopping of convection, the rise and fall of the cold Labrador water—may all have different typical timescales. Sometimes they could beat in sync and sometimes they could cancel each other out. That would fit with the fact that the NAO's variability varies. In the 19th century, most of the variation was on timescales of between two and three years. Today the variability is measured in decades.

## Greenhouse Connection

The NAO's oscillations are a hot topic in global warming. McCartney remembers that when The New York Times ran a piece last year reporting that the NAO effects and those from other natural cycles might account for most of the observed warming in the northern hemisphere, he got two calls. One was from a noted critic of ideas about greenhouse warming as a result of human activity. He was "very happy", McCartney recalls. The other, he says, was from a group representing automobile manufacturers, who "'wanted to sign me up as a consultant." McCartney admits that natural shifts in the NAO may be responsible for a fair-sized chunk of the warming previously ascribed to greenhouse gases. But he prefers a different view: that the greenhouse effect may be changing the way the NAO and other natural climate variations actually vary.

The idea is that if the greenhouse effect is going to manifest itself, it will have to do so via existing climate patterns such as the NAO. Changes in the stratosphere's winds and structures could reflect changes in the NAO's behaviour, and the high indexes in the early 1990s may be an example of just that. The eruption of Mount Pinatubo in the Philippines threw a girdle of dust around the world which cooled down the surface but heated up the tropical stratosphere. As a result, stratospheric winds, driven by the temperature difference between equator and poles, got stronger,

especially those around the poles. Chris Folland of the Hadley Centre points out that strong circumpolar winds in the stratosphere mean strong westerlies in the troposphere below, just like those seen in the early 1990s.

McCartney observes that global warming and ozone depletion also cause changes in the stratosphere—both are thought to cool it. No clear mechanism for this has been worked out yet—as Folland stresses, no models capture the details of the interaction between troposphere and stratosphere very well. But until we know if global warming is strengthening or changing the NAO in some way, it's worth being cautious. As McCartney says, "No matter what you do to the atmosphere, it will get mapped on the same systems." It's just that, as yet, no one knows exactly how the mapping works. To find out will require rather better modelling of the subtleties of the NAO, the excesses of El Niño and the intricacies of other naturally varying climate systems—and a lot more messing about in boats.

## Questions:

1. What path does the Gulf Stream and the North Atlantic Current follow?
2. What does a high index signify for the NAO?
3. What does ENSO stand for?

Answers are at the back of the book.

## Activity:

Go to the Website: **http://www.pbs.org/wgbh/nova/elnino/**
Select: the Global Weather Machine.
Explain what a convection cell is and what the jetstream is.

You can also select: Anatomy of El Niño.
If your computer has the appropriate plug-ins, try selecting: Make the Earth's Weather.
You can also download a stand-alone version of make the Earth's weather. Hopefully you can play around with this program and have some fun with it.

# PART 6
## The History of Life—Origins, Evolution, and Extinction

# 32

One of the most intriguing questions to even the most unscientific of minds is "How did life begin?" Over time different theories have been presented. One theory has its basis in the primitive atmosphere, which was quite different from today's atmosphere. Stanley Miller ran an experiment in 1953 based on the volatile gases of the oxygen-free primitive atmosphere and created organic molecules.

The 1977 discovery of hydrothermal vents along volcanic ridges on the ocean floor has sparked a new theory of the origins of life. An area once thought to be void of life is teeming with organisms in scalding water. A growing number of scientists believe that these vents could hold the key to the first steps of life.

# Life's First Scalding Steps

By Sarah Simpson

In a back room at the Carnegie Institution of Washington (D.C.), Jay A. Brandes packs water and powdered rock into a 24-carat-gold capsule not much bigger than a daily vitamin. He injects gases like those spewed by a volcano and then seals the receptacle. Finally he places the shiny amulet inside "the bomb," a device isolated from the rest of the world by steel panels salvaged from a scrapped battleship. Within, the mixture is besieged with scorching temperatures and bonecrushing pressures, forces like those at work at seafloor geysers. Brandes and his colleagues then look inside the tortured capsule for evidence of the chemical steps that sparked the beginnings of life.

Using no more than the cocktail of chemicals discharged at undersea hydrothermal vents, the researchers are trying to mimic reactions that occur within living cells.

Oceanographer Jack Corliss first discovered these vents in 1977, while exploring a volcanic ridge at the bottom of the Pacific Ocean. Before then, oceanographers had considered the ocean floor cold and mostly barren. Corliss' view outside the tiny research submersible revealed a different world.

Peering out a porthole, Corliss became the first person to witness the biological wonderland of shoe-sized clams, 6-foot tube worms, and blizzards of strange microbes thriving at the vents, which spew out a hellish mix of shimmering brines. If these creatures can subsist in a bath of scorching chemicals and heat from the planet's interior, he reasoned, then perhaps this is where life got its start.

Corliss' proposal that life sprung from water, gas, and rocks far out of the sun's reach inspired grandiose theories but remained virtually untested for 20 years. Most origins-of-life researchers were still busy pondering the long-held notion that life's precursor chemicals linked up at the surface of a sun-drenched pond or ocean. In this more traditional scenario, the sun simmered a prebiotic soup for millions of years to cook up the first cellular organisms.

Only in the past few years have scientists such as those at the Carnegie Institution begun to roll up their shirtsleeves and get down to the business of determining what biochemical reactions are possible at hydrothermal vents. In a series of recent experiments, researchers have found that the harsh vent conditions can foster some of the chemical steps thought necessary for early life. Their results are capturing the attention of a growing group of scientists—and raising belief in the chance of finding life elsewhere in the universe.

The most detailed step-by-step blueprint for how Earth's oldest raw materials could have given rise to the stuff of life came out of the imagination of Günter Wächtershäuser, an organic chemist at the University of Regensberg in Germany. Ten years ago, Wächtershäuser conceived of an assembly-line process at the ocean floor that transforms basic inorganic chemicals into organic chains, the biological molecules that are the building blocks of life.

Wächtershäuser's factory enlists the elements of modern industry—all readily available at vents. The conveyor belt is the flat surface of metal sulfide minerals, such as iron pyrite, abundant in seafloor rocks. The raw materials are carbon- and hydrogen-rich gases from volcanic belches dissolved in the seawater. The workers that drive the assembly line—the keys to the whole process—are metallic ions in the sulfides.

In living cells, complex proteins called enzymes play the role of factory laborers, bringing certain molecules together and splitting others apart. Before enzymes appeared on the planet, Wächtershäuser says that metallic ions filled that catalytic role. Without these mediators, reactions might take months or years, or never happen at all, he adds. New components would never get added to the molecules passing by on the conveyor.

In Wächtershäuser's theory, the first organic molecule put together on the conveyor belt was acetic acid, a simple combination of carbon, hydrogen, and oxygen that is best known for giving vinegar its pungent odor. Formation of acetic acid is a primary step in metabolism, the series of chemical reactions that provides the energy that cells use to manufacture all the biological ingredients an organism needs.

According to the theory, metabolism came before all else. Once a primitive metabolism evolved, it began to run on its own, and only later were cells other basic elements, such as a genetic code, invented.

Wächtershäuser focuses on the heart of modern metabolism, the citric acid cycle. All living cells use this series of reactions to extract energy from food. The cycle makes changes in several chemical compounds, but it always begins with acetic acid. Inside a cell, the two carbon atoms in each acetic acid molecule are eventually expelled as carbon dioxide in a reaction that gives off a packet of energy.

Because the citric acid cycle is intrinsic to all modern life, Wächtershäuser guesses that its basic reactions are close to the chemistry with which life began—with one significant variation. In the oxygen-deficient world at hydrothermal vents, heat-loving bacteria operate the cycle backward (SN: 3/29/97, p. 192). Instead of giving off carbon dioxide to make energy, they incorporate carbon atoms to build a succession of more complex organic molecules. Wächtershäuser says life's first chemicals were built the same way.

Around the vents, he theorizes, catalytic metallic ions first enabled the materials around them to fashion acetic acid. In the next step, the ions catalyzed the addition of a carbon molecule to the acetic acid to get three-carbon pyruvic acid, which is another key chemical in the citric acid cycle and also reacts with ammonia to form amino acids, which themselves link up to form proteins.

After writing the blueprint, Wächtershäuser set out to prove each step. He produced the first important component of his assembly process 2 years ago. He and fellow German chemist Claudia Huber, of the Technical University of Munich, reported in the April 11, 1997 Science that they generated large quantities of activated acetic acid from basic raw materials at 100°C. This is the temperature typical of the fringe of a hydrothermal vent, where the volcanic brines mix with near-freezing ocean water.

Critics complain that Wächtershäuser's experimental temperature represents a limited zone in the vent environment. Toward the scalding core of a vent system, temperatures are nearer 350°C.

A longtime supporter of the primordial soup hypothesis, Jeffrey L. Bada of the Scripps Institution of Oceanography in La Jolla, Calif., has conducted several experiments showing that certain life-critical chemicals could not survive in their scorching birthplace for more than a few minutes or days. By testing reactions at only a single temperature, Bada says, Wächtershäuser is "not playing with a full deck of cards."

Such a rebuke doesn't apply to other experimentalists, such as the Carnegie Institution team, who are joining the game. The Carnegie researchers at times closely follow Wächtershäuser's blueprint, but they're aiming for results much broader, and perhaps more convincing, than the German chemist has achieved to date.

They are trumping Bada's criticism by testing chemical reactions over a much wider range of temperatures. What's more, with their bomb apparatus, the team can perform experiments at the extreme pressures that are typical under thousands of meters of seawater, a factor Wächtershäuser never explored.

Christopher Chyba of the Search for Extraterrestrial Intelligence (SETI) Institute in Mountain View, Calif., is encouraged by the growing interest in this research. A variety of ideas, many of them flowing out of Wächtershäuser's hypothesis, are now leading to a kind of renaissance of experiments in the origins of life," he says.

It was another origins-of-life theorist who got the Carnegie team involved. About the same time that Wächtershäuser began considering metabolism as the root of life, a similar idea came to biologist Harold J. Morowitz of George Mason University in Fairfax, Va. He also was drawn to the primitive power of the citric acid cycle. Unlike Wächtershäuser, Morowitz's first ponderings were still steeped in the primordial soup.

It took prodding from his friend Corliss, now at the Central European University in Budapest, for Morowitz to move his envisioned birthplace of prebiotic metabolism out of the light and into the ocean depths. That's when he turned to his George Mason colleague Robert M. Hazen, who also holds a position at Carnegie's Geophysical Laboratory. Hazen had long studied what happens to the structures of mineral crystals buried deep inside the Earth, so his high-pressure expertise translated easily into methods for testing what might happen to chemicals at vents.

Hazen and Morowitz assembled specialists, each capable of attacking the origins-of-life question from a different area of expertise, from several institutions. What's more, they are backed by NASA's new Astrobiology Institute, which is exploring where and how life may exist throughout the universe. "We're trying to use a systematic approach to what is a huge and complicated field," Hazen says.

Their approach recently took advantage of Brandes' focus on nitrogen, another important life-building element. Organisms gain nitrogen through reactions involving ammonia, a simple combination of nitrogen and hydrogen. In Wächtershäuser's theoretical assembly line, ammonia is a key player: It helps convert compounds from the citric acid cycle into amino acids. Yet few had expected ammonia to survive the vent inferno.

Brandes led the way to proving that expectation false. In the tiny gold capsules, he mixed water with nitrogen oxides presumed to be present in ancient oceans and added Wächtershäuser's sulfide minerals to jump-start any reactions. Inside the bomb, heating elements and pneumatic pistons subjected the capsules to conditions typical of hydrothermal vents. After only 15 minutes at 500°C and a pressure 500 times that at the planet's surface, the experiment created ammonia—and lots of it. What's more, the ammonia was stable up to a fiery 800°C. The team reported their findings in the Sept. 24, 1998 NATURE.

Once the Carnegie researchers knew that the hydrothermal mix could create ammonia, they turned to the next step, in which ammonia combines with pyruvic acid to build the amino acid alanine. Again, the scientists mixed their ingredients in a gold capsule and heated and squeezed them. Even without a catalyst, as much as 40 percent of the pyruvic acid converted to alanine.

Amino acids made at hydrothermal vent conditions don't impress the researchers who have most enthusiastically promoted the primordial soup hypothesis. They've had amino acids in hand for decades. Back in 1953, chemist Stanley L. Miller of the University of California, San Diego shot a streak of electricity through a laboratory mixture of methane, hydrogen, and ammonia—a replica of the primordial atmosphere. This imitation lightning sent chemicals raining down into a flask of oceanlike water below, which grew red and yellow with amino acids in a week's time.

Research has since drawn Miller's hypothetical atmosphere into question, causing many scientists to doubt the relevance of his findings. Recently, scientists have focused on an even more exotic amino acid source: meteorites. Chyba is one of several researchers who have evidence that extraterrestrial amino acids may have hitched a ride to Earth on farflung space rocks.

Amino acids from a variety of sources almost certainly seasoned a broth on the planet's surface 4 billion years ago, Chyba says, but he points out that no one has ever satisfactorily explained how the widely distributed ingredients linked up into proteins. Presumed conditions of primordial Earth would have driven the amino acids toward lonely isolation. That's one of the strongest reasons that Wächtershäuser, Morowitz, and other hydrothermal vent theorists want to move the kitchen to the ocean floor. If the process starts down deep at discrete vents, they say, it can build amino acids and link them up—right there.

Last year, Wächtershäuser and Huber did just that. They reported in the July 31, 1998 SCIENCE that at 100°C, they got amino acids to connect into short proteinlike chains called peptides. Even so, Bada won't budge from his surface-soup stance. He sees no reason why amino acids simmering in pools of water on rocky shorelines couldn't link up through just the same reaction. "I guarantee that experiment would work just as well at 25°C," says Bada. "It would take longer, but you'd make the same product."

That may be so, Wächtershäuser says, but the molecular links are broken nearly as fast as they're created in both situations. All the chances lie in the blindingly quick activity at the vents, he says. "You throw the dice much more often."

Busily throwing the dice in his laboratory, Wächtershäuser is trying to find the right metal catalysts that will get his short peptide chains to lengthen and reproduce themselves, the ultimate criterion of life. Everyone digging around for the origin of life would like to discover the first molecule that learned to make copies of itself.

"That's really what the struggle is all about," Wächtershäuser says. "and so far, it hasn't been found."

Articles appearing regularly in scientific journals claim to have generated self-replicating peptides or RNA strands (SN: 8/10/96, p. 87), but they fail to provide a natural source for their compounds or an explanation for what fuels them. Brandes, now at the University of Texas Marine Sciences Institute in Port Aransas, compares this top-down approach to a caveman coming across a modern car and trying to figure out how to make it.

"It would be like taking the engine out of the car, starting it up, and trying to see how that engine works," but Carnegie's approach is from the bottom up, he says. "What we do is start with the block of iron and try to make a car out of that."

The Carnegie team's next refinement to their iron block is to build a laboratory model of a hydrothermal vent. A maze of tubes and flasks will route hot, pressurized gases and water through vent minerals to see whether reactions that take place inside the gold capsules also happen in a more realistic environment. Brandes guesses that one day bottom-up and top-down researchers will together forge a complete chain of events—from the simplest raw materials to a collection of molecules that qualifies as being alive.

Yet even if scientists ultimately find a set of steps for the origin of life in the laboratory, everyone agrees that it may not replicate what occurred on Earth 4 billion years ago.

"I'm optimistic, not that we will ever really know with certainty what happened on Earth but that we will have a plausible account to give," Chyba says.

An account based on hydrothermal forces deep in the sea opens up the possibility of life elsewhere in the solar system. While few places may have had sunlit ponds, more have evidence of geothermal forces. Prime candidates are Mars and Jupiter's ice-covered moon Europa (SN: 11/7/98, p. 296). An upcoming mission to Europa will search for a liquid ocean overlying chemical-spewing geysers similar to Earth's hydrothermal vents.

There's an interesting synergy between the experimental results suggesting that life originated on Earth's seafloor and speculation about an ocean on Europa, Chyba says. "To some extent Europa will provide a test of the deep-origins hypothesis."

**Questions:**

1. What is the theory for the origin of life that is now being challenged by the discovery of life around the hydrothermal vents?

2. According to Wächtershäuser's theory, what must evolve first?

3. What are some of the criticisms of Wächtershäuser's experiment?

Answers are at the back of the book.

**Activity:**

Search the web for pictures of the organisms surrounding hydrothermal vents.
Go to **http://www.neaq.org/learn/kidspace/vent.html**
Follow the directions to make a model of a hydrothermal vent.

# 33

There are two ways to study the evolution of life on Earth. One way is to study fossils to determine when different species evolved and when they went extinct. The other way is to assess the relatedness of the organisms that live on the Earth today. Until recently, only fossils from well-dated rocks could indicate at what point in the Earth's history a particular evolutionary event took place. More recently, molecular biologists have been able to look at mutations in species' genetic code to tell how closely it is related to another. Studying rates of genetic mutation, molecular biologist can actually determine the time when two groups or organisms diverged from species or genera, all the way up to kingdoms. When scientists compare the results of the molecular studies with the fossils, they should get similar results. Right?

# Digging Up the Roots of Life

By Arne Ø. Mooers and Rosemary J. Redfield

When did life arise? What did it look like? These questions were once the province of theologians, but in recent years palaeontologists have been better guides. Molecular systematists have now joined the game, and the latest molecular evidence, published last month in *Science*,[1] points to a date almost 2 billion years (Gyr) younger than that implied by the fossil record.

Until now, the fossil evidence that life arose and bacteria diversified more than 3.5 billion years ago, soon after the Earth first developed a crust and ocean, has been virtually uncontested. This evidence consists of both structurally complex microfossils with a striking resemblance to modern photosynthetic cyanobacteria, and stromatolites, layered structures resembling modern cyanobacterial mats.[2]

Searching for deep roots in the tree of life has lately become a hot topic for molecular systematists. Comparisons of slowly evolving RNA and protein sequences have identified bacteria, Archaea ('archaebacteria') and eukaryotes as the three main domains of life, and it seems more and more likely that the first divergence was between bacteria and (Archaea + eukaryotes), with all of the modern bacterial groups diversifying after this initial split.[3] Although most deep phylogenetic trees have been uncalibrated (that is, based on sequence divergence, not actual time), the resemblance of the oldest fossils to modern bacteria implies that this initial divergence occurred more than 3.5 Gyr ago. But Doolittle and colleagues' now present an analysis placing the first branch much more recently, at about 1.8 Gyr ago. This means that our early history might be in need of serious re-evaluation.

Although squeezing evidence from billion-year old stones is difficult business, the evidence for 3.5-Gyr-old bacteria seems strong. Because many reported 'fossils' turn out to be bubbles of non-biogenic matter, contaminants from later eras or even spores floating about the laboratory,[4] micropalaeontologists have agreed upon a set of stringent criteria addressing provenance and protocol, as well as biogenic origin. These include morphological relationships to modern groups, and evidence of plausible morphological variation and developmental stages. Based on these criteria, there are now two sites that have produced convincing fossils reliably dated at about 3.5 Gyr old. These sites have produced a wide array of forms, often attributed to different families and genera of modern cyanobacteria, a group with complex morphologies allowing tentative assignment to known taxa. Micropalaeontologists go so far as to speak of diversified communities and ecosystems represented in the Archaean rocks.[5] Such a flowering of life arising so soon after it was physically possible to do so bears directly on the question of the difficulty or inevitability of life beginning at all.[6]

Although the fossil date seems rock solid, Doolittle and colleagues' contradictory evidence is hard to dismiss outright. The usual procedure in molecular analyses compares the different versions of a protein found in different organisms. The number of differences between the amino-acid sequences of each pair of species gives an 'evolutionary distance' between them, and a matrix of such distances is used to assign branch points on a phylogenetic tree. Because rates of changes may vary, different proteins may give very different trees. Doolittle et al. therefore took the unprecedented step of pooling data from genes for 57 different metabolic enzymes, using a total of 531 sequences representing 15 major phylogenetic groups. The trees built with the resulting evolutionary distances[7,8] allowed fast-and slowly-evolving groups to be identified, and appropriate corrections to be made. The data are internally consistent, in that different proteins tend to give the same tree, and the pattern of branching agrees with that estimated with other molecules and other techniques.[9-11]

The crucial (and controversial) step was calibrating the tree. Doolittle et al. began by plotting the dates when seven key animal lineages first appeared in the fossil record against adjusted sequence divergence at the seven corresponding branchpoints. For example, the earliest fossil echinoderms and chordates both date from mid-Cambrian,[12] which sets the latest possible date for the split between them at 550 million years. A straight line was then fitted to these points and extrapolated back another 2 billion years, allowing the dates associated with the more ancient divergences to be estimated. The discordance between these divergence dates and the early fossil record is obvious. To restore our previous view of early evolution we must either discard the Archaean fossil evidence or move the first divergence of the Doolittle tree back almost 2 billion years. Could either of these be justified?

Neither the age nor the biological origin of the 3.5-Gyr-old microfossils has recently been seriously questioned, and the similarities between modern and fossil forms supports their common biological origin. Furthermore, the occurrence of stromatolites of different ages in the Archaean and early Proterozoic is consistent with a continuous flow of life from its inception 3.5 Gyr ago to the present. Still, the stromatolites are not unquestionably considered to be of biological origin,[5] suggesting we might do well to reconsider the oldest microfossils.

On the other hand, the molecular data may be misleading. Although the extrapolation seems unreliable (a line extended back to more than three times its original length), the discrepancy with the fossil data is so extreme that only drastic measures could reconcile the two. It could be that the reference fossils used by Doolittle et al. are misdated, being older than they seem. However, we should have as much or more faith in these more recent fossils as in the ancient microfossils. Perhaps we are simply missing much earlier fossils for each of the groups used in the calibration. Or, it might be that the observed rate of amino-acid substitution, which seems to be fairly constant over the time span used for calibration ($r = 0.94$ for $n = 7$ points) actually slows down far into the past, due to a slower rate of amino-acid substitution during early times. This idea has no basis in theory.

Alternatively, the model of substitution[8] used by Doolittle and colleagues might not be sophisticated enough to adjust fully for several substitutions at the same site, and so is underestimating divergence times. Doolittle et al. tested for one form of departure from the standard model, keeping certain amino-acids constant, and found that this had little effect. But there is some evidence[13,14] that the positions within proteins that cannot support a substitution may change through time, as other amino-acids around them change, and this is a much harder bias to contend with. Although the effect would be most serious for very divergent lineages, evidence, in the form of too-recent extrapolated dates, should be present throughout the tree. Without fossils, this is hard to test, but there is no glaring evidence for such a bias. Indeed, the estimated split between plants and (animals + fungi) at 1.0 Gyr may strike some as too old.

What if both dates are correct? Perhaps, given the measurement errors associated with both fossils and molecules, we should be happy with a factor-of two discrepancy. Or, if there was a diverse assemblage of cyanobacteria-like bacteria around 3.5 Gyr ago, and all living organisms looked at by Doolittle *et al.* have a common ancestor at 1.8 Gyr, then we could well be but footnotes to some particular cyanobacterial lineage. Or, if the resemblance between fossil and modern cyanobacterial forms is only convergent, perhaps we are descended from some other, unnamed member of the early biota. Might the original cyanobacterial-like lineage (or another branch from this time) persist? If so, it should root in the tree well below the rest of us.

In any case, we must account for the extinction of all other lineages found in the oldest rocks. If this seems far-fetched, then consider that, if the resemblance between the fossil and modern cyanobacteria is truly convergent, it is also possible that we are the descendants of a completely different diversification. The fossil record between the early Archaen and the early Proterozoic (around 2 Gyr ago) is very sparse indeed.[4] If life arose and diversified with alarming speed once, three-and-a-half billion years ago, why not again a billion-and-a-half years later? Unfortunately, this hypothesis would be hard to test.

The new study[1] is the first broadly based molecular sally into the debate on the timing of life's origin. But it is surely not the final word. Analyses of molecular evolution continue to improve, and fossils continue to be both found and re-evaluated. Only time will tell who is right—only, however, if time can.

# References

1. Doolittle, R.F., Feng, D.-F., Tsang, S., Cho, G. & Little, E. *Science* 271, 470-477 (1996).

2. Schopf, J.W. *Proc. Natl. Acad. Sci. U.S.A.* 91, 6735-6742 (1994).

3. Iwabe, N., Kurna, K., Hasegawa, M., Osawa, S. & Miya, T. *Proc. Natl. Acad. Sci. U.S.A.* 86, 9355-9359 (1989).

4. Schopf, J.W. & Walter, M.R. in *Earth's Earliest Biosphere* (ed. Schopf, J.W.) 214-238 (Princeton Univ. Press, 1983).

5. Schopf, JW. *Science* 260, 640-646 (1993).

6. Thaxton, C.B. The *Mystery of Life's Origin: Reassessing Current Theories* (Lewis & Stanley, Dallas, 1992).

7. Feng, D.-F. & Doolittle, R. D. *J. Molec. Evol.* 25, 351-360 (1987).

8. Dayhoff, M. O., Schwartz, R. M. & Orcutt, B. C. in *Atlas of Protein Sequence and Structure*, Suppl. 3 (ed. Dayhoff, M. O.) 345-352 (National Biomedical Research Foundation, Washington, D.C., 1978).

9. Knoll, A. H. *Science* 256, 622-627 (1992).

10. Wainright, P. O. *et al. Science* 260, 340-342 (1993).

11. Jukes, T.H. *Space Life Sci.* 1, 469-490 (1969).

12. Benton, M. J. (ed.) *The Fossil Record* 2 (Chapman & Hall, London, 1993).

13. Fitch, W. M. & Markowitz, E. *Biochem. Genet.* 4, 579-593 (1970).

14. Fitch, W. M. & Ayala, F. J. *Proc. Natl. Acad. Sci. U.S.A.* 91, 6802-6807 (1994).

## Questions:

1. How old are the oldest fossils, and what are they?

2. What is the age of the split between the bacteria, the archea, and the eukaryotes, based on molecular analysis?

3. What did the molecular biologists use for their study?

Answers are at the back of the book.

## Activity:

Draw a timeline beginning 4.5 billion years ago and running to the present. Record major biological events. Be sure to include the organisms mentioned in this article. Also, make sure to draw the timeline to scale. It is all right to break the scale apart for clarity.

# 34

Many people may not know that for almost the first four billion years of Earth's history, life on Earth was mostly microscopic bacteria and algae, with a few simple multicellular organisms also present. When complex, multicellular life did make an appearance, it did so in a geologic instant throughout the world. Within this short period of time, often referred to as the Cambrian explosion, the ancestors to almost every phylum of animals evolved. If you could compare the history of the Earth to the length of a human life, the Cambrian explosion would be like one season of a year. Watching the divergence of animal life would be like watching the flowers bloom in the spring. This amazingly rapid period of evolution brings with it some of the most interesting questions in evolution: why did so many animal phyla evolve so rapidly, why did they evolve when they did, and why have there been no more recent periods of evolution at the phylum level? These are questions that scientists are trying to answer as more fossils from this critical period in Earth history are found and studied.

# When Life Exploded

By J. Madeleine Nash

***For billions of years, simple creatures like plankton, bacteria and algae ruled the earth. Then, suddenly, life got very complicated.***

An hour later and he might not have noticed the rock, much less stooped to pick it up. But the early morning sunlight slanting across the Namibian desert in southwestern Africa happened to illuminate momentarily some strange squiggles on a chunk of sandstone. At first, Douglas Erwin, a paleobiologist at the Smithsonian Institution in Washington, wondered if the meandering markings might be dried-up curls of prehistoric sea mud. But no, he decided after studying the patterns for a while, these were burrows carved by a small, wormlike creature that arose in long-vanished subtropical seas—an archaic organism that, as Erwin later confirmed, lived about 550 million years ago, just before the geological period known as the Cambrian.

As such, the innocuous-seeming creature and its curvy spoor mark the threshold of a critical interlude in the history of life. For the Cambrian is a period distinguished by the abrupt appearance of an astonishing array of multicelled animals—animals that are the ancestors of virtually all the creatures that now swim, fly and crawl through the visible world.

Indeed, while most people cling to the notion that evolution works its magic over millions of years, scientists are realizing that biological change often occurs in sudden fits and starts. And none of those fitful starts was more dramatic, more productive or more mysterious than the one that occurred shortly after Erwin's wormlike creature slithered through the primordial seas. All around the world, in layers of rock just slightly younger than that which Erwin discovered, scientists have found the mineralized remains of organisms that represent the emergence of nearly every major branch in the zoological tree. Among them: bristle worms and roundworms, lamp shells and mollusks, sea cucumbers and jellyfish, not to mention an endless parade of arthropods, those spindly legged, hard shelled ancient cousins of crabs and lobsters, spiders and flies. There are even occasional glimpses—in rock laid down not long after Erwin's Namibian sandstone—of small, ribbony swimmers with a rodlike spine that are unprepossessing progenitors of the chordate line which leads to fish, to amphibians and eventually to humans.

Where did this extraordinary bestiary come from, and why did it emerge so quickly? In recent years, no question has stirred the imagination of more evolutionary experts, spawned more novel theories or spurred more far-flung expeditions. Life has occupied the planet for nearly 4 billion of its 4.5 billion years. But until about 600 million years ago, there were no organisms more complex than bacteria, multicelled algae and single-celled plankton. The first hint of biological ferment was a plethora of mysterious palm-shape, frondlike creatures that vanished as inexplicably as they appeared. Then, 543 million years ago, in the early Cambrian, within the span of no more than 10 million years, creatures with teeth and tentacles and claws and jaws materialized with the suddenness of apparitions. In a burst of creativity like nothing before or since, nature appears to have sketched out the blueprints for virtually the whole of the animal kingdom. This explosion of biological diversity is described by scientists as biology's Big Bang.

Over the decades, evolutionary theorists beginning with Charles Darwin have tried to argue that the appearance of multicelled animals during the Cambrian merely seems sudden, and in fact had been preceded by a lengthy period of evolution for which the geological record was missing. But this explanation, while it patched over a hole in an otherwise masterly theory, now seems increasingly unsatisfactory. Since 1987, discoveries of major fossil beds in Greenland, in China, in Siberia, and now in Namibia have shown that the period of biological innovation occurred at virtually the same instant in geological time all around the world.

What could possibly have powered such a radical advance? Was it something in the organisms themselves or the environment in which they lived? Today an unprecedented effort to answer these questions is under way. Geologists and geochemists are reconstructing the Precambrian planet, looking for changes in the atmosphere and ocean that might have put evolution into sudden overdrive. Developmental

biologists are teasing apart the genetic toolbox needed to assemble animals as disparate as worms and flies, mice and fish. And paleontologists are exploring deeper reaches of the fossil record, searching for organisms that might have primed the evolutionary pump. "We're getting data", says Harvard University paleontologist Andrew Knoll, "almost faster than we can digest it."

Every few weeks, it seems, a new piece of the puzzle falls into place. Just last month, in an article published by the journal *Nature*, an international team of scientists reported finding the exquisitely preserved remains of a 1-in.-to 2-in.-long animal that flourished in the Cambrian oceans 525 million years ago. From its flexible but sturdy spinal rod, the scientists deduced that this animal—dubbed *Yunnanozoon lividum*, after the Chinese province in which it was found—was a primitive chordate, the oldest ancestor yet discovered of the vertebrate branch of the animal kingdom, which includes *Homo sapiens*.

Even more tantalizing, paleontologists are gleaning insights into the enigmatic years that immediately preceded the Cambrian explosion. Until last spring, when John Grotzinger, a sedimentologist from M.I.T., led Erwin and two dozen other scientists on an expedition to the Namibian desert, this fateful period was obscured by a 20 million-year gap in the fossil record. But with the find in Namibia, as Grotzinger and three colleagues reported in the Oct. 27 issue of *Science*, the gap suddenly filled with complex life. In layer after layer of late Precambrian rock, heaved up in the rugged outcroppings the Namibians call *kopfs* (after the German word for "head"), Grotzinger's team has documented the existence of a flourishing biological community on the cusp of a startling transformation, a community in which small wormlike somethings, small shelly somethings— perhaps even large frondlike somethings—were in the process of crossing over a shadow line into uninhabited ecospace.

Here, then, are highlights from the tale that scientists are piecing together of a unique and dynamic time in the history of the earth, when continents were rifting apart, genetic programs were in flux, and tiny organisms in vast oceans dreamed of growing large.

# The Weird Wonders

Inside locked cabinets at the Smithsonian Institution nestle snapshots in stone as vivid as any photograph. There, engraved on slices of ink-black shale, are the myriad inhabitants of a vanished world, from plump *Aysheaia* prancing on caterpillar-like legs to crafty *Ottoia*, lurking in a burrow and extending its predatory proboscis. Excavated in the early 1900s from a geological formation in the Canadian Rockies known as the Burgess Shale, these relics of the earliest animals to appear on earth are now revered as priceless treasures. Yet for half a century after their discovery, the Burgess Shale fossils attracted little scientific attention as researchers concentrated on creatures that were larger and easier to understand—like the dinosaurs that roamed the earth nearly 300 million years later.

Then, starting in the late 1960s, three paleontologists—Harry Whittington of the University of Cambridge in England and his two students, Derek Briggs and Simon Conway Morris—embarked on a methodical re-examination of the Burgess Shale fossils. Under bright lights and powerful microscopes, they coaxed fine-grain anatomical detail from the shale's stony secrets: the remains of small but substantial animals that were overtaken by a roaring underwater mudslide 515 million years ago and swept into water so deep and oxygen-free that the bacteria that should have decayed their tissues couldn't survive. Preserved were not just the hard-shelled creatures familiar to Darwin and his contemporaries but also the fossilized remains of soft-bodied beasts like *Aysheaia* and *Ottoia*. More astonishing still were remnants of delicate interior structures, like *Ottoia's* gut with its last, partly digested meal.

Soon, inspired reconstructions of the Cambrian bestiary began to create a stir at paleontologial gatherings. Startled laughter greeted the unveiling of oddball *Opabinia*, with its five eyes and fire-hoselike proboscis. Credibility was strained by *Hallucigenia*, when Conway Morris depicted it as dancing along on needle-sharp legs, and also by *Wiwaxia*, a whimsical armored slug with two rows of upright scales. And then there was *Anomalocaris*, a fearsome predator that caught its victims with spiny appendages and crushed them between jaws that closed like the shutter of a camera. "Weird wonders," Harvard University paleontologist Stephen Jay Gould called them in his 1989 book, *Wonderful Life*, which celebrated the strangeness of the Burgess Shale animals.

But even as *Wonderful Life* was being published, the discovery of new Cambrian-era fossil beds in Sirius Passet, Greenland, and Yunnan, China, was stripping some of the weirdness from the wonders. *Hallucigenia's* impossibly pointed legs, for example, were unmasked as the upside-down spines of a prehistoric velvet worm. In similar fashion, *Wiwaxia*, some scientists think, is probably allied with living bristle worms. And the anomalocarids whose variety is rapidly expanding with further research—appear to be cousins, if not sisters, of the amazingly diverse arthropods.

The real marvel, says Conway Morris, is how familiar so many of these animals seem. For it was during the Cambrian (and perhaps only during the Cambrian) that nature invented the animal body plans that define the broad biological groupings known as phyla, which encompass everything from classes and orders to families, genera and species. For example, the chordate phylum includes mammals, birds and fish. The class Mammalia, in turn, covers the primate order, the hominid family, the genus *Homo* and our own species, *Homo sapiens*.

## Evolving at Supersonic Speed

Scientists used to think that the evolution of phyla took place over a period of 75 million years, and even that seemed impossibly short. Then two years ago, a group of researchers led by Grotzinger, Samuel Bowring from M.I.T. and Harvard's Knoll took this long-standing problem and escalated it into a crisis. First they recalibrated the geological clock, chopping the Cambrian period to about half its former length. Then they announced that the interval of major evolutionary innovation did not span the entire 30 million years, but rather was concentrated in the first third. "Fast," Harvard's Gould observes, "is now a lot faster than we thought, and that's extraordinarily interesting."

What Knoll, Grotzinger and colleagues had done was travel to a remote region of northeastern Siberia where millenniums of relentless erosion had uncovered a dramatic ledger of rock more than half a mile thick. In ancient seabeds near the mouth of the Lena river, they spotted numerous small, shelly fossils characteristic of the early Cambrian. Even better, they found cobbles of volcanic ash containing minuscule crystals of a mineral known as zircon, possibly the most sensitive timepiece nature has yet invented.

Zircon dating, which calculates a fossil's age by measuring the relative amounts of uranium and lead within the crystals, had been whittling away at the Cambrian for some time. By 1990, for example, new dates obtained from early Cambrian sites around the world were telescoping the start of biology's Big Bang from 600 million years ago to less than 560 million years ago. Now, information based on the lead content of zircons from Siberia, virtually everyone agrees that the Cambrian started almost exactly 543 million years ago and, even more startling, that all but one of the phyla in the fossil record appeared within the first 5 million to 10 million years. "We now know how fast fast is," grins Bowring. "And what I like to ask my biologist friends is, How fast can evolution get before they start feeling uncomfortable?"

## Freaks or Ancestors?

The key to the Cambrian explosion, researchers are now convinced, lies in the Vendian, the geological period that immediately preceded it. But because of the frustrating gap in the fossil record, efforts to explore this critical time interval have been hampered. For this reason, no one knows quite what to make of the singular frond-shape organisms that appeared tens of millions of years before the beginning of the Cambrian, then seemingly died out. Are these puzzling life-forms—which Yale University paleobiologist Adolf Seilacher dubbed the "vendobionts"—linked somehow to the creatures that appeared later on, or do they represent a totally separate chapter in the history of life?

Seilacher has energetically championed the latter explanation, speculating that the vendobionts represent a radically different architectural solution to the problem of growing large. These "creatures" which reached an adult size of 3 ft. or more across did not divide their bodies into cells, believes Seilacher, but into compartments so plumped with protoplasm that they resembled air mattresses. They appear to have had no predators, says Seilacher, and led a placid existence on the ocean floor, absorbing nutrients from seawater or manufacturing them with the help of symbiotic bacteria.

UCLA paleontologist Bruce Runnegar, however, disagrees with Seilacher. Runnegar

argues that the fossil known as *Ernietta*, which resembles a pouch made of wide-wale corduroy, may be some sort of seaweed that generated food through photosynthesis. *Charniodiscus*, a frond with a disklike base, he classifies as a colonial cnidarian, the phylum that includes jellyfish, sea anemones and sea pens. And *Dickinsonia*, which appears to have a clearly segmented body, Runnegar tentatively places in an ancestral group that later gave rise to roundworms and arthropods. The Cambrian explosion did not erupt out of the blue, argues Runnegar. "It's the continuation of a process that began long before."

The debate between Runnegar and Seilacher is about to get even more heated. For, as pictures that accompany the *Science* article reveal, researchers have returned from Namibia with hard evidence that a diverse community of organisms flourished in the oceans at the end of the Vendian, just before nature was gripped by creative frenzy. Runnegar, for instance, is currently studying the fossil of a puzzling conical creature that appears to be an early sponge. M.I.T.'s Beverly Saylor is sorting through sandstones that contain a menagerie of small, shelly things, some shaped like wine goblets, others like miniature curtain rods. And Guy Narbonne of Queen's University in Ontario, Canada, is trying to make sense of *Dickinsonia*-like creatures found just beneath the layer of rock where the Cambrian officially begins. What used to be a gap in the fossil record has turned out to be teeming with life, and this single, stunning insight into late-Precambrian ecology, believes Grotzinger, is bound to reframe the old argument over vendobionts. For whether they are animal ancestors or evolutionary dead ends, says Grotzinger, *Dickinsonia* and its cousins can no longer be thought of as sideshow freaks. Along with the multitudes of small, shelly organisms and enigmatic burrowers that riddled the sea floor with tunnels and trails, the vendobionts have emerged as important clues to the Cambrian explosion. "We now know," says Grotzinger, "that evolution did not proceed in two unrelated pulses but in two pulses that beat together as one."

## Breaking Through the Algae

To human eyes, the world on the eve of the Cambrian explosion would have seemed an exceedingly hostile place. Tectonic forces unleashed huge earthquakes that broke continental land masses apart, then slammed them back together. Mountains the size of the Himalayas shot skyward, hurling avalanches of rock, sand and mud down their flanks. The climate was in turmoil. Great ice ages came and went as the chemistry of the atmosphere and oceans endured some of the most spectacular shifts in the planet's history. And in one way or another, says Knoll, these dramatic upheavals helped midwife complex animal life by infusing the primordial oceans with oxygen.

Without oxygen to aerate tissues and make vital structural components like collagen, notes Knoll, animals simply cannot grow large. But for most of earth's history, the production of oxygen through photosynthesis—the metabolic alchemy that allowed primordial algae to turn carbon dioxide, water and sunlight into energy—was almost perfectly balanced by oxygen-depleting processes, especially organic decay. Indeed, the vast populations of algae that smothered the Precambrian oceans generated tons of vegetative debris, and as bacteria decomposed this slimy detritus, they performed photosynthesis in reverse, consuming oxygen and releasing carbon dioxide, the greenhouse gas that traps heat and helps warm the planet.

For oxygen to rise, then, the planet's burden of decaying organic matter had to decline. And around 600 million years ago, that appears to be what happened. The change is reflected in the chemical composition of rocks like limestone, which incorporate two isotopes of carbon in proportion to their abundance in seawater—carbon 12, which is preferentially taken up by algae during photosynthesis, and carbon 13, its slightly heavier cousin. By sampling ancient limestones, Knoll and his colleagues have determined that the ratio of carbon 12 to carbon 13 remained stable for most of the Proterozoic Eon, a boggling expanse of time that stretched from 2.5 billion years ago to the end of Vendian. But at the close of the Proterozoic, just prior to the Cambrian explosion, they pick up a dramatic rise in carbon 13 levels, suggesting that carbon 12 in the form of organic material was being removed from the oceans.

One mechanism, speculates Knoll, could have been erosion from steep mountain slopes. Over time, he notes, tons of sediment and rock that poured into the sea could have buried algal remains that fell to the sea floor. In addition, he says, rifting continents very likely changed the geometry of ocean basins so that water could not circulate as vigorously as before. The organic carbon that fell to the sea floor, then, would have stayed there, never cycling back to the ocean surface and into the atmosphere. As levels of atmospheric carbon dioxide dropped, the earth would have cooled. Sure enough, says Knoll, a major ice age ensued around 600 million years ago—yet another link in a complex chain that connects geological and geochemical events to a momentous advance in biology.

Biology also influenced geochemistry, says Indiana University biochemist John Hayes. In fact, in a paper published in *Nature* earlier this year, Hayes and his colleagues argue that guts, those simple conduits that take food in at one end and expel wastes at the other, may be the key to the Cambrian explosion. Their reasoning goes something like this: animals grazed on the algae, packaging the leftover organic material into fecal pellets. These pellets dropped into the ocean depths, depriving oxygen depleting bacteria of their principal food source. The evidence? Organic lipids in ancient rocks, notes Hayes, underwent a striking change in carbon-isotope ratios around 550 million years ago. Again, the change suggests that food sources rich in carbon 12, like algae, were being "express mailed" to the ocean floor.

## The Genetic Tool Kit

The animals that aerated the Precambrian oceans could have resembled the wormlike something that left its meandering marks on the rock Erwin lugged back from Namibia. More advanced than a flatworm, which was not rigid enough to burrow through sand, this creature would have a sturdy, fluid-filled body cavity. It would have had musculature capable of strong contractions. It probably had a heart, a well-defined head with an eye for sensing light and, last but not least, a gastrointestinal tract with an opening at each end. What kind of genetic machinery, Erwin wondered, did nature need in order to patch together such a creature?

Over the summer, Erwin pondered this problem with two paleontologist friends, David Jablonski of the University of Chicago and James Valentine of the University of California, Berkeley. Primitive multicelled organisms like jellyfish, they reasoned, have three so-called homeotic homeobox genes, or *Hox* genes, which serve as the master controllers of embryonic development. Flatworms have four, arthropods like fruit flies have eight, and the primitive chordate *Branchiostoma* (formerly known as *Amphioxus*) has 10. So around 550 million years ago, Erwin and the others believe, some wormlike creature expanded its *Hox* cluster, bringing the number of genes up to six. Then, "Boom!" shouts Jablonski. "At that point, perhaps, life crossed some sort of critical threshold." Result: the Cambrian explosion.

The proliferation of wildly varying body plans during the Cambrian, scientists reason, therefore must have something to do with *Hox* genes. But what? To find out, developmental biologist Sean Carroll's lab on the University of Wisconsin's Madison campus has begun importing tiny velvet worms that inhabit rotting logs in the dry forests of Australia. Blowing bubbles of spittle and waving their fat legs in the air, they look, he marvels, virtually identical to their Cambrian cousin *Aysheaia*, whose evocative portrait appears in the pages of the Burgess Shale. Soon Carroll hopes to answer a pivotal question: Is the genetic tool kit needed to construct a velvet worm smaller than the one the arthropods use?

Already Carroll suspects that the Cambrian explosion was powered by more than a simple expansion in the number of *Hox* genes. Far more important, he believes, were changes in the vast regulatory networks that link each Hox gene to hundreds of other genes. Think of these genes, suggests Carroll, as the chips that run a computer. The Cambrian explosion, then, may mark not the invention of new hardware, but rather the elaboration of new software that allowed existing genes to perform new tricks. Unusual-looking arthropods, for example, might be cobbled together through variations of the genetic software that codes for legs. "Arthropods," observes paleoentomologist Jarmila Kukalovd-Peck of Canada's Carleton University, "are all legs"—including the "legs" that evolved into jaws, claws and even sex organs.

## Beyond Darwinism

Of course, understanding what made the Cambrian explosion possible doesn't address the larger question of what made it happen so fast. Here scientists delicately slide across data-thin ice, suggesting scenarios that are based on intuition rather than solid evidence. One favorite is the so-called empty barrel, or open spaces, hypothesis, which compares the Cambrian organisms to homesteaders on the prairies. The biosphere in which the Cambrian explosion occurred, in other words, was like the American West, a huge tract of vacant property that suddenly opened up for settlement. After the initial land rush subsided, it became more and more difficult for naive newcomers to establish footholds.

Predation is another popular explanation. Once multicelled grazers appeared, say paleontologists, it was only a matter of time before multicelled predators evolved to eat them. And, right on cue, the first signs of predation appear in the fossil record exactly at the transition between the Vendian and the Cambrian, in the form of bore holes drilled through shelly organisms that resemble stacks of miniature ice-cream cones. Seilacher, among others, speculates that the appearance of protective shells and hard, sharp parts in the late Precambrian signaled the start of a biological arms race that did in the poor, defenseless vendobionts.

Even more speculative are scientists' attempts to address the flip side of the Cambrian mystery: why this evolutionary burst, so stunning in speed and scope, has never been equaled. With just one possible exception—the *Bryozoa*, whose first traces turn up shortly after the Cambrian—there is no record of new phyla emerging later on, not even in the wake of the mass extinction that occurred 250 million years ago, at the end of the Permian period.

Why no new phyla? Some scientists suggest that the evolutionary barrel still contained plenty of organisms that could quickly diversify and fill all available ecological niches. Others, however, believe that in the surviving organisms, the genetic software that controls early development had become too inflexible to create new life-forms after the Permian extinction. The intricate networks of developmental genes were not so rigid as to forbid elaborate tinkering with details; otherwise, marvels like winged flight and the human brain could never have arisen. But very early on, some developmental biologists believe, the linkages between multiple genes made it difficult to change important features without lethal effect. "There must be limits to change," says Indiana University developmental biologist Rudolf Raff. "After all, we've had these same old body plans for half a billion years."

The more scientists struggle to explain the Cambrian explosion, the more singular it seems. And just as the peculiar behavior of light forced physicists to conclude that Newton's laws were incomplete, so the Cambrian explosion has caused experts to wonder if the twin Darwinian imperatives of genetic variation and natural selection provide an adequate framework for understanding evolution. "What Darwin described in the *Origin of Species*," observes Queen's University paleontologist Narbonne, "was the steady background kind of evolution. But there also seems to a non-Darwinian kind of evolution that functions over extremely short time periods—and that's where all the action is."

In a new book, *At Home in the Universe* (Oxford University Press; $25), theoretical biologist Stuart Kauffman of the Santa Fe Institute argues that underlying the creative commotion during the Cambrian are laws that we have only dimly glimpsed—laws that govern not just biological evolution but also the evolution of physical, chemical and technological systems. The fanciful animals that first appeared on nature's sketchpad remind Kauffman of early bicycles, with their odd-size wheels and strangely angled handlebars. "Soon after a major innovation," he writes, "discovery of profoundly different varia-tions is easy. Later innovation is limited to modest improvements on increasingly optimized designs."

Biological evolution, says Kauffman, is just one example of a self-organizing system that teeter-totters on the knife edge between order and chaos, "a grand compromise between structure and surprise." Too much order makes change impossible; too much chaos and there can be no continuity. But since balancing acts are necessarily precarious, even the most adroit tightrope walkers, sometimes make one move too many. Mass extinctions, chaos theory suggests, do not require comets or volcanoes to trigger them. They arise naturally from the intrinsic instability of the evolving system, and superior fitness provides no safety net.

In fact, some of prehistory's worst mass extinctions took place during the Cambrian itself, and they probably occurred for no obvious reason. Rather, just as the tiniest touch can cause a steeply angled sand pile to slide, so may a small evolutionary advance that gives one species a temporary advantage over another be enough to bring down an entire ecosystem. "These patterns of speciations and extinctions, avalanching across ecosystems and time," warns Kauffman, are to be found in every chaotic system—human and biological. "We are all part of the same pageant," as he puts it. Thus, even in this technological age, we may have more in common than we care to believe with the weird—and ultimately doomed—wonders that radiated so hopefully out of the Cambrian explosion.

---

## Questions:

1. When was the beginning of the Cambrian period, and how long did the period of rapid animal evolution last?
2. What is the evidence that the amount of organic matter decaying in the oceans was decreasing prior to the beginning of the Cambrian?
3. What genetic features could allow for rapid evolution of new animal phyla?

Answers are at the back of the book.

## Activity:

For each kingdom recognized today, give an example of a prehistoric ancestor which evolved during the Cambrian explosion.

**35**

There are two important questions raised about Australia's ancient past. First, did *Homo sapiens* kill off Australia's megamarsupials? And second, when did the first Australians venture inland, away from the bountiful coastal regions. These questions may be answered by the discovery of bones from extinct megafauna alongside bloodstained stone tools, charcoal and other signs of human activity found at an inland area in New South Wales called Cuddie Springs.

Scientists are currently testing minute traces of blood found on these ancient tools and perfecting more reliable ways of identifying the sources of the blood. One such way is DNA sequencing. Sequencing fragments of DNA from residues and replicating them many times over with a polymerase chain reaction (PCR). These results will determine which type of species or megamarsupials these people were feasting on and possibly if it led to their unfortunate demise.

# Cooking Up a Storm
## The fossilized remains of a lakeside supper are fueling the debate over the demise of Australia's gentle giants.

By Stephanie Pain

There is a place in New South Wales called Cuddie Springs. You won't find it on any road map but you can see it from the plane as you fly over the continent's interior towards Sydney. If it has rained, you should spot a shallow lake surrounded by saltbush scrub. If it's dry, all you'll see is a claypan in a sea of coolabah and blackbox trees.

More than 30,000 years ago, when the lake was a more permanent feature of the landscape, a group of people sat down on the shore to eat. Nothing strange about that, except that what they were eating could help to answer two of the most important questions about Australia's ancient past. Did people have anything to do with the disappearance of the gigantic animals that once roamed the continent? And when did the early Australians head inland from the bountiful coastal regions to colonise the less hospitable interior?

The main course at that lakeside dinner has provided the first incontrovertible evidence that there were people around at the same time as some of Australia's extinct giants. The discovery of bones from long-dead megafauna alongside bloodstained stone tools, charcoal and other signs of human activity may provide the best clue yet to the fate of megamarsupials—the Australian equivalent of Europe's mammoths and giant elk, and the great bison and ground sloths of the Americas.

As a side dish, the diners ate porridge or "cakes" made from expertly milled grass seeds, a type of food processing that is not supposed to have developed for another 20,000 years—and then on the other side of the world. It shows clearly that these people were able to make a reasonable living in apparently hostile country from a very early stage. "Cuddie Springs is arguably the most brilliant site ever found in

Australia," says Richard Fullagar, an expert on stone tools at the Australian Museum in Sydney and one of the team excavating at the lake. "It is the linchpin in lots of different arguments."

The question of what happened to the world's megafauna is a global one. By the end of the last ice age, around 10,000 years ago, most of the giant species of reptiles, birds and mammals had disappeared. After almost a century of debate over their fate, there are two main schools of thought. One maintains that the great cooling and drying of the climate that accompanied the last ice age made most of the world uninhabitable for animals that needed prodigious amounts of drinking water. The other argues that the arrival of that most effective of predators, Homo sapiens, heralded an age of slaughter that soon saw off the gentle giants.

## Shrinking Monsters

There are several variations on this theme: that diseases spread by the incomers and their animals wiped out the megafauna, or that the deliberate burning of overgrown scrub to promote new growth destroyed their habitats. Some researchers believe that a combination of factors was probably to blame. In fact, not all the great Australian beasts disappeared. Some shrank. Today's red kangaroos and Tasmanian devils, for example, are smaller versions of prehistoric giants.

None of these explanations is wholly convincing. While climate change hit all the continents at the same time, there are huge variations in the timing of extinctions. In Australia, almost all the biggest animals appear to have gone by about 35,000 years ago—well before the ice age reached its maximum. In the Americas, they persisted for another 20,000 years or more—until well after the ice retreated to the poles. Not only that, the last ice age was merely one of 17 over the past 2 million years, and none of the others triggered such a rash of extinctions.

The idea that people wiped out the great beasts within a few hundred years of arriving on the scene poses an even bigger problem: there is no archaeological evidence that humans systematically slaughtered the biggest animals. And in Australia, there is no sign of hunting at all. Indeed, until Cuddie Springs was excavated there was not one piece of unequivocal evidence

that put people and megafauna in the same place at the same time.

"Most of the discussion about people and megafauna so far has been based on supposition from a few bits of bone and a few artefacts, but there have been no complete bones and stone artefacts together," says Judith Field, the archaeologist at the University of Sydney who is in charge of excavations at the site. "Cuddie Springs is it—where you have both together in undisturbed sediments."

An expedition by the Australian Museum early this century retrieved a collection of megafaunal bones from Cuddie Springs, but the archaeological significance of the site only became clear in 1990, when a preliminary dig revealed signs of human presence. In 1991 and 1994, Field carried out further excavations with a team from the University of New South Wales, the museum and the local Aboriginal community.

They uncovered a series of layers of sediment that records the comings and goings of animals and people, and documents with rare precision the changing environment around the lake.

At the deepest, prehuman level lie the bones of a large number of species of long extinct animals, from the monstrous marsupial "tree feller" Palorchestes, to the giant emu-like Genyornis. There were huge reptiles too: horned turtles, crocodiles and a formidable predatory goanna. These remains probably date back more than 100,000 years.

In a distinct layer above this ancient graveyard are the first signs of a human presence, where animal bones are mixed with stone tools and charcoal from camp fires. This layer, dated to between 30,000 and 35,000 years, includes Genyornis and the wombat-like Diprotodon, the largest marsupial ever. Above this is a layer 30,000 years old which is jam-packed with animal remains and tools, including the first of many seed-grinding stones.

These levels are capped by a stone pavement that represents a long dry period around 28,000 to 19,000 years ago when the sediments were blown away. "The pavement is so dense with bones and stones that you can't get a trowel between them," says Field. After this, the megafauna disappear, but grinding stones are present all the way to the surface.

Pollen preserved in the layers, along with the nature of the mud and silt particles, document the changing climate and vegetation. When the first people arrived, Cuddie Springs was a swamp with abundant freshwater fern that grows only in still, shallow waters. By 30,000 years ago, the climate had begun to dry out and people were camping on the claypan floor of the lake, and would probably have had to dig for water in dry years. The lake then entered a prolonged dry phase as the ice age tightened its grip on the world, and by 19,000 years ago, at the peak of the ice age, the saltbush had more or less given way to grassland.

The finds at Cuddie Springs put people and megafauna in the same place at the same time, but they don't prove that people were killing the animals. The bones show no obvious signs of butchering. But as Field points out, recent experiments have shown that it is possible to cut up a carcass with stone tools and leave no mark.

## Butcher's Tools

The tools have yielded more information. Their design generally provides some clue to their function: hammerstones were for smashing joints, sharp-edged flakes could be used for cutting or scraping meat from bones. Field and Fullagar's analyses of the microscopic patterns of wear on the tools from Cuddie Springs—and the residues of blood, hair and other tissues—lead them to believe that those from the lowest levels were used mainly for stripping meat off bones and processing soft tissue. "This is compelling evidence for a relationship between people and megafauna—because if they weren't butchering these animals, what were they butchering?" says Field.

Tom Loy, a molecular archaeologist at the University of Queensland in Brisbane, is convinced that some of the traces of blood on the tools from Cuddie Springs are from megafauna, or at least from *Diprotodon*. Loy has spent more than a decade studying blood, fat and other remains on dirty tools, borrowing techniques from pathology, immunology, molecular genetics and forensic science. Red blood cells are often visible under the microscope. Hairs have a characteristic pattern on the outside and collagen looks like a rope unravelling. "They are all pretty distinctive," says Loy.

There is surprisingly little degradation of blood, says Loy. "We don't know how much of it disappears over time but what's left is very recognisable as blood." Anything that looks like a bloodstain is tested: first with an Ames Hemastix, a dipstick that turns from yellow to green in the presence of minute traces of the blood protein haemoglobin. The test, more familiar in hospital labs where it is used to check for blood in the urine, is a cheap and easy way of deciding whether further investigation is worthwhile. Field and Fullagar have used Loy's techniques to test for blood on hundreds of artefacts from Cuddie Springs, confirming that most of the greasy marks and spots that look like blood under the microscope really are what they seem.

The next step is to narrow down the origin of the blood with an antibody test. The antibodies staphylococcal protein A and streptococcal protein G bind to immunoglobulin G, an important protein in mammalian blood. If the test is positive, the blood came from a mammal. And if it was a mammal, the million-dollar question is which one? One way to find out is to grow haemoglobin crystals from the blood sample. The size and shape of the crystals are suggestive of a species, says Loy. The technique is difficult and the crystals don't last long, yet it provided what Loy believes was the first, fleeting glimpse of megafauna on a tool from Cuddie Springs. "We tested a tool taken from among the bones and found a few *Diprotodon*-like crystals," he says. "Then they disappeared."

Loy is close to perfecting a more reliable means of identifying the source of the blood—by sequencing fragments of DNA retrieved from the residues. With a PCR machine, which can accurately replicate the fragments millions of times over, Loy can generate enough DNA to work with. He looks at a stretch of DNA about 100 base pairs long from a gene called *28S rRNA*, which has a role in assembling ribosomes, the protein factories found in every animal cell.

The first section in the sequence is common to all animals. "Then there is a bit that's diagnostic of mammals," says Loy "And then there's a variable bit that depends on which mammal it is." Before you can identify an animal, however, you need something to match it to—in this case sequences from known species of extinct animals. So Loy is building up a reference library of sequences with DNA extracted from bones held in collections and museums. "Fortunately there are heaps of diprotodons," he says.

Loy has tried to sequence DNA extracted from residues on two of the Cuddie Springs tools. One produced a match with the hairy-nosed wombat, a species that is not present in New South Wales today. The second produced something partway between a wombat and a *Diprotodon*.

"These two tools were relatively small and unspecialised and were unlikely to show much, yet we turned up these sequences," says Loy. So the chances of finding a definite *Diprotodon* look pretty good. "'We have concentrated on diprotodons because people want to know about them. They were so weird and so big that they capture the public imagination," he says. "But now we need to look at giant kangaroos such as *Sthenurus* and get some *Genyornis* samples to work on."

The DNA might prove beyond a doubt that people were butchering *Diprotodon* and its contemporaries, but it won't show whether they hunted them or scavenged from a dead or dying animal. Some Palaeontologists believe that the first hunters did not pursue large animals because their weapons weren't good enough or because they were not prepared to take the risk.

Direct evidence for hunting is hard to find. The early Australians made their spears and arrows from wood, which is rarely preserved. But, as Loy points out, the risks of attacking *Diprotodon* might not have been that high. Although it was enormous, weighing in at around 2 tonnes, *Diprotodon* had a tiny brain and was probably a slow and shambling beast. "They weren't very bright," says Loy. "And the arrangement of their shoulder blades meant they would have had a cumbersome gait." If Loy finds the blood of the faster-moving and more agile giant kangaroos, the team can be more confident that people were hunting them.

Whether the lake people were hunters remains unproven, but they were certainly millers. Field has excavated 33 fragments of grinding stones from Cuddie Springs, 21 of them dating from the same time as the megafauna. Fullagar suggests that the polished surfaces of these stones, and plant material still sticking to them, show that they were not used to process energy-rich roots and tubers, as you would expect at that time, but grass seeds, a foodstuff that requires a lot of hard work for far less reward. "Seed grinding was thought to be a late development in human history linked to the development of agriculture and settled societies," says Fullagar. Before this, the earliest known seed grinding stones came from the Middle East and date from around 12,000 years ago. "It pushes the date for this technology way back," he says.

## Shiny Stones

Some of the Cuddie Springs grindstones show evidence of "wet milling" of grass seeds, a technique in which seeds are first ground dry, and then water is added and the mixture ground again to make a floury paste which might be eaten raw or cooked in the fire. "Wet milling produces a distinctive gloss on stone," says Fullagar. "It's the effect of silica [from plant tissues] on stone. It smooths the stone at the molecular level to produce a mirrorlike shine." Wet milling was previously thought to have developed only 3000 years ago.

As with butchery, processing plants leaves telltale residues on stone tools, including starch grains and phytoliths, the silica skeletons of plants. "Phytoliths are the key to identifying which plants were being processed. Tubers don't have them. Grasses do," says Fullagar. Starch grains and phytoliths have such distinctive forms that they should eventually help in identifying the species of grass.

Some archaeologists argue that seed grinding is a very recent innovation developed after the ice age as part of a "dietary revolution." Once they had adopted grasses as foods, people would be better equipped both to live in marginal country and to deal with changes in climate and food supply. In Australia, there has been much debate about whether people had to begin milling grass before they could inhabit the arid interior— which would have put a late date on their movement inland. The evidence from Cuddie Springs suggests that early Australians had the means to survive in an arid landscape 30,000 years ago.

Field argues that people may have milled grains occasionally but only began to rely on seeds as the climate dried out in the last ice age. "Perhaps Cuddie Springs is where seed grinding developed. It's difficult to say," she says. There may have been compelling reasons why people wanted to stay despite the worsening climate. And if they wanted to stay they needed to switch to new types of food. The lake might have been an important ceremonial site: the team has found a conical stone or "cylcon," which possibly had some ritual function. Cuddie Springs is part of an Aboriginal dreaming track, and there is a dream time story that provides an explanation for the presence of the bone deposits there.

Whatever their reasons for staying, the residents of Cuddie Springs do not fit the traditional image of a small band of people scraping a meagre living from the inhospitable land of Australia's interior. Even in the face of dramatic climate changes and the dwindling supply of large animals, these people had the means to support a thriving group. But did they—and their counterparts across the continent—have any impact on the megafauna?

"Cuddie Springs won't answer the question of whether there was a Pleistocene overkill. Nor will any one site do that," admits Loy. If people hunted *Diprotodon*, then it was probably just one of many animals they killed. But it is possible that people played some part in the demise of the large animals in the immediate area, suggests Field. "'Only a handful of megafaunal species are found in association with artefacts at Cuddie Springs. So maybe the megafauna were already in decline." It might have been a case of hitting a large animal when it was already down.

---

## Questions:

1. What does a PCR machine do?

2. What are the two main schools of thought as to why the world's megafauna disappeared?

3. Name two examples, that are smaller versions of the ancient Australian giants.

Answers are at the back of the book.

## Activity:

Write a short paper (1-2 pages) saying whether or not this is a plausible explanation (if the demise of Australia's megamarsupials was due to man) and give some examples of man's destruction of other species in the recent past.

# 36

Former inhabitants of the Earth can leave various kinds of evidence of their lives behind. The type that most professional and amateur paleontologists like to find is remains of the organism itself—the shells, bones, and teeth, or the cast, molds, and impressions of the organism. These are the kinds of evidence that we normally think of then we think of fossils. Other types of evidence of an organism's existence are what we call trace fossils: tracks, burrows, and other marks indicating the activities of a long-dead organism. Included in the category of trace fossils are the droppings of animals, which paleontologists refer to by the rather sanitized term "coprolites." At present there is only one person we know of who actually specializes in coprolites, particularly the coprolites of dinosaurs. Wouldn't you like to know how someone would get started in this line of work?

# What the Dinosaurs Left Us

By Karen Wright

*To learn what dinosaurs ate and how they ate it, a brave (and good-humored) researcher has dedicated herself to analyzing the most humble of relics: the fossilized droppings of yesteryear.*

The sun has just set over the tranquil Santa Barbara campus of the University of California, and the crisp evening air is redolent of warm sand and eucalyptus. Scores of students are jogging or cycling under the rosy gold autumn sky; a few stroll back from the beach with surfboards under their arms.

But in a low white building on the east side of campus, in a cavelike room that smells of wet stone, Karen Chin is hard at work. Chin is hunched over a cluttered bench, her dark hair fanning halfway down her lab coat, her slender fingers holding a small gray rock against the motionless blade of a circular saw. She has repositioned the rock several times, in search of the right cut, when her concentration is shattered by a colleague entering the lab.

"Hey, Karen," calls the colleague in greeting. "You still messing around with poop?"

The short answer is yes. Karen Chin was, is, and probably always will be messing around with poop—petrified, prehistoric poop, the poop of ages past. She's a pioneer in a specialty so peculiar it's not taught in any university. It doesn't even have a formal name, though one does suggest itself: paleoscatology. It is safe to say that Chin is the world's leading paleoscatologist. Also the world's only paleoscatologist.

For the past six years this doctoral student has been analyzing and categorizing hundreds of the fossilized leavings that go by the polite name of coprolites. The specimens come from around the world and across the epochs. They include 300 million-year-old fish feces; dinosaur dung from the Triassic, Jurassic, and Cretaceous; and a sloth stool issued during the last ice age. Some of the fossils have been ravaged by time and are nearly unrecognizable. But others have survived more or less intact, their humble morphologies uncannily familiar in spite of their antiquity.

It is Chin's dream to bring order to this exocolonic chaos. In coprolites, she hopes to find evidence of feeding habits and behavior available from no other fossil source. Most important, she expects to discover the diets of ancient creatures so that paleontologists may one day reconstruct ecological webs from the very bowels of prehistory.

So far, however, Chin's results aren't much more impressive than her subject matter. If she is to tease out the secrets of coprolites, Chin must first devise a way of grappling with the daunting variety and anonymity of her specimens. On this particular evening, she has set out to section a fragment of putative *T. rex* turd from the Royal Saskatchewan Museum. "The whole specimen was 15 inches long and this big around," she says, putting the tips of her thumbs and forefingers together in a disconcertingly large O. The fragment, which was cleaved from its fecal parent with a pair of wire cutters, resembles a chunk of light-colored concrete with darker, elongate inclusions that Chin recognizes as bone. Pieces of bone are common in carnivore coprolites, she says; she's cutting open the fossil to see what else she can find out.

"Karen's the first person in the history of coprolites who's had the technology and the will to treat them to such a detailed analysis," says paleobotanist Bruce Tiffhey, Chin's doctoral adviser. "And she's only at the very beginning of that. It'll take her lifetime's worth of work, plus some other people's, to make a reasonable picture out of this."

But no one else has volunteered.

• • •

"I don't like to be associated with the nasty aspects of defecation," Chin declares some time later from the carpeted floor of her office, where an assortment of aged scat surrounds her like a most unappetizing picnic lunch. "I'm interested in coprolites from a biological standpoint. I'm trying to develop an overview of what they can tell us about the past. And that means that I need to look at all different types of coprolites and all different types of preservation."

Hence the protean display on the office floor. There are fractured loaves of dark gray rocks, chalky palm-size crescents, thumb-shaped orange nuggets, Irregular pebbles in stunning aquamarine, and numerous variations on the common brown sausage. Time and geology have bestowed upon these specimens a flamboyance they undoubtedly lacked when they first saw daylight. Yet many still bear a signature of their provenance in the form of

faint longitudinal striations. "Those are probably marks from the anal sphincter," Chin explains.

Coprolites form much the way bone fossils do, when minerals invade the microscopic interstices in organic matter and grow into crystals there. Sometimes mineralization helps preserve the living material itself; other times the crystals replace the organic template. In either case, the more readily a substance decomposes, the less likely it is to remain intact long enough to become fossilized. Dung is at a definite disadvantage there, and because of that disadvantage coprolites are rarer than bone fossils. But they are still plentiful: hundreds have been collected from the field, and millions more languish in fossil beds the world over.

Until Chin came along, nobody much cared. "Coprolites haven't gotten much respect," says Chin. "A lot has been published on them, but mostly the authors just described the appearance of coprolites from a given locality, then put them on a shelf in a museum." Paleontologists considered the information coprolites might provide far too dubious to warrant the time it would take to decipher it. And there was a certain stigma attached to the enterprise.

Trained as a botanist and ecologist, Chin doesn't share her colleagues' prejudice. Where others see ambiguity and guffaws, she sees opportunity. Chin thinks the main problem is that there is no context in which to evaluate coprolites. So she has decided to provide that context, by devising schemes for identifying and classifying the phenomena of interest.

"In the past, most coprolites have been identified on the basis of shape," she says. "Here's one from Nebraska." Chin offers a pale, slightly bowed cylinder with rounded ends. "This is probably mammal, about 31 million years old. It looks like feces, right? I mean, you look at this and go, yeah, this is fecal material. Right?"

Definitely.

"This is dinosaur, from Montana," says Chin, handing over an innocuous gray-brown individual with blocky edges. Darker speckles in the rock give it a heathered texture, but mostly it's just evocative of other rocks. "Now, this lacks an identifiable shape," Chin notes. "It's because I've been collecting these for so long that I recognized it easily."

"These two samples are very different. And that is a problem when you're trying to interpret paleobiological information from coprolites. How can we recognize coprolites with atypical shapes? How can we compare coprolites from different ages, from different depositional environments, from different animals? Those are the kinds of questions I'm trying to address in my research."

In an ideal world, of course, coprolites would be classified and compared according to their organism of origin, just as fossil bones are. Unfortunately, because of what Chin calls the "detached nature of feces," it's almost impossible to match droppings with droppers.

You can, however, narrow the range of possible culprits. The criterion of shape does sometimes say a little bit about who did what. Spiral coprolites, for example, are thought to be the exclusive province of primitive fish, which include sharks and lungfish as well as many extinct taxa. Because their intestinal valves are spiral shaped, these fish produce (and produced) distinctive oblong coils, many of which have turned up in sediments from the Paleozoic (570 million to 245 million years ago) and Mesozoic (245 million to 65 million years ago).

But shape isn't usually a reliable indicator of source. Anyone who's emptied a litter box or flushed a toilet knows that the issue of a single individual can change shape dramatically over time. Conversely, the droppings of different species can look very similar. Logs, pellets, piles, pinched ends—these morphologies are generously distributed by and among all manner of vertebrate life-forms.

The size of a fecal deposit may also suggest something about its maker: many large Mesozoic coprolites are attributed to dinosaurs because paleontologists assume that nothing else alive at the time could have manufactured mounds of such breadth or slugs of such girth. But size, too, can confound. A 1,000-pound moose leaves morsels no more than an inch long. Rodents that share a community latrine can generate heaps of fused waste several feet high.

"Even pristine dung in modern ecosystems can be ambiguous," says Chin. And then it gets rained on, stepped on, decomposed—eaten, even. Imagine, then, what confusion can be wrought by a few million years of geologic pressures. In Yorkshire, England, one paleontologist found Jurassic droppings that were nearly two-dimensional. "We're talking soft material," Chin says, "and soft materials is

subject to deformation even under the best circumstances."

So, with a characteristic disregard for appearances, Chin is exploring some of the less obvious features of her specimens. She's cut them open, sliced them up, and pulverized them. She's examined their insides with electron microscopes and made exhaustive inventories of their contents—animal, vegetable, and mineral. She's run geochemical analyses to characterize organic matter in the fossils, and elemental and mineral profiles to examine the processes by which they became fossilized. Her subject matter may lack sophistication, but her methods have it in spades.

• • •

Chin's strange scatological journey began in 1989 in Bozeman, Montana, where she had a job making thin sections of fossil bones for paleontologist Jack Horner at the Museum of the Rockies. Chin had recently turned to dinosaurs after more than a decade studying extant ecosystems. For 15 summers she'd worked as a Parks Service ranger and naturalist. On the job, in nearby Glacier National Park, she'd come to appreciate the informational value of feces. Though she might see elk, mountain lions, and grizzlies only rarely, their stools were comparatively easy to find and much more approachable. From such samples, researchers could deduce the animals' numbers, territory, and diet. Chin bought field guides to Rocky Mountain scat and began assembling her own personal photo collection. When Horner told her that he'd found some suspected dinosaur coprolites, it seemed fated that she and they should meet.

The coprolites came from a site in northwest Montana called the Two Medicine Formation, where Horner was unearthing the bones and nesting grounds of *Maiasaura*, a duck-billed dinosaur. The first thing Chin noticed about these fossils was that they were not the discrete orbs, pellets, or cylinders commonly described in the literature. These were more indiscreet: vast and seemingly formless, like massive cow pies turned to stone and then broken by a nasty fall. And the broken pieces were big: some of the blocks measured more than a foot on a side.

The Two Medicine specimens were full of fibrous bodies large enough to be seen with the naked eye. Chin sectioned them, just as she had Horner's bone fossils. Under the microscope the dark fibers revealed themselves to be the mineralized remnants of shredded wood. The fossils, Chin decided, were the by-product of a large vegetarian—most likely the herbivorous *Maiasaura*. They didn't look like the coprolites she'd seen in the literature, because the majority of those were the more firm and robust doings of carnivores.

Chin visited the Two Medicine sites again and again, gathering samples with various compositions from different locations. She wanted to find a way to characterize the members of her 76-million-year-old suite, to tease them apart, expose their contents, and force them to yield information. Her preoccupation soon took on the proportions of a doctoral dissertation. She found a sponsor in Tiffney, whose theories about the role of herbivory in the evolution of plants were in need of empirical substantiation.

"When she first came to me with the idea of this project," Tiffney remembers, "I said, 'Coprolites?' My basic response was 'Prove it.'"

It seemed like a reasonable request. But as it happened, very few dinosaur coprolites had ever been demonstrated unequivocally to be in fact what they appeared to be. Chin carefully amassed the evidence for the scatological nature of her Two Medicine specimens. First, she argued, the presumed coprolites occurred as scattered aggregations rather than as one continuous layer, a distribution inconsistent with geologic deposition but consistent with the cow-pie hypothesis. Second, they occurred in the same sediments in which the duckbills had been preserved—so a connection between the two was chronologically kosher. Third, angular breaks in the woody fibers suggested that the plant material had been chewed up rather than stepped on or weathered by water.

And then there were the dung beetle burrows. Chin had noticed that several of her fossils were riddled with smooth channels, some measuring more than an inch in diameter. On a hunch, she'd shown these specimens to a noted dung beetle specialist in Ontario.

"I had a look at these things and it was like, 'Wow. I don't know anything about paleontology or fossil rocks, but these look like the perfect soil trace of a modern-day dung beetle,'" says Bruce Gill, an entomologist at Agriculture Canada who became her collaborator. What clinched the ID, says Gill, was the presence of "backfilled" burrows: tunnels in the shadowy black rock that had been plugged with sand-colored sediments. Any number of invertebrates can dig through a dung pile, but only dung beetles fill their tunnels back up, using the soil displaced from brood cells hollowed out in the ground beneath the flop. And when you find a dung beetle backfilling, you know you've found dung.

"They wouldn't go to just any old rotting plants," says Gill. "They'd go to the rotting plants that had passed through the gut. Very rich. Very enticing."

The presence of dung beetle burrows also revealed something of the perpetrator's toilet habits: the animal had simply left its waste where it landed, rather than burying it as a cat would.

"It just kept getting better and better—I mean, if you're into this kind of thing," says Chin. "We were able to say that what we were looking at in the Two Medicine Formation was the largest verifiable dinosaur coprolite. And now we had evidence for an ecological relationship between dinosaurs and some of the insects that lived at the time. We had the beginnings of a Mesozoic food web."

Chin had proved that her coprolites could live up to their name. She had discovered the oldest evidence of dung beetle activity. She had found the only evidence of dinosaur-insect interactions. In her Two Medicine fossils, Chin could even look forward to finding some dung beetle dung.

• • •

After the beetle breakthrough, Chin broadened her mission. She scouted museums for coprolites from other animals and other eras. She talked up her vision of a grand classification scheme at diggers' meetings. As word got around, other researchers started sending their suspects to Chin. As always, she would do the dirty work.

Truth be told, the work isn't all that dirty. The thin-section lab is a bit dusty, but that's because it's full of rocks getting cut up and ground down. Chin uses saws with diamond-studded blades and grinding wheels with abrasive slurries to reduce her samples to slices thousandths of an inch thick. Mounted on square glass slides, the sections project the semi-opaque jumble of a kaleidoscope image.

"What I'm trying to do now is classify the fossils in terms of fabrics," says Chin. "I've looked at so many of these slides, I'm beginning to recognize patterns. You can have fabrics full of bioclasts—fragments of organic origin, such as bits of plants or clamshells—and you can classify the kinds of mineral grains, their size and distribution." Chin picks up a slide in comely tones of gray, brown, and off-white. The victim was one of her Two Medicine fossils. " I could say, for example, that this type of fabric has larger bits of woody material in a fine ground mass. Now, that fine-ground mass, under magnification, is a series of disaggregated tracheids—water-conducting plant cells."

Chin has been able to group the coprolites from 14 Two Medicine sites into four distinct categories based on the proportion of woody fibers and the type and quantity of plant cells they contain. Those categories, she believes, represent diets with different fiber contents and probably different nutritional values as well. A high-fiber diet, full of stems and bark, might keep an animal regular, but the animal would have to eat a lot to meet nutritional needs. The low fiber samples may correspond to nutritionally rich diets of ferns, young leaves, and flowering plants, which would require more selective foraging. At one time Chin thought that her low-fiber coprolites might be the work of juvenile duckbills, because they seemed to be concentrated in a *Maiasaura* nesting area. Later excavations turned up a specimen too big to have come from a juvenile, and Chin went back to the drawing board.

While the Two Medicine investigation is still in progress, Chin has been examining coprofabrics from other locations. She's found that coprolites offer a cornucopia of intestinal itinerants in addition to wood and shellfish. Sections of specimens have revealed whole and fragmented teeth, bones, seeds, leaves, stems, spores, fish scales, snail shells, and shards of volcanic ash. Some of the inclusions represent the animals' intended diet; others were probably passengers therein.

"Some people say, well, you wouldn't get all that coming through the digestive tract," says Chin. "But we all know that some things go through us, and you can see them if you look: tomato seeds, corn, and lettuce. In bear scat, whole berries will come out."

Still, Chin concedes, some of the items she finds in her specimens may have arrived ex post facto. She's begun examining some of the bioclasts in her samples for signs of gastric etching, evidence that they had passed through the gut.

Other techniques have helped Chin reconstruct the original composition. With organic geochemist Simon Brassell at Indiana University, Chin has used biogeochemical analysis to detect the carbon skeletons of organic compounds that can persist in coprolites even when gross structures such as plant cells have degraded. Some of the compounds isolated in this way implicate a particular source. Oleanane, for example, indicates the presence of angiosperms, or flowering plants; certain diterpanes are peculiar to plants called gymnosperms (which include conifers, such as pine trees). Biogeochemical analysis has provided the only evidence that *Maiasaura* ate flowering plants, since all the plant material in the Two Medicine coprolites that could be identified under the microscope was coniferous.

Chin also relies on X-ray diffraction to determine the mineral content of her fossils. Mineral analysis can reveal the preservational environment of a coprolite as well as what manner of manure was preserved. Carnivore leavings, for example, tend to be high in calcium phosphate, or apatite, a principal constituent of bone. Meat eaters' dung fossilizes more readily than plant feeders' because it is richer in minerals from the get-go.

"I still can't say for sure which animal did what," says Chin. "But these techniques release you from the confines of size and morphology as criteria for classification. You're not dependent on having an intact specimen anymore. It's a place to start."

In time, Chin says, she may try to develop bile acid markers that could help her match species with feces, and imaging techniques that would disclose a coprolite's contents without its having to be sliced up first. She might even discover unknown species among the bioclastic melange of her specimens. But in spite of the progress she's made and her ambitions for the future, Chin hasn't yet inspired a larger movement. And there's only so much one woman can do.

"There's no renaissance in paleoscatology at the moment," says Adrian Hunt, director of the Mesalands Dinosaur Museum in Tucumcari, New Mexico, and a former collaborator of Chin's. "But maybe there will be. The laughter's subsided."

**Questions:**

1. What would be a piece of evidence that a particular coprolite came from a carnivore?
2. Spiral coprolites are characteristic of what organisms?
3. What does the presence of dung-beetle burrows in a coprolite indicate?

Answers are at the back of the book.

**Activity:**

Read the information on coprolites and answer the following questions:

What different types of remains are found in coprolites from herbivores, carnivores, and omnivores?

Why are the coprolites of carnivores more likely to be preserved than the coprolites of herbivores?

What are good environments to preserve "scat"?

# 37

Sometimes a single fossil discovery can radically change our ideas about the past. This is especially true with the dinosaurs, which have accumulated many long-standing misconceptions because of their popularity. For example, a number of recent discoveries of well-preserved dinosaur embryos and juveniles have greatly changed our understanding of dinosaur developmental processes. This article describes yet another striking fossil find, that of a dinosaur possibly in the act of brooding its eggs. Birds, of course, do this in order to keep the eggs warm and protect them. Many paleontologists have suspected that birds evolved from dinosaurs, based on anatomical and other information, but this is the first direct evidence of birdlike brooding behavior.

# Fossil Indicates Dinosaurs Nested on Eggs

By Robert Lee Hotz

*Discoverers say the find shows a behavioral link to birds.*

In the sands of Mongolia's Gobi Desert, scientists have discovered a unique fossil of a carnivorous dinosaur nesting on its eggs like a brooding bird, revealing for the first time how Earth's most fearsome parents may have tenderly cared for their young.

The 80-million-year-old fossil is graphic testimony that the nesting behavior so common among birds today actually originated long before modern feathers and wings, reinforcing the intimate evolutionary link between birds and the long-extinct dinosaurs. It proves they share complex behavior, several dinosaur experts said, in addition to important anatomical features.

Indeed, the sandstone fossil of a 9-foot-long, beaked carnivorous dinosaur called an oviraptor, preserved with a nest and a brood of unhatched young, is the sole direct evidence of *any* dinosaur behavior, experts said.

Until now, scientists could only make educated guesses about parental care among dinosaurs, by studying fossil nests and juvenile dinosaurs. The fossil of the oviraptor on its eggs offers the first concrete proof that dinosaurs

actively protected and cared for their young, said researchers from the American Museum of Natural History and the Mongolian Academy of Sciences who made their find public Wednesday.

"What makes this specimen so spectacular is that, while there is a lot said and a lot written about dinosaur behavior, there is very little real evidence," said Mark A. Norell, associate curator of vertebrate paleontology at the American museum, who led the team that discovered the bones. "This is about the only piece of hard evidence we have."

David B. Weishampel, a dinosaur expert at Johns Hopkins University, called the discovery "astonishing and incontrovertible evidence." Jack Horner, curator of paleontology at the Museum of the Rockies in Bozeman, Mont., said it is "the strongest evidence of some kind of parental attention." Jacques Gauthier, curator of reptiles at the California Academy of Science, said the find "opens all kinds of possibilities."

Norell and his colleagues, like many dinosaur experts, are convinced that all modern birds are direct descendants of a group of meat-eating, bipedal dinosaurs called theropods, a group that includes the oviraptor, the rapacious velociraptors made famous in the novel

"Jurassic Park" and the *tyrannosaurus rex*—the largest carnivore to walk the planet. Many ornithologists do not subscribe to that theory. Several said the new fossil is unconvincing, circumstantial evidence.

Alan Feducci, an authority at the University of North Carolina on bird evolution, said the fossil "makes no sense" and challenged the way its discoverers have linked their find to the development of birds.

"I have no faith in their conclusions whatsoever," he said. "It is a stretch of credulity. There are numerous animals preserved in bizarre poses. Maybe it was laying an egg instead of brooding the eggs. That would make more sense."

As presented by the American Museum of Natural History on Wednesday, however, the fossilized rock captures a moment more than 80 million years ago when a bipedal, meat-eating dinosaur roughly the size of an ostrich, with a whip tail and six-inch talons, warmed its clutch of about 15 large eggs. The eggs, shaped like baking potatoes, are neatly arranged in a circle.

Overtaken perhaps by a sudden sandstorm, the dinosaur was preserved as it sat on the eggs.

Its arms are turned back to encircle the nest and its legs are tucked tightly against its body, identical to the nesting posture of birds, such as chickens and pigeons, living today.

It was the sight of a few daggerlike talons protruding from the earth at Ukhaa Tolgod Mongolia that first drew the scientists' attention. They excavated only enough of the skeleton to identify the specimen and discover the nest. Then they bundled the fossil in a 400-pound plaster cast for shipment to New York. It was not until the fossil was examined at the museum in New York that the researchers realized the importance of what they had found.

Scientists do not know why the creature was sitting on the eggs. Perhaps, Gauthier suggested, it was incubating the eggs to warm them as birds do, shading them, or simply protecting them. The researchers do not know when the brooding behavior first developed, but suggested it was extremely ancient even among dinosaurs.

"Because oviraptors are so closely related to birds, our best bet is that it was present in the common ancestor of birds and oviraptors, so this behavior predates the origin of modern birds," Norell said.

## Questions:

1. What kind of dinosaur was found? What did it eat?

2. How are the eggs arranged? How many were found?

3. What happened to preserve the fossil parent in this way?

Answers are at the back of the book.

## Activity:

Using the Internet, research the oviraptor. What was the misconception about this dinosaur? What is the general consensus about the oviraptor now?

# 38

Dinosaur bones that were discovered in China showed evidence that feathers were attached to the arms and tails of some dinosaurs. These were long feathers, much like modern flight feathers. This could mean that feathers cloaked dinosaurs before birds arose, possibly functioning as an insulator or for sexual courtship. But, many dispute this proposed bird-dinosaur link.

It has long been thought that birds and dinosaurs branched off separately from primitive reptiles over 250 million years ago. This discovery raises the question: are birds the direct descendants of such dinosaurs?

# Feathers Don't Make the Bird

By Josh Fischman

The dead do not all rest in peace at the Royal Tyrrell Museum in Alberta. One of them rests on an artist's easel. The three-foot dinosaur has its head pulled back in the rictus of death, the neck and tail bowed toward each other. Extending from the tail is a peacocklike fan of feathers; other feathers form fringes under the arms. The figure is actually a model of the real fossil, which remains in China, where it was discovered and where Tyrrell paleontologist Phil Currie had the privilege of examining it. "It's a beautiful specimen," he says. "You look at the anatomy on that thing, and everything that's been used to separate birds from dinosaurs is there. But it's too primitive to be a bird. It didn't fly. The only thing truly birdlike about it is its feathers."

And that means, Currie says, that feathers don't make the bird—they could make the dinosaur just as easily. The animal, named *Caudipteryx*, came to light this June when Currie, geologist Ji Qiang of the National Geological Museum of China in Beijing, and their colleagues published a description of it and another feathered fossil named *Protarchaeopteryx*. For many paleontologists, these overlapping bird and dinosaur features

sealed the 30-year argument that birds are the direct descendants of such dinosaurs. And the discovery seemed to push some intractable opponents of this idea farther out on a limb.

For most of this century, scientists thought that birds and dinosaurs had branched off on their separate paths from primitive reptiles more than 250 million years ago. But in 1970, Yale paleontologist John Ostrom was struck by the similarities between the oldest known bird, the 150-million-year-old *Archaeopteryx* and a two-legged carnivorous dinosaur named *Deinonychus*. After Ostrom's insight, scientists began piling-up the anatomic similarities between the two groups until, by the mid-1980s, they had a list of about 150 shared features. One they didn't have was feathers, and given the ravages of time and the delicate nature of plumage, finding some seemed unlikely.

Then, in 1996, Ji presented Currie with a gift-wrapped box containing a chicken-size dinosaur dubbed *Sinosauropteryx*. The fossil was surrounded by a featherlike fringe. Paleontologists are still arguing over just what it is—protofeathers, long scales, or something more exotic—but a year later Ji produced another dinosaur so similar to the earliest bird that he called it *Protarchaeopteryx*. And he had not just one but three, all from Liaoning Province in northeastern China, that were between 150 and 120 million years old.

Currie hotfooted it back to China to examine the new fossils, and thanks to the fine-grained sediments in which the animals were buried, he could make out their anatomy in stunning detail. "You could definitely see there were feathers attached to the arms and the tail," says Currie. These feathers were long, with tightly interlaced strands splitting off from a central spine, like modern flight feathers; the bodies were covered with fuzzy, downy structures. On close examination, the second and third specimens turned out to be yet another species, which Currie and Ji named *Caudipteryx*, which means "tail feather."

And then the scientists did what many a birder would do when faced with homeless avians: put them in a tree. Ji, Currie, and their colleagues compared the two new fossils with *Velociraptor*, a sleek dinosaur considered by many researchers to be a close relative of birds, as well as *Archaeopteryx* and other primitive birds. The analysis produced a sort of family tree, with its first split coming between the lineage leading to *Velociraptor* and the line leading to all the other dinosaurs and birds. *Protarchaeopteryx* branched off this line soon after that, and then *Caudipteryx's* ancestry split away from that of *Archaeopteryx* and all modern birds. In other words, the Chinese fossils were dinosaurs. That meant that feathers, apparently, cloaked dinosaurs long before birds arose, perhaps functioning as a layer of insulation or as courtship display. Only later did protobirds adapt them for flight.

Skeptics of the bird-dinosaur link have long pointed out that what appear to be the closest relatives to birds lived 80 million years after *Archaeopteryx*. "The timing is a real problem," says ornithologist Alan Feduccia of the University of North Carolina. The new fossils don't resolve the gap. Feduccia suggests that the pre-Ostrom idea still holds and that the Chinese fossils are early, flightless birds. "*Caudipteryx* is really just an old kiwi."

Currie admits at this stage there are still shaky branches on his tree, but the classification of the Chinese finds as feathered dinosaurs is not one of them. Feduccia's scenario creates an even bigger disparity with the fossil record and requires that the Chinese animals independently evolved many dinosaurian traits. "It's good to have a dissenting voice that says, 'Hey, there's more than one possibility here,' because it makes you look at the specimens harder," says Currie. "But some of these arguments are simply drifting into the realm of silliness."

---

**Questions:**

1. How old were the dinosaur fossils of *Protarchaeopteryx* which Ji Qiang discovered?
2. Before the discovery of the feathers, how many shared features were listed to compare birds and dinoaurs?
3. What's the name of the feathered fossil in question (it's name means 'tail feather')?

Answers are at the back of the book.

**Activity:**

Write a short essay explaining either side of the argument raised in this article. Hint: Do all birds have feathers? And if something has feathers is it necessarily a bird? If something can fly does it have to be a bird? Give some examples. Try using the computer as a resource.

# 39

Jurassic Park captured the imaginations of millions, but its portrayal of *Tyrannosaurus rex* was not that accurate. The small forearms and weak leg bones of *T. rex* indicate that the famous jeep race would most likely have not been possible. Other studies are pointing towards a more social behavior and possibly scavenging for food. The ferocious *T. rex* may be beginning to be seen as somewhat different from what was previously thought.

# The Dinosaur Detectives

By Stephen Young

With a crack of his bullwhip and a touch of computer wizardry, Nathan Myhrvold conjures up a prehistoric sonic boom. It is an echo from more than 100 million years ago when plant-eating giants called apatosaurs were alive and generating 200 decibels with each swipe of their tails. A tonne and a half of flesh and bone moving from side to side must have created quite a spectacle, but what use was it? Myhrvold particularly likes the idea that it might have been some form of mating display. Of course, there's no knowing for sure. "Behaviours don't fossilise all that well," he admits.

This lack of evidence has put the study of dinosaur behaviour beyond the pale for some palaeontologists. But Myhrvold, who holds down a day job as chief technology officer at Microsoft in Redmond, Washington, relishes a challenge. In fact, his ambitious goal of recreating the lives of long-dead animals is shared by a growing number of researchers. Ingenuity, lateral thinking and a forensic scientist's eye for detail are the tools of their trade as they search for clues among the scant remains of another age.

Take *Tyrannosaurus rex*, "king of the tyrant lizards." Dinosaur detectives who want to get under this monster's skin have meagre evidence to work with. The first specimen was found in 1900, and since then excavations have turned up only three animals that are more than half complete. The same sites in America's Midwest have yielded another 20 or so skeletons that are at least 15 per cent intact. Even so, clever dino-sleuths have turned up plenty of clues upon which to work their magic. Jurassic Park it isn't, but this work paints a pretty good picture of what a day in the life of a *T. rex* might have been like.

## Rise and Shine

Were tyrannosaurs and their ilk coldblooded, relying on outside heat to warm their bodies and get them going? Or could they remain active whatever the temperature of their surroundings, by burning metabolic fuel to generate their own body heat? The long-running controversy over whether dinosaurs were ectothermic, like modern lizards, or endothermic, like birds and mammals, is central to our understanding of their lifestyles.

In one strand of research, John Ruben and his colleagues at Oregon State University in Corvallis have been using CAT scanning to probe the secret recesses of dinosaurs' noses. The team found that an assortment of dinosaurs—including a member of the tyrannosaur family—had relatively narrow nasal passages which would have had little space for

From "The Dinosaur Detectives," by S. Young, *New Scientist*, 1998, pp. 24-29. Reprinted with permission. : From "What's This Volcano Trying to Tell Us?" D. Pendick, *New Scientist*, February 20, 1999, pp. 26-30. Reprinted with permission.

special scroll-shaped structures called respiratory turbinates. Existing endotherms use these as air conditioners to moisten and warm incoming air and recover some heat and moisture from outgoing air—essential processes when breathing rates are high, as they are in endotherms. The implication is that the dinosaurs had low breathing rates and so wouldn't have inhaled enough oxygen to fuel the high metabolic rate needed for life as an endotherm.

In another study, the Oregon team focused on the lung structure of the famous Chinese "feathered dinosaur", *Sinosauropteryx* (This Week, 19 April 1997, p 6). A distant relative of tyrannosaurs and a fellow member of the Theropoda group, this fossil comes from the Yixian formation in northeast China, where a treasure trove of exquisitely preserved prehistoric remains is being unearthed. Astonishingly, signs of soft tissues are visible in the fossil, and these have helped the team to conclude that *Sinosauropteryx* had relatively simple, crocodilelike lungs that would be incapable of achieving the rates of gas exchange most endotherms need.

What does this all mean for dinosaurs? According to Ruben they didn't necessarily have the same rate of oxygen consumption or activity as living lizards. "They could have had something intermediate," he says. "But the evidence from this seems to preclude the possibility that they would have been warm-blooded in the sense that we'd ordinarily think of warm-blooded animals.'"

Ruben points out that this doesn't necessarily mean that dinosaurs were sluggish and dozy. After all, they lived at a time when the world was warm, and their large bulk would have helped them to keep a steady body temperature. Today's Komodo dragon offers a clue to what they might have been like. "If we were to reconstruct theropod dinosaurs with the same sort of metabolic physiology we see in some of these very active lizards we would come up with a very active, dangerous animal that would have been very mammallike in its behaviour,'" says Ruben.

## First Brush Your Teeth

One palaeontologist memorably described the huge, curved teeth of *T. rex* as "lethal bananas." Their serrated edges and the slots between the serrations were extremely good at trapping meat fibres, according to Chicagobased researcher William Abler. As a result, *T. rex* almost certainly had terrible breath and its mouth would have been a dental hygienist's nightmare. Abler believes its bite may have caused serious infections in any prey that survived an attack.

Even so, until recently some experts believed that this impressive oral weaponry was rather fragile. Then, a few years ago, Gregory Erickson was studying for his master's degree at the Museum of the Rockies in Montana when Ken Olson, a fossil collector, appeared with a Triceratops pelvis. The pelvis bore some extraordinary bite marks that looked like the work of *T. rex*. "If you took your thumb and pushed it down into clay—that was the depth of the holes," says Erickson.

Erickson and Olson's studies on the pelvis suggested that *T. rex* fed by "puncture-and-pull" biting. "They'd bite very deeply into flesh and bone and once they stopped they'd pull straight back," says Erickson, "and that would rip out a big chunk of flesh." With a group of colleagues at Stanford University, Erickson staged further tests to try to measure the bite force of a *T. rex*. The team simulated bites on a cow's pelvis using a life-size replica tooth and a hydraulic press, and measured the force needed to produce holes like the ones on the *Triceratops* specimen. The results showed that *T. rex* would have chomped like a champion, producing a force of at least 13.4 kilonewtons—outperforming wolves and lions, and biting in the same league as alligators.

## Off to Work

In Hollywood, *T. rex* runs races with Jeeps, but some researchers doubt it was so fleet. McNeill Alexander of Leeds University has looked for clues about the running speeds of various dinosaurs by studying the structure of their leg bones. He calculates a "strength indicator" which represents the strength of the bones in relation to the animal's weight. "If you do that for *Tyrannosaurus* you find that the leg bones were relatively weak for an animal that was that heavy," he says. Faster animals need stronger leg bones and his approach suggests that *T. rex*, which could have weighed 6 tonnes or more, moved more like an elephant than a rhino or a gazelle. "We're not talking about the sort of speeds that are good if you're going to chase jeeps," says Alexander. He offers a tentative figure of about 25 kilometres an hour. That's less than half the speed of one modern top predator, the lion.

However, Theagarten Lingham-Soliar of the Russian Academy of Sciences in Moscow argues that tyrannosaurs might have been faster than their bone structure alone implies, because of other factors such as the presence of large, shockabsorbing chunks of cartilage in their legs and their highly flexed knees.

With a little lateral thinking, James Farlow and his colleagues at Indiana-Purdue University at Fort Wayne have come up with another ingenious way to deduce tyrannosaurs' top speed. They asked themselves: what damage might a fully grown *T. rex* do to itself if it came a cropper at speed? A fall at 72 kilometres an hour, they calculated, could have been fatal. They came up with a top speed of about 36 kilometres an hour. "For an animal its size, I think it was pretty fast," says Farlow. "I suspect it could have caught any other large dinosaur in its environment."

In theory, fossil footprints could provide decisive evidence, but there are very few known tyrannosaur tracks, says Martin Lockley of the University of Colorado, Denver. The best, in New Mexico, shows just one footprint with about 3 metres of untrampled surface in front of it. It's difficult to draw conclusions from a single footprint, says Lockley, but if this distance is taken as the distance between steps, then the calculations show that the animal was travelling at around 11 or 12 kilometres an hour. Of course, this need not represent its top speed. It may well have put on a spurt when in pursuit of its dinner.

## Dinner Time

The idea that *T. rex* was a mighty hunter looks like an open-and-shut case. But is it? According to some researchers, including palaeontologist Jack Horner of Montana's Museum of the Rockies, there is evidence that this terrifying beast lived by scavenging. Horner points out that *T. rex*'s femur was longer than its tibia, whereas bipedal animals that run fast have the opposite arrangement. *T. rex* appears to be adapted for long-distance walking, he says. Its brain had a huge olfactory lobe like that of a turkey vulture, which depends on smelling carrion from a long way off. And those extraordinarily small arms look hopeless for grappling with prey. What's more, its physique would have been ideal for scaring hunters away from a fresh carcass.

Most researchers still see *T. rex* as the most ferocious predator of its day, although it might have scavenged when it got the chance, just as modern hunters do. "I would be astonished if tyrannosaurs had not been hunters as well as scavengers," says Farlow. Hunting was their main method of procuring food, according to Lingham-Soliar. He bases his conclusion on studies of tyrannosaur anatomy and comparisons with modern animals. In particular, he points to the huge skulls of tyrannosaurs, which were immensely strong in critical areas and clearly designed to resist large stresses such as those that might be encountered in hunting and dismembering large prey.

The same imposing anatomy has prompted David Norman of the University of Cambridge to suggest that *T. rex* might have charged at its victims with mouth agape. Lingham-Soliar, on the other hand, believes that this would have been highly damaging to their teeth: "Like running at a brick wall with one's mouth open." He offers a different rationale for the stoutly built skull, which also explains how tyrannosaurs managed to break up their prey despite their lack of effective grasping forelimbs. He believes they tore their victims apart by seizing them in the mouth and shaking them violently rather like many of today's marine predators that also feed without the help of limbs, such as sharks and killer whales. Enormous forces would have acted upon the skull and neck, putting a premium on size and strength. Larger prey wouldn't have been shaken, but chunks of flesh would have been gouged out with the teeth, he says.

## Family Life

Female tyrannosaurs were larger than males. That's the conclusion of Peter Larson from the Black Hills Institute of Geological Research in South Dakota, following his discovery of two distinct body types in *T. rex*. "The pelvis in the robust form is wider inside, which might be an indication that this form is female," says Larson. Further evidence comes from a living descendent of dinosaurs. Male crocodiles have an extra bone known as a chevron at the base of their tail where the muscle that retracts the penis is attached. The same is true of saurornithoides, a group of dinosaurs from China that are closely related to tyrannosaurs. Larson is looking for a similar pattern. "I haven't got absolute proof on *T. rex* yet," he admits. The clincher will come later this year when the robust skeleton of Sue, the most complete *T. rex* ever found, goes on display at Denver Museum of Natural History. "I believe we're going to find that Sue has one less chevron than the males," says Larson.

Finding that female tyrannosaurs were more stoutly built than males is not as surprising as it sounds. Larson points out that, contrary to most people's expectations, throughout the animal kingdom females tend to be larger than males, because of the obvious advantages for laying eggs or carrying young. "The only time you see males larger is where they have a harem and have to compete for females," he says. "*T. rex* was not a herding animal in that sense." He goes even further, pointing out that in birds of prey—which are probably among dinosaurs' closest living relatives—outsized females and monogamy go hand in hand. "Tyrannosaurs may have pair-bonded," concludes Larson.

Another palaeontologist who wants to dispel the image of *T. rex* as a loner is Tom Holtz from the University of Maryland, College Park. He points out that Sue was found in what looks like a family group, with a male and two juveniles. And this is not an isolated example. "There are multiple occurrences of multiple rexes" says Holtz. "Whether or not they hunted together or had division of labour within the group is difficult to tell." Pact hunting in *T. rex* society is Myhrvold's current area of interest, but he is yet to publish his results.

"*T. rex* probably organised into highly social, protective and cooperative family groups," concludes Larson. Even if this harmonious picture is correct, other clues suggest that there were outbreaks of violence between individuals. For example, tyrannosaur teeth sometimes bear telltale marks made by the teeth of their fellows. These could have been made during feeding, fighting or courtship, according to Abler. Then there are the numerous broken bones found in fossilised specimens. Many of these were healed by the time of the animal's death, suggesting that struggles with prey and competitors were common. Bite marks on Sue's skull show that she died after the left side of her face was almost torn off by another *T. rex*. Even before this brutal attack Sue's life must have been agony. Research last year at the Denver Museum of Natural History revealed that Sue suffered from gout.

---

## Questions:

1. Why is it so hard to study *Tyrannosaurus rex*?
2. What are ectotherms and endotherms?
3. What evidence supports a more social animal rather than the previously thought loner?

Answers are at the back of the book.

## Activity:

Make a list of all the things that come to your mind when thinking of dinosaurs. Which ones do you think may be myths?

Go to **http://www.ucmp.berkeley.edu/diapsids/dinosaur.html** How many myths about dinosaurs were you aware of, and were any on your own list? Feel free to check out the *T. rex* expo, too.

# 40

Movies such as *Armageddon* and *Deep Impact* have caught the attention of millions. People leave the movie wondering if something like that can ever happen. It has happened, and it killed off the dinosaurs. Surfaces of other planets and the moon prove the beatings that meteors can give, and earth, too, has received these beatings. However, earth's atmosphere and its erosional processes hide the scars. Though most meteors that have come to earth are not as big as the dinosaur killer, many have had their impact on civilization.

# Meteors that Changed the World

By Bradley Schaefer

The fabulous Leonid meteor storms have produced some of the most spectacular astronomical sights in history. These blizzards of light struck terror in people's hearts as the sky appeared to be falling. At other times, the glare of a bright bolide—perhaps followed by a tremendous explosion—can provoke anything from momentary surprise to vast destruction.

Most meteoroids from space burn up as meteors—streaks of light—in the upper atmosphere. Only rarely does a meteorite survive the plunge to reach the ground. When they do, they become objects that pass from the heavens to Earth. As such, they have left a bigger imprint on human history that is generally realized.

## The Iron Age

One characteristic that separates humans from animals is that we make lots of tools. Our early ancestors fashioned their first tools from bone, wood, and stone. The Stone Age takes it name from when stone implements represented the highest technology available. Later, copper and its alloy with tin (bronze) were discovered, yielding more durable weapons and tools that offered significant advantage to their possessors. The Bronze Age started during the third millennium B.C. for much of Eurasia, though its

beginning and end varied widely around the world. Then around 1400 B.C., the Hittites of Asia Minor discovered that iron could be smelted from common ores to produce even more superior tools and armory, thus giving birth to the Iron Age.

But how was iron discovered? Copper's melting point (1,980° Fahrenheit) is low enough that simple fires can both reveal and smelt the ore, while iron's melting point (2,795°F) requires intentional discovery and special methods for processing. What gave the clue that iron should be sought and developed?

Many late Bronze Age archaeological sites actually contain artifacts made of 90 percent iron. A famous example is the dagger recovered from the tomb of the 14th-century B.C. Egyptian pharaoh Tutankhamen. Chemical analyses show the dagger's "impurities" to be largely nickel, a sure sign that the iron came from a meteorite. So early metalsmiths found and used naturally smelted iron. They would have quickly realized its superiority. The Hittites and Sumerians acknowledged this connection by calling iron "fire from heaven." The Egyptian word for it means "thunderbolt of heaven," and the Assyrian term was "metal of heaven." With meteorites as an inspiration and a direct guide, the recognition of Earthly iron ores was probably inevitable. Meteorites jump-started the Iron Age.

When European explorers encountered a tribe of Inuits in northwestern Greenland in 1818, they were astounded to find knife blades, harpoon points, and engraving tools made of meteoritic iron. Tools from the fabled Greenland meteorite had been found as far as 1,400 miles away, having been transported as treasured trade goods. The area has no natural metal deposits, yet the abundant availability of meteoritic iron allowed the polar hunters to skip to the Iron Age and helped them survive in an extremely harsh and frigid land.

Five expeditions from 1818 to 1883 failed to find the "Iron Mountain" until Robert E. Peary was led by a local guide to the site on Saviksoah Island off northern Greenland's Cape York in 1894. The meteorite was found in three primary masses, named the Tent or "Ahnighito" (34 tons), the Woman (3 tons), and the Dog (1/2 ton). Over the next three years Peary's expeditions managed to load them onto ships despite savage weather, engineering problems, and having to build Greenland's only railway for transporting the behemoths. Upon arrival in New York City, the sources of Greenland's Iron Age were sold to the American Museum of Natural History for $40,000, where they are now on display at the Arthur Ross Hall of Meteorites.

Ironically, even today, 27 percent of the world's nickel comes from mines in the large Sudbury meteor crater in Ontario, Canada.

## Sacred Stones

In 205 B.C., Hannibal's Carthaginian armies had marauded Italy for more than a dozen years, threatening the very existence of the Roman republic. After witnessing a frightening meteor shower, the Roman magistrates consulted the Sibylline books, which prophesied that Hannibal could be defeated if the Idaean Mother was brought to Rome. This "Mother," a large, conical meteorite thought to represent the Great Mother of Gods, was enshrined in a temple at Pessinus in central Turkey. The resplendent Roman delegation that was sent to King Attalus gained permission to remove the meteorite only after an earthquake changed the monarch's mind. A sacred ship was built and the meteorite sailed up the Tiber River to Rome. The leading citizen in the city performed the rituals of hospitality for the rock, then a procession carried it to the Temple of Victory. With this divine morale boost, the Romans expelled Hannibal from Italy within the year and soon conquered Carthage.

In gratitude, the Romans built a special temple on the Palatine hill where the meteorite was worshiped for at least 500 years. However, the stone eventually fell into oblivion. In A.D. 1730 it was apparently excavated from its chapel, only to be discarded for lack of recognition.

Many meteorites have been worshiped as gods through the ages. One such sacred rock was uncovered in 1808, when a party of explorers found a 1,635-pound hunk of metal near the Red River in central Texas. The Pawnee Indians had worshiped this meteorite for its curing powers and made regular pilgrimages to its site. The explorers, who thought the metal was platinum, returned to the city to get provisions and equipment. They eventually split into two rival groups as they tried to outrace one another in cashing in on the treasure. The first party, which had left hastily without the gear for transporting the hefty load, could only hide the "platinum" under a flat stone for safekeeping while they searched for horses.

It took several days for the second party to find the prize and drag it toward the Red River. In their long overland journey they were repeatedly attacked by the desperate Indians. The meteorite was transported by boat down the river to New Orleans, and finally to New York. There, Benjamin Silliman of Yale University recognized its true nature based on its high nickel content, after which the disappointed owners sold the object to Colonel George Gibbs. After Gibbs's death in 1833 his wife rescued the meteorite from Irish laborers seeking to bury it; she then donated the rock to Yale as the largest meteorite in any collection at the time. Currently, the Red River iron is on display near my office at Yale in a hallway of the Gibbs building, where I have the weird feeling of often passing by an out-of-work god.

In 1853 a meteorite that fell in eastern Africa just north of Zanzibar was declared to be a god by the Wanika tribe until famine and a massacre by the Masai rendered its owners skeptical of its powers. On December 2, 1880, a 6-pound meteorite fell at the feet of two Brahmins near Andhra, India, who immediately proclaimed themselves as ministers of the "Miraculous God" and attracted up to 10,000 pilgrims a day. On August 14, 1992, dozens of rocks fell on the African town of Mbale in Uganda (S&T: June 1993, page 96). Local residents ground up some of the fragments and ingested the powder as medicine. They believed the rocks had been sent by their god to cure AIDS.

It's easy to understand that a fiery descent from heaven can be interpreted as a god, or a god's gift, come to Earth.

## Elagabalus

The slow fall of the Roman empire was a chaotic drama in which a wide variety of religions coexisted and competed, often with the emperor's worship playing a key role. In addition to the traditional Roman pantheon, gods with Greek, Hebrew, and Egyptian origins enjoyed widespread followings.

Popular among soldiers was the worship of Mithras, the god of truth and light. The idolization of Dionysus, the god of fertility, wine, and drama known for orgiastic festivals of merriment, dominated the urban middle class. And the Egyptian goddess of fertility, Isis, was patronized by the lower middle class and courtesans. To many Romans of the time, Judaism and Christianity seemed like just two more examples of the eastern flair for mysticism. Local cults from around the Mediterranean enjoyed periods of favor, including the Great Mother of Gods (from Asia Minor) and Sol Invictus (from Syria).

In this marathon of competing faiths a steady front-runner was the Sun god, variously called Apollo, Helios, Sol, Sol Invictus, or Elagabalus. The overwhelming power and importance of our day star makes a Sun god a natural chief deity; its worship is deeply embedded in many societies. The cult of the Syrian Sun god Elagabalus was known in Rome from before the time of Julius Caesar, but it only came to dominance with the ascension of the Roman emperor Marcus Aurelius Antoninus.

As a young boy, this emperor was made the high priest at the magnificent temple in Emesa (now Homs) in Syria, as a hereditary responsibility through his mother's family. He took his religious duties very seriously and even changed his name to that of his god, Elagabalus. Following a military plot arranged by his grandmother, the 14-year old boy was proclaimed emperor on June 8, 218. Elagabalus had no interest in affairs of state, as he was entirely devoted to his fanatical worship and sensual orgies. In the name of religion, the new emperor ravaged traditional Roman gods, depleted the treasury, and wore women's clothes.

The worship of Elagabalus was centered on the Black Stone of Emesa, a large, cone-shaped meteorite. When its high priest moved to Rome the stone came with him. Two magnificent temples were built for it, one on the Palatine hill and the other in the suburbs. The emperor personally led daily worship of the meteorite while dressed in silk robes, a lofty tiara, and cheeks painted red and white. The bejeweled Elagabalus directed choruses, sacrificed numerous bulls and sheep, and danced with Syrian damsels around the Black Stone. A special chariot decked with gold and gems and drawn by large, spotless white horses was used to move the meteorite between temples. No mortal was permitted in the chariot, while Elagabalus himself held the reins and ran backward to keep his face toward his god, with gold dust strewn under his feet supposedly to prevent him from stumbling. Thus meteorite worship became the official religion of the Roman empire.

Elagabalus's cruelty, extravagance, neglect, and depravity alienated him from all segments of society, so he was doomed to a short reign. On March 6, 222, the emperor and his mother were assassinated by the Praetorian Guards, and their mutilated corpses were dragged through the streets of Rome.

As for the meteorite, it was quietly returned to Syria and reinstalled in the temple of Emesa. The rock was likely broken into pieces when the temple was converted to a Christian church sometime in the 4th century. The site, now a mosque, has never been excavated.

Even after the disastrous reign of Elagabalus, the Sun god remained as the most important deity. Many cities throughout the empire had instituted games in its honor by 240. The emperor Aurelian elevated Sol to the highest rank among all gods in 274, and this practice was followed by all rulers until the time of Constantine, who reigned from 306 to 337. Even Constantine wavered between Sol and Jesus Christ, and his coins celebrated the Sun god as the grantor of imperial power as late as 324, after which he converted both himself and his empire to Christianity.

The spread of Christianity throughout Europe occurred for many reasons; still, it is amusing to wonder if it was only by the whim of an emperor that the Western world now is predominantly Christian instead of meteorite-worshipers.

## The Black Stone of the Kaaba

The most famous sacred meteorite is the Black Stone of the Kaaba. The Kaaba is a cubical building in Mecca, Saudi Arabia, toward which Moslems pray five times daily. The Black Stone, set in the northeastern outside corner of the Kaaba, is considered to be the most sacred treasure of Islam.

The Kaaba served as a center of worship for pre-Islamic Arabs and was reputed to contain 360 idols. In 630 the triumphant prophet Mohammed returned to Mecca and cleansed the temple of the idols after honoring the Black Stone. The heretical Qarmatian sect stole the stone in 930, but it was recovered 21 years later, positively identified because of its ability to float on water. In 1050 a mad Egyptian caliph sent a man to destroy the relic. The Kaaba was twice burned down and was flooded in 1626. During these trials the original stone was broken up into about 15 pieces. It was finally set in cement surrounded by a silver frame.

Islamic tradition variously describes the stone as coming from heaven and as originally hyacinth in color before it turned black because of humanity's sins. Reports of its visual appearance mention a dark, reddish black color, smooth surface, some apparent banding, and small crystal inclusions. Although generally regarded as a meteorite, it cannot be an iron-type meteorite since it fractures, nor can it be a stony meteorite since it would not have withstood the handling and would sink in water. Similar arguments may reject nonmeteoritic origins, such as basalt or agate.

In 1980 Elsebeth Thomsen proposed that the Black Stone is an impactite (fused sand mixed with meteoritic material) from the Wabar Craters in Saudi Arabia's Empty Quarter (S&T: November 1997, page 44). The Wabar impactite is a hard glass (so it is tough enough for repeated fondling yet can shatter) with a porous structure (so it can float) and having inclusions of white glass (the crystals) and sandstone (the banding). The change from its original lighter color could be due to the accumulated oils from frequent kissing and handling. A critical problem confronting this proposal is that several measurements suggest the Wabar Crater complex is only a few centuries old, though other analyses suggest it was formed 6,400 years ±2,500 years ago. Whether or not Wabar is the source, the Black Stone still fits well with a desert impactite and a meteorite tradition.

## The End of the Dinosaurs

What ended the age of dinosaurs 65 million years ago? Until 1980, paleontologists would answer their biggest mystery by mumbling something about climate, mammals, or volcanoes. Then Luis Alvarez and his coworkers hypothesized that a giant meteorite hit the Earth and caused global environmental damage, leading to the reptiles' extinction. They tested their idea by seeking the element iridium in sediments around the world laid down at the correct time. Iridium is often abundant in meteorites but rare from Earthly sources. Alvarez and his team examined clay from the distinct K-T boundary (named after the Cretaceous-Tertiary geological epochs) at three widely separated sites. They indeed found a thin layer of iridium-rich sediment coinciding with the time of death of the dinosaurs.

The startling discovery of iridium at a few sites was intriguing, but the scientific community wasn't won over to the new idea. Various groups rushed to check many sites widely spread around the globe and confirmed that there really is an iridium layer and it really is worldwide. Then, the groups searched for and found other evidence, such as the presence of the element osmium and right-handed amino acids, which are relatively abundant in meteorites but rare on Earth. As the excitement mounted, yet more researchers found microtektites in the K-T boundary layer.

But most geologists withheld judgment until the unique signature of shocked quartz was also found to be widespread and sharply confined to the K-T boundary. Finally the 110-mile-diameter Chicxulub crater was discovered under Mexico's Yucatán Peninsula. Its age of 64.98 million years ±60,000 years matches that of the K-T boundary (64.3 million years ± 1.2 million years).

This progression from intriguing proposal to certainty of a gigantic meteor impact of worldwide consequences was made in only a few years and is a classic example of the repeated tests of critical predictions that is the hallmark of the scientific method.

Such "catastrophism" was in direct opposition to a strong tradition among geologists, who tend to believe in "uniformitarianism." Most of the landforms around us are fashioned by slow, fairly steady processes, but now it appears that there are also sharp catastrophes of great importance. After some rearguard proposals concerning volcanism, the geological community has now converted to the Alvarez hypothesis.

"Death by meteor" was also a hard sell to paleontologists at first. They pointed to the complex patterns of mass die-offs, but these were readily explained by the many deadly aspects of a major impact and by the necessarily complex responses of intricate ecological systems. The paleontologists then claimed that the dinosaurs were already dying out by the time of the K-T boundary and lived a bit past the time of impact, but these notions were soon revealed to be caused by inadequate sampling of fossil bones. Subsequent large, systematic studies showed that big dinosaurs were flourishing up to the K-T layer and not past, while ammonites and other small fossil fauna show the same extinction pattern with impressive statistics and fine time resolution. Finally, narrow iridium layers have since been identified at many of the other geological boundaries that mark mass extinctions through Earth's long history.

The research and debate has now shifted to understanding the consequences of the Chicxulub event. The immediate shock and heat from the impact would annihilate all life within perhaps a 600-mile radius from ground zero, while tsunamis would devastate anything near sea level in the Atlantic basin. Indeed, layers of tsunami-wave rubble have been found more than 30 feet thick all around the Caribbean, and the deposits extend up to 430 miles inland. The initial blast would also send out large masses of ejecta, which would reenter the Earth's atmosphere over the entire globe; the combined heat from all these simultaneous secondary meteors lighting up the sky would kindle worldwide forest fires. Indeed, the volume of soot in the K-T layer indicates that the majority of the world's biomass was burned to a crisp. In the months and years that followed, a dust veil would have cut off most sunlight, stopping photosynthesis and creating a "meteor winter." Together with other forced and deadly effects, entire food chains collapsed. The dinosaurs, unable to adapt, died out. Only about half of the world's species, including mammals, managed to have some breeding population survive through the holocaust.

Here we have a scientific revolution fought and finished in a decade, where the scientific community was won over by the great strength of the evidence, despite various deep prejudices. This is science at its best, and it makes me proud.

## Target Earth

NASA's robotic missions to other planets have shown that all old, solid surfaces in the solar system are heavily cratered. Earth too must have been heavily bombarded, but the evidence has been largely hidden by the erosion acting on the surface of our dynamic planet. Even so, the many meteorite finds, Chicxulub, and the approximately 140 known craters around the world demonstrate that Earth has suffered massive hits.

In modern times, the best known destructive impact occurred in a remote region of taiga forests near the Tunguska River in eastern Siberia at about 11:30 a.m. on June 30, 1908. The explosive energy released by the event was equivalent to roughly 15 million tons of TNT—a thousand times more powerful than the Hiroshima bomb and matching a large hydrogen bomb. The meteorite, likely of stony composition with a diameter of 200 feet, exploded at an altitude of 5 miles, creating an air burst that leveled more than 1,200 square miles of forest. But this famous event is by no means unique even in the current century.

Early on the morning of August 13, 1930, a large meteorite exploded over the Amazon jungles in an isolated area on the Curuça River with a force estimated to be a tenth of the Tunguska event. The bolide was heard as a shriek of artillery shells followed by great balls of fire that fell from the sky like thunderbolts. Three massive explosions and three shock waves ripped through the jungle, followed by a very light rain of ash that veiled the Sun until midday. The blasts were heard up to 150 miles away, while the resulting magnitude 7 earthquake was recorded 1,300 miles away in La Paz, Bolivia. This massive meteor explosion would not have been known to the outside world if not for a Capuchin monk, Father Fedele d'Alviano, who had visited the terrified population during his yearly apostolic mission and then written about the event for the papal newspaper.

The collision of Comet Shoemaker-Levy 9 with Jupiter in July 1994 was a widely publicized event. Most of the educated population in the world heard about the dark impact scars on the Jovian atmosphere, each larger than Earth. If such a comet were to hit Earth, mass extinction would be the most likely scenario. The moral of the Amazonian Tunguska and Shoemaker-Levy 9 events is that such occurrences are not unique even in modern times. Collisions of catastrophic proportions and worldwide scale can happen anytime within our lifetimes.

Over the last decade or so, there has been a big change of consciousness among both scientists and the general public. The paradigm now is the realization that the Earth is a target in a giant shooting gallery, with the stakes as large as civilization itself. Countless articles, books, and movies on the threat of meteorite impacts have already appeared. Thankfully, the larger the explosion, the rarer the event. Metropolis destroyers, with an explosive energy on the order of 100 million tons of TNT, happen roughly once per millennium. Regional destroyers, about 100 billion tons of TNT, have an event rate of around once per hundred millenniums. Civilization destroyers, about 100 trillion tons of TNT, average once every 10 million years or so.

So how can we defend ourselves? Most ideas center on finding any objects on a collision course and shifting their orbits perhaps using nuclear warheads. Any such solutions must have tradeoffs between the degree of advance notice and the size of the diversion. A tradeoff for society is to decide how many resources should be put into the meteorite threat instead of into more immediate possible catastrophes, such as overpopulation, environmental collapse, or nuclear war.

Astronomical events have changed history and affected the lives of ordinary people in a surprising number of ways, but only meteorites actually come down to Earth. They have had a great impact on humanity—as a resource for building civilization, as a god come to ground, and as the cause of mass die-offs. But as you are watching the fiery trails of Leonid bolides, please remember only the majesty and not the terror.

## Questions:

1. How often does a meteor actually reach the ground without burning up in the atmosphere?
2. What is a sure sign that iron came from a meteorite?
3. What is the evidence proving a meteor killed off the dinosaurs?

Answers are at the back of the book.

## Activity:

Got to **http://seds.lpl.arizona.edu/nineplanets/nineplanets/meteorites.html**
Read the webpage and the links that interest you. Write a page summary.

# 41

In 1972, Niles Eldredge and Stephen Gould formulated a model of evolutionary change called "punctuated equilibrium," in which most speciation occurs during short intervals, with long periods of stability in between. This model contradicted the traditional Darwinian notion of continuous, gradual change. Eldredge and Gould's proposal touched off one of paleontology's biggest controversies: what is the rate and pattern of evolutionary change? Over the last twenty-plus years, some studies have purported to show evidence for punctuated equilibrium while others purported to show evidence for phyletic gradualism. Still other studies showed evidence of some combination of the two processes, with these evolutionary patterns being dubbed "punctuated gradualism."

One problem that has plagued these studies has been a concern over how accurately fossil species, which can only be defined based on their mineralized shells and skeletons, represent actual biological species which may have had distinct differences in the soft parts of their bodies that are not preserved. Recent work by Alan Cheetham and Jeremy Jackson has helped to solve this problem with regard to one group of organisms: the bryozoa. They have shown that the criteria used to distinguish fossil species of bryozoa also accurately distinguish between modern species. These findings have given great weight to their studies of evolution in fossil bryozoa. And their studies indicate that punctuated equilibrium is the predominant mode of speciation. While this study will probably not end the debate, it does lend more support to Eldredge and Gould's 1972 proposal.

# Did Darwin Get It All Right?

By Richard A. Kerr

*The most thorough study yet of species formation in the fossil record confirms that new species appear with a most un-Darwinian abruptness after long periods of stability.*

In a 20-year debate about the pace of evolution, paleontologist Alan Cheetham had always known exactly where he stood. Since 1972, when Niles Eldredge of the American Museum of Natural History and Steven Gould of Harvard University first proposed their theory of punctuated equilibrium, some paleontologists have argued that new species appear suddenly in the geological record, after millions of years of evolutionary stasis. But Cheetham, like many of his colleagues, thought differently. As a student

of the renowned evolutionary paleontologist George Gaylord Simpson, Cheetham had learned that a species changes gradually, through millions of years of natural selection—Darwin's survival of the fittest—until it is so different that it constitutes a new species.

That's the pattern he expected to confirm when he began an exhaustive study of the filigreed remains of corallike animals known as bryozoa, hoping to determine the pace at which new species had appeared during the past 15 million years. But, Cheetham, who works at the Smithsonian Institution's National Museum of Natural History, was in for a surprise. "I came reluctantly to the conclusion," he says, "that I wasn't finding evidence for gradualism." What

Cheetham did see, again and again, was individual species persisting virtually unchanged for millions of years and then, in a geologic moment lasting only 100,000 years or so, giving rise to a new species.

This just completed study isn't the first to confirm punctuated equilibrium in the fossil record. But it is the strongest yet, other researchers agree. "Theirs is by far the most complete," says Dana Geary of the University of Wisconsin. Recognizing new species based only on their fossils can be problematic, as critics of earlier studies have emphasized. Cheetham and his collaborator Jeremy Jackson of the Smithsonian Tropical Research Institute in Panama seem to have defused that criticism, at least for the bryozoa, by testing their methods for distinguishing fossil species on living bryozoa. With their study, some paleontologists are now leaning toward punctuated equilibrium as the dominant mode of speciation. "Those who have looked hard, and that's not a large number, have tended to find punctuation," says Geerat Vermjij of the University of California, Davis.

Eldredge and Gould made their original proposal as graduate students, after they had been sent off in search of fossils—Eldredge to upstate New York for trilobites and Gould to Bermuda for land snails—to document the gradual, pervasive evolution that the textbooks said was there. Neither could find it. Instead they saw species that had gone unchanged for millions of years suddenly give rise to new ones.

Since Darwin, paleontologists have attributed such findings to flaws in the geologic record: The stratum recording the gradual change that led to speciation must simply be missing. But Gould and Eldredge decided to take the fossil record at face value. They proposed that a long-standing mechanism for generating new species—the geographic isolation of a small population of one species for tens of thousands of years—could produce geologically abrupt speciation when the isolation broke down and the new species spread into the rest of the world.

Cheetham, though, regarded punctuated equilibrium as an unnecessary complication, and in 1986 he set out to demonstrate gradualism among the fossil bryozoa he had already spent decades studying. It was awkward for his views that the bryozoan species he was acquainted with did seem to have appeared abruptly. But he was confident that when he made a detailed study of their skeletal features to identify as many different species as he could, gradual speciation would prove to be the norm.

First, with the help of colleagues, Cheetham amassed a large sample of bryozoan fossils of the genus *Metrarabdotos* from the Caribbean and adjacent regions. He meticulously classified them into 17 species using 46 microscopic characteristics of their skeletons such as the length of the individual zooids (the animals that make up bryozoan colonies) and the detailed dimensions of the pores and larger orifices that dot the zooids. Then he arranged them in a *Metrarabdotos* family tree. Yet even though Cheetham's analysis often allowed him to split what had seemed to be a single species into several, the abruptness was stronger than ever. Through 15 million years of the geologic record, these species would persist unchanged for 2 to 6 million years, then, in less than 160,000 years, split off a new species that would continue to coexist with its ancestor species. Cheetham the gradualist "was amazed when he saw the punctuated result that he got," says Jackson.

## A Biological Test

Jackson too was impressed, but he wasn't convinced. What if the subtle morphological differences Cheetham was using to split his fossil species really didn't mark separate species at all but rather, say, variants within a species? "Clearly, the strength of any discovery of punctuated equilibrium—a model of speciation—depends on our ability to recognize species," says Jackson. "So I challenged him to submit his methods to biological examination."

As test material, Jackson gathered many different modern bryozoa that are native to the Caribbean. He and Cheetham then tried to distinguish among modern species by applying the same kinds of morphological measurements Cheetham had used for the fossils. The first part of the exam tested consistency: Would the classification depend on how many morphological features they applied? No, the morphological differences that defined 22 species in three distantly related genera of modern bryozoa held up whether Cheetham and Jackson used 20 or 40 morphological characters.

Then came reliability: Do the skeletal details accurately distinguish species? One worry was that the immediate environment might affect skeletal morphology, making populations of the same species living in different environments look like separate species. But when the researchers transplanted bryozoan colonies from different reefs to a single spot, the skeletons of the descendant colonies still closely resembled those of their parents in spite of the changed environment.

Another concern was that morphology might not be a fine enough scalpel: It might lump several different real species into a single apparent "species." So Jackson resorted to genetics. Using protein electrophoresis, he analyzed enzymes extracted from specimens of each of eight morphologically defined species. In each case, all the specimens from each morphological species had much the same enzymes, indicating that they belonged to the same genetically defined species. Cheetham's fossil species had passed the biological test with flying colors.

"Morphology still seems to be a way to say something about the way evolution occurred," says Cheetham. With this confirmation of Cheetham's method of identifying species, Jackson says he "became a believer." He and Cheetham have now extended the earlier work with *Metrarabdotos* to the genus *Stylopoma*. And once again, the 19 different fossil species they traced revealed textbook cases of punctuated equilibrium.

Because of their fastidious identification of species, Cheetham and Jackson's work is widely regarded as the strongest such evidence so far, but it has some competition. Timothy Collins of the University of Michigan and his colleagues, for example, recently took the same biologically based approach as Cheetham and Jackson when they studied a genus of coastal snails called *Nucella*. Although *Nucella* has fewer distinctive characteristics than the bryozoa, Collins and his colleagues also found punctuated equilibrium in the evolution of these snails in California over the past 20 million years.

Those who doubt the importance of punctuated equilibrium, however, can still take heart from earlier studies of fossil freshwater snails by Geary, who documented gradual change within two snail species over periods as long as 2 million years, along with six cases of punctuated speciation. Another verdict of gradual change came from Peter Sheldon of the Open University in Milton Keynes, United Kingdom, who studied morphological change among trilobites from Wales.

Faced with a welter of evidence, some paleontologists are sticking to a middle ground. In *New Approaches to Speciation in the Fossil Record*, the soon-to-be-published proceedings of a 1992 symposium, editors Douglas Erwin of the National Museum of Natural History and Robert Anstey of Michigan State University survey 58 studies published since 1972. They conceded that many of these studies have their weaknesses, but they still conclude that "paleontological evidence overwhelmingly supports a view that speciation is sometimes gradual and sometimes punctuated, and that no one mode characterizes this very complicated process in the history of life."

But Jackson, for one, thinks most of the studies supporting gradualism are flawed—for example, because the researchers relied on only a single characteristic to monitor evolutionary change and couldn't be sure they had identified all the species. "I'm imposing pretty strict criteria," says Jackson, "but in the few cases I know [that meet those criteria], it's perhaps 10-to-1 punctuated." Geary, whose work has been used to buttress both sides of the argument, tends to agree. "Gould was my adviser," she says, "but I don't think I have a stake in it. I think that a whole host of patterns is possible, but it does seem to me punctuated patterns predominate."

## How to Punctuate Evolution

If so, evolutionary biologists will feel new pressure to explain how punctuated equilibrium could actually work, a topic about which "there are a lot of hypotheses and not many facts," says evolutionary theorist Mark Ridley of Emory University in Atlanta. One mystery is what would maintain the equilibrium in punctuated equilibrium, keeping new species from evolving in spite of environmental vagaries.

One much-discussed possibility is that species become caught in what Vermeij calls "an adaptive gridlock." Called stabilizing selection, this gridlock results because "there's so much [natural] selection pushing at a species from different directions," Vermeij explains. "It can't go anywhere because moving in one direction has implications for its other competing functions." If a shellfish could reduce the weight of its shell, for example, it might have a better chance of escaping from some fastmoving predators. But that evolutionary route could be closed because a lighter, thinner shell would also decrease its resistance to other predators that bore into their victims. So the species remains unchanged for millions of years until a small population, isolated in a new environment, quickly evolves into a new species.

Stabilizing selection gets some new support in Cheetham and Jackson's chapter in *New Approaches to Speciation in the Fossil Record*. Over millions of years, they point out, any species would be expected to change slightly because of random genetic drift, but their analysis of the *Metrarabdotos* and *Stylopoma* bryozoa suggests something more like evolutionary paralysis. "Our tests strongly favor stabilizing selection" as an explanation of long-term species stasis, says Cheetham.

But that explanation only deepens another mystery: "If stability is the rule, how do you get large-scale shifts in morphology" over many successive species? asks paleontologist David Jablonski of the University of Chicago. "How do you get from funny little Mesozoic mammals to horses and whales? From *Archaeopteryx* to hummingbirds?" One possibility is species selection, a process analogous to Darwin's natural selection but acting at a higher level. A species might be especially likely to spawn new species because of some characteristic of that species that could never appear in an individual, such as having a broad geographic range. As a species wins out in this higher level evolutionary game, Jablonski explains, "all sorts of things get swept along." Body characteristics of individual members of the species, which might have nothing to do with the success of the species as a whole, would turn up in an increasing number of descendant species.

To finally resolve how common such processes are, and how many of his teacher's lessons Cheetham will eventually have to reject, researchers will have to apply a paleontological scalpel as sharp as Cheetham and Jackson's to a variety of organisms, living in many different environments. As Eldredge and Gould have written, "Only the punctuational and unpredictable future can tell."

## Questions:

1. Alan Cheetham originally thought that his studies would support which theory?

2. In Cheetham's study, what was the length of time that individual species persisted, and how long were speciation events?

3. What is one theory for why species persist for long periods of time, with only short intervals in which new species evolve?

Answers are at the back of the book.

## Activity:

To understand punctuated equilibrium and speciation better, draw three evolutionary trees. There should be one tree to represent punctuated equilibrium, one for punctuated gradualism, and another for Darwin's continuous, gradual change.

# 42

How big a role does climate play in shaping the tempo of evolution? Some scientists see climatic change as a primary force shaping evolution, while others see climate change as one of many more equal factors. Often, two studies using similar data will reach entirely different conclusions. Understanding the role of climate in evolution has become an important goal for paleontology, because in the last ten years many scientists have seen climate change as the major force shaping the evolution of hominids, the lineage leading straight to us. A better understanding of the role of climate change in the recent geologic past may also help us to understand the role of climate in the entire evolution of life, as well as what biological changes might be caused by human-induced global warming.

# New Mammal Data Challenge Evolutionary Pulse Theory

by Richard A. Kerr

Paleontologists anxious to make sense of the rise and fall of species in the fossil record have long invoked climate change as a prime mover in evolution, a force that triggers the evolution of new species while condemning others to extinction. But although there are plenty of rough correlations between climate change and evolution, proving a causal link has been difficult, given the imperfect preservation of the geologic record.

In the 1980s, however, paleontologist Elisabeth Vrba of Yale University documented a striking coincidence in the African geologic record about 2.6 million years ago, when a major climatic step toward the ice-age world occurred just as African antelopes underwent a burst of evolution and extinction. Adding popular appeal to the work, the human family tree branched out at about the same time, giving rise to the lineage that eventually led to *Homo sapiens*. Vrba proposed that a single climate-driven "turnover pulse" involving antelopes, hominids, and other animals had in a geologic moment turned evolution in a fateful new direction.

The idea attracted much attention, but few paleontologists managed to test it. Now new data reported at the North American Paleontological Convention (NAPC) In Washington, D.C., last month raise doubts about the theory. One of the richest, best dated African fossil records—which includes some of the same species Vrba studied—shows no sign of a turnover pulse. Rather, it shows "a more sustained shift" over a million years or more from woodland species toward grassland species, says Anna K. Behrensmeyer of the Smithsonian Institution's National Museum of Natural History, who led the study. "There was global change," she says, "but its effect on the fauna was not punctuated."

This and other new work could provide new ammunition to those who see a limited role for climate change in evolution. "I'm a real skeptic" about the effects of climate change, says mammal paleontologist Richard Stucky of the Denver Museum of Natural History. "When you look at the whole range of species, very seldom is there a climate event that changes the course of mammalian evolution." But even as new data come in, it's clear that the subject of how changing climate affects mammalian evolution continues to spark a range of opinions, with Stucky's minimal role at one extreme, Vrba's turnover pulse at another, and Behrensmeyer's prolonged shift somewhere in between.

Vrba wasn't at the meeting to defend the turnover pulse idea—she's on sabbatical in South Africa. (She also could not be reached for this article, despite repeated attempts to locate her through colleagues in the United States and Africa.) But her latest data were published late last year in two conference proceedings chapters. She compiled her own and published records of the first and last appearances of 147 species of African antelopes, most from eastern and southern Africa, during the past 7 million years. That analysis showed that from 3 million to 2 million years ago, the total number of species doubled, and 90% of all species recorded in that interval either first appeared or went extinct during that time. Furthermore, almost all of this considerable turnover was concentrated between 2.7 million and 2.5 million years ago. Meanwhile, although the exact timing is in dispute, the genus *Homo* also appeared between 3 million and 2 million years ago, possibly right about 2.5 million years ago.

Climatic data are consistent with Vrba's theory too. After about 3 million years ago, Earth was gradually cooling, as the climate system headed toward glaciation in the Northern Hemisphere. But Africa didn't slide smoothly toward the ice ages—it jumped, according to Peter deMenocal of Columbia University's Lamont-Doherty Earth Observatory (*Science*, January 14 1994, p. 173). By analyzing climate indicators in marine muds off the African coasts, he showed that between 2.8 million and 2.6 million years ago, subtropical Africa's climate abruptly shifted from one mode of operation to another, switching from a 20,000-year beat controlled by Earth's wobbling on its rotation axis to a more intense, 40,000-year beat driven by the changing tilt of the axis. This new regime left tropical Africa oscillating between a warmer, wetter climate and a cooler, drier one.

Vrba suggests that the longer, cooler episodes drove antelope evolution by means of a classic mechanism—breaking up the antelope's preferred woodland habitat into isolated ecological islands scattered among grasslands. The small woodland populations then spawned new species better adapted to the grasslands. Her hypothesis predicts that other species, including hominids, would respond the same way. As might be expected, such a sweeping generalization drew strong reactions. Those who didn't see pulses in their data were doubtful, while those whose world view includes abrupt evolutionary steps were enthusiastic. "The idea is wonderful," says Niles Eldredge of the American Museum of Natural History in New York City, co-creator of the theory of punctuated equilibrium.

But testing Vrba's idea requires an unusually rich and well-dated fossil record. One such record is a new computerized database developed under the Evolution of Terrestrial Ecosystems program run by the National Museum of Natural History. This includes the first and last appearances of 510 mammal taxa ranging from antelopes to baboons for the past 6 million years. For their test, the group focused on the fossiliferous and well-studied Lake Turkana region of East Africa, which has yielded a variety of animals, including hominids. What's more, the Turkana fossils are the best dated in Africa for the period from 1 million to 4 million years ago, thanks to repeated volcanic eruptions that blanketed the region with radiometrically datable ash layers.

When the Smithsonian team plotted the pace of evolution in the Turkana fauna about 2.5 million years ago, the turnover pulse theory "just didn't seem to hold up," says Behrensmeyer. "Clearly, there was a shift going on, but I think we can show the event was occurring over at least a million years and doesn't qualify as a pulse." Instead of a 90% turnover in a few hundred thousand years, the team found a 50% to 60% turnover spread between 3 million and 2 million years ago. Diversity during the period rose 30% rather than doubling, as Vrba reported for the antelopes. Even for the 53 species common to both studies, there is little sign of a Turkana pulse, says Behrensmeyer.

Slowing the pace of the shift toward grassland-adapted animals and starting it earlier blurs Vrba's link between evolutionary change and Africa's jump to a new climate mode. Instead, the Turkana data suggest that the fauna was steadily nudged toward grassland-adapted species by a global cooling and related African drying. "There isn't a pulse," says paleontologist David Pilbeam of Harvard

University, who has seen the Smithsonian data. "I had considered that maybe around 2.5 million years ago there was sufficient environmental change that you would get a turnover pulse, but the evidence would now suggest that you didn't."

Exactly why Vrba's record for African antelopes is punctuated and the Turkana record isn't remains unclear. One possibility is that variations in fossil abundance through time skewed Vrba's data, creating a false peak. Another is that the Turkana rift valley—which held a river bounded by woodland at this time—was buffered from the dramatic climatic shifts, suggests paleontologist Steven Stanley of Johns Hopkins University. Testing whether the Turkana region was typical of Africa isn't yet practicable, says Pilbeam, noting that only in the Turkana basin is the African mammalian record detailed enough to offer a more or less complete documentation of the changing fauna. "If you really want to know what happened in Africa over the past 2 to 3 million years, you need many such [records]," he says. "The quality of record that we would need [to test the turnover pulse hypothesis] is way beyond what we currently have, and it may indeed be beyond what we are ever likely to have."

Detailed comparisons of methodology may eventually sort out why these African studies differ, but they are not likely to settle the broader question of how climate influences mammalian evolution. On that the record is mixed. In addition to Vrba's pulse and Behrensmeyer's slow drift, there are also reports of no mammal response at all to abrupt climate change. At the NAPC meeting, Donald Prothero of Occidental College in Los Angeles argued that two major cooling events 37 million and 33 million years ago failed to affect North American mammals, although these cold spells apparently triggered extinctions in the sea and among terrestrial nonmammals. "The mammal response is negligible," Prothero says. "There is no turnover pulse, at least in North America."

Yet previous studies have shown that climate can have at least an indirect effect on mammal evolution. For example, 33 million years ago, when North American mammals were blithely ignoring climate change, European mammals were suffering through "La Grande Coupure," or the great break. It was a brief but momentous evolutionary event in which up to 60% of European mammals went extinct, to be replaced by more modern forms (*Science*, September 18 1992, p. 1622). But researchers think climate's role was indirect: A burst of glaciation created a land bridge to Asia, and the European mammals lost out to Asian invaders.

The dearth of evidence that climate change has forced mammalian extinction and speciation has Prothero and others questioning traditional assumptions. "We've oversold the idea that animals, especially land mammals, are responsive to environmental change," he says. "Animals seem to be remarkably resistant to a lot more change than we thought." All of which leaves open the question of why our favorite mammals, our ancestors, emerged in Africa as Earth was entering its ice age.

---

## Questions:

1. What group of organisms did paleontologist Elisabeth Vrba study in developing her turnover pulse hypothesis?

2. When did she find a pulse of extinctions?

3. What may have caused the difference in results between Vrba's study and the study of Behrensmeyer?

Answers are at the back of the book.

## Activity:

Make a timeline beginning with the Ice Ages and ending at the present. Record the major climate changes and biological events.

# ANSWERS

## PART ONE: THE ORIGIN OF THE EARTH AND ITS INTERNAL PROCESSES

### 1. Dynamics of Earth's Interior

1. Plate tectonics and improved observational and experimental constraints on the rheologic properties of mantle materials.
2. There are no phase transitions at depths greater than ~900 kilometers.
3. The core system shows that it is able to generate predominantly dipolar magnetic fields that undergo spontaneous reversals.

### 2. Mantle Plumes and Mountains

1. The islands are progressively older, less elevated, and more eroded to the northwest along the length of the chain. All these characteristics point toward a westward motion of the plate over the plume. After the newly formed island passed over the plume it began to subside and erode.
2. The model calls for the Yellowstone hot spot to have existed for 75 million years.
3. The Laramide Orogeny has widespread deformation from vast portions of continental crust tectonically sliced and heaved on top of one another. Also, there is no volcanic activity.

### 3. The Globe Inside Our Planet

1. The studies on the rotation of the inner core present different rates of rotation that lead some to question the possibility of this theory.
2. According to geophysicists, gravity should lock the inner core into place with the rest of the earth.
3. Anistrophy is the property of waves traveling north-south across the inner core faster than those going east-west.

### 4. Travels of America

1. The Appalachian Mountains and the Ouachita Mountains were once connected.
2. An "exotic terrane" is a landmass that has become attached to the leading edge of a continent. It could have originated as an island, a small continent, or a piece of another continent that has been torn loose.
3. The fossils of trilobites and brachiopods were like the ones found in North American rocks of the same age, and not like the ones found in South American rocks of that age. This indicates that the Precordillera was not near South America at that time.

## PART TWO: EARTHQUAKES AND VOLCANOES

### 5. What's This Volcano Trying to Tell Us?

1. The two most abundant gases in magma are water vapor and carbon dioxide.
2. Volcanic 'belches' are exploding bubbles of gas and rock, which throws magma and other debris into the air.
3. The first step in deciphering a volcano is to model its physical structure mathematically, using such parameters as the size of the conduit, the rigidity of the bedrock and the gas content of the magma.

### 6. Faulty Premise

1. It is on the border of one of the Earth's tectonic plates, lying above a subduction zone.
2. The only successsfully predicted earthquake occurred near Heicheng, China, on February 4, 1975.
3. The last large earthquake to strike Cascadia occurred on January 26, 1700 at 9:00 P.M., according to tsunami records from Japan.

## 7. Clues from a Village: Dating a Volcanic Eruption

1. The eruption occurred between 500 and 600 A.D. based on radiocarbon dating of pottery found at the site.
2. The eruption probably occurred during the rainy season, probably in September, based on the maturity of corn found in local fields.
3. Open blossoms of cacao trees indicate that the eruption occurred during the night, and artifacts in houses suggest that the eruption occurred in the evening, after dinner but before people went to bed.

## 8. Like a Bolt from the Blue

1. Dilantancy is the network of microscopic fractures that fluff up and expand a rock when it is being stressed to near its breaking point. Dilantancy was found in the field to apply only to small portions of the crust, and this portion is too small to explain precursors.
2. The black box approach was based on the philosophy that it was not needed to understand very much about earthquakes to be able to predict them.
3. Premonitory creep is the slippage of nearby rock before the fault ruptures in a granite model. It is possibly a predictor of earthquakes.

## PART THREE: EXTERNAL EARTH PROCESSES AND EXTRATERRESTRIAL GEOLOGY

## 9. Ancient Sea-Level Swings Confirmed

1. Unconformities (erosional gaps) in different parts of the world that have the same age.
2. There are large margins of error in the dating which makes correlation imprecise.
3. The amount of ice locked up in ice sheets may have controlled sea-level change as far back as 49 million years ago, but before that global temperatures were too high for ice sheets to exist.

## 10. The Seafloor Laid Bare

1. Satellites determined the elevation of features on the sea floor by measuring small variations in the elevation of the sea surface (caused by the pull of gravity) from those features on the ocean floor.
2. The satellite has a resolution of about 3000 feet in elevation and six miles across, which is much better than previous data.
3. Stretching and thinning of the Pacific Plate may have allowed magma to rise to the surface, forming a volcanic ridge.

## 11. Rocks at the Mars Pathfinder Landing Site

1. Petrological investigations aid in the investigation of terrestrial rocks' histories, the timing of various events, and processes such as melting.
2. *Barnacle Bill* has the composition of an andesite which may imply a planet with a wet interior and/or plate tectonics. It also could imply repeated melting events of the rock is only representative of a small portion of the Martian crust.
3. The Pathfinder had limited spatial resolution due to cameras and dust that obscures the rocks' textural and spectral features and contaminates the analyses of the rocks' chemical composition.

## 12. A Martian Chronicle

1. The new ambulatory robot is better than *Sojourner* because it will be able to travel ~100 meters a day (*Sojourner* travelled just a few meters a day), it's equipped with more sophisticated photographic and chemical equipment, it can take core samples of rock and collect ~100 grams of sample and this is all done by remote control.
2. McKay's 4 lines of evidence for ancient life on Mars are: the precipitated carbonate minerals, distinct grains of the mineral magnetite (resembling the crystals formed within some terrestrial bacterial cells), complex organic molecules (evidence fo the breakdown of biomolecules), and miniscule rodlike structures (interpreted as microfossils).
3. True.

## 13. Pluto Reconsidered

1. The discovery of the Kuiper Belt filled with sizable somethings.
2. An object massive enough to fuse deuterium in its core it widely considered to be "substellar," a brown dwarf.
3. The first is anything more than 1,000 kilometers across. The second is anything large enough to shape itself into a sphere.

## PART FOUR: RESOURCES AND POLLUTION

## 14. Sharing the Planet: Can Humans and Nature Coexist?

1. A 'paper park' is a park established on paper to protect its inhabitants, but in reality, or on the ground, it is completely unprotected.
2. Invasive species are highly adaptable plants and animals that spread outside their native ecological ranges and thrive in human-disturbed habitats. They are usually introduced by humans.
3. If top predators or dominant herbivores disappear, this could trigger a cascade of disruptions in the ecological relationships among species that maintain an ecosystem's diversity and function.

## 15. Sharing the Wealth

1. The oil and gas industry, the timber industry, and mining industry, development, sport-utility vehicles are all subsidized by our economy.
2. The tax lost by two votes due to confusion generated by an oil and gas campaign about the tax.
3. The fiscal environment tax placed on wastes general waste and pollution. Specific cost-covering charges are placed upon specific activities or things, such as batteries.

## 16. Water for Food Production: Will There Be Enough in 2025?

1. Evapotranspiration is simply the joint processes of evaporation and transpiration.
2. The harvest index is the edible portion of a crop that contributes to food supplies, the portion of a crop's dry matter that's actually harvested.
3. Estimates suggest that salinization affects 20 % of irrigated lands worldwide.

## 17. Paradise Lost: America's Disappearing Wetlands

1. Wetlands are being destroyed at a rate of an acre a minute.
2. Under the "quick permits" builders will not always have to notify the U.S. Army Corps of Engineers of building on a wetland. Also, the requirement to avoid wetlands when possible will not apply, and public input will not be allowed.
3. Man-made wetlands are usually not as diverse and many dry up or die because of a poor understanding of wetland function.

## 18. Hydrologic Manipulations of the Channelized Kissimmee River

1. Within two years after reflooding, the demonstration project impoundment had the highest wading bird and waterfowl density in the channelized pools.
2. 90% of the fish found in a floodplain marsh were less than 100 mm long, in the 1957 survey.
3. The Kissimmee River basin is located in southcentral Florida.

## 19. Do Soils Need Our Protection?

1. Pedology is the study of soil properties and processes *in situ* on the landscape.
2. Pedology states that a change in any of the factors of soil formation inevitably leads to a new or different soil. Human activities change many of these factors.
3. Soils are geological bodies that take thousands to millions of years to develop. Also, soils can not be recreated, and any type of disturbance can change its properties.

## 20. Waste Not

1. Sea dumping was outlawed in N. America in 1993.
2. Pasteurization is used for top-grade sludge in the U.S.
3. It would cost 3 to 6 times the cost of spreading sludge to incinerate it.

## 21. The Greening of Plastics

1. 'Green' composite is a new biodegradable material from natural plant fibers.
2. Green composites could be a substitute for wood in many products, like crates and packaging materials and it would be beneficial because composites would help preserve forests. It takes about 25 years for wood to generate, but it only takes a year for the plants to generate.
3. Netravali believes it may be difficult to convince industries and consumers that using green composites will be useful and practical.

## 22. Is Combustion of Plastics Desirable?

1. No more than 50% of the plastic gets recycled.
2. Burning waste generates less sulfur oxides than coal and oil and less nitrogen oxides than coal, oil, and natural gas. Both sulfur and nitrogen oxides are precursors to acid rain.
3. Concern is focused on the emissions of heavy metals and chlorine from the combustion of plastics.

## 23. Acid Rain Control: Success on the Cheap

1. There is 8.95 million tons of sulfur dioxide expected to be released in 2010.
2. The chief benefit of the U.S. emissions trading system is that it puts free market forces to work. Emitters have a 'checking account' system and the EPA limits the amount of 'currency' in the system. Anyone can find the best buy in emissions reduction as long as they don't 'overspend' what they are allowed. This can broaden a power plant operator's options.
3. The latest cost estimate for U.S. acid rain control is about $1 billion per year.

## 24. The Complexity of Urban Stone Decay

1. Stone decay occurs twice as fast in urban areas.
2. The average pH of rainfall is 5.6 which is slightly acidic.
3. Surface water and groundwater can bring corrosive substances into contact with the stone. Also, the capillary action of the stone can transport water long distances bringing along any corrosive substance, too.

## PART FIVE: GLOBAL CLIMATE CHANGE—PAST, PRESENT, AND FUTURE

## 25. Spending Our Great Inheritance—Then What?

1. The U.S. reached peak oil production in 1970.
2. The U.S. consumes 19 million barrels a day. Worldwide, 72 million barrels a day are consumed.
3. Solar and wind powers are undependable, intermittent, and produce electricity for which there is no battery pack to store the amount of energy needed.

## 26. Deep Freeze

1. There are three Milankovitch cycles. One concerns the shape of the earth's orbit, and has a cycle of 100,000 years. Another cycle concerns the tilt away from the vertical of the earth's spin axis, and has a cycle of 41,000 years. The third cycle concerns the time of year in which the earth is closest to the sun, and has a cycle of 23,000 years.
2. Milankovitch cycles run into a problem with explaining the ice ages because the 100,000 year cycle matches the time scale of the ice ages, but the effects are too small. Also, the other two cycles don't match the ice age cycles that are ten times the size of these two Milankovitch cycles.
3. Some believe that the ocean circulation changes in the Southern Hemisphere. Also, the south could act as an amplifier by changing the amount of carbon dioxide in the atmosphere, and, therefore, change the amount of warming caused by the natural greenhouse effect.

## 27. Underground Temperatures Reveal Changing Climate

1. There are 500 years worth of surface temperature history imprinted on and contained within the upper 500 meters of the Earth's crust.
2. Borehole drilling allows one to obtain a temperature versus depth profile.
3. There were 50,000 years of climate change obtained through boreholes in the ice of central Greenland.

## 28. Verdict (Almost) In

1. To exclude natural suspects, researchers have been looking not just at average global temperature but at the geographic pattern.
2. We release six billion tons of carbon and 23 million tons of sulfur, mostly from fossil fuels, into the atmosphere each year.
3. Carbon dioxide spreads around the earth.

## 29. The Heat Is On

1. Glaciers are good indicators of global warming because they only reflect long term changes in climate
2. A doubling of carbon dioxide will raise the global temperature by another 3.5 degrees by 2100.
3. Ecosystem characteristics being studied include stream temperature, glacial movement, and fire vulnerability. The parks were chosen for their ecological health, historical records, and their representation of various ecosystems.

## 30. The Day the Sea Stood Still

1. Foraminifera are good indicators of warming water because as water warms foraminifera more easily absorb the isotope Oxygen 16 than Oxygen 18. The Oxygen 16 molecule vibrates faster in warmer water and is more easily absorbed by forminifera.
2. The circulation of the ocean is weakened because temperature increases are much greater in higher latitutdes decreasing the temperature gradient between the tropics and high latitudes.
3. A volcanic eruption would have cooled the area around the lower latitudes pushing Earth over a threshold and halting ocean circulation. With the ocean at a standstill, greenhouse gases could bubble to the surface and cause the spike in global warming.

## 31. The Storm in the Machine

1. The Gulf Stream and North Atlantic Current follow a west to east path.
2. For the NAO, a high index signifies low pressure around Iceland and high pressure of Portugal, yielding strong westerly winds and relatively warm temperatures in Northern Europe.
3. ENSO means, El Niño/Southern Oscillation.

## PART SIX: THE HISTORY OF LIFE— ORIGINS, EVOLUTION, AND EXTINCTION

## 32. Life's First Scalding Steps

1. The theory being challenged is life's precursor chemicals linked up at the surface of a sun-drenched pond or ocean. Also, known as the sun simmering a prebiotic soup for millions of years to cook up the first cellular organisms.
2. According to Wächtershäuser, metabolism had to evolve first and begin to run on its own before other basic elements could evolve.
3. Wächtershäuser is criticized for only testing materials at one temperature and for not adding a pressure variable that is present around these vents.

## 33. Digging Up the Roots of Life

1. The oldest fossils are 3.5 billion years (Gyr) old and they are similar to modern bacteria, including the photosynthetic cyanobacteria.
2. The split between bacteria, archea, and eukaryotes occurred about 2 billion years ago.
3. The molecular biologists used genes from 57 metabolic proteins which were obtained from 15 different groups of organisms.

## 34. When Life Exploded

1. The beginning of the Cambrian is now set at 543 million years ago and the period of rapid animal evolution was about 5 to 10 million years.
2. The amount of carbon 13 relative to carbon 12 increased, which indicated that organic matter, which has high levels of carbon 12, was probably being buried before it could decay.
3. The increase in the number of Hox genes, and possibly changes in their linkage to other genes may have led to major changes during embryonic development.

## 35. Cooking Up a Storm

1. A PCR machine can accurately replicate DNA fragments millions of times over, so scientists can generate enough DNA to work with.
2. The two main schools of thought describing how the World's megafauna disappeared are first, the great cooling and drying of the climate that accompanied the last ice age made most of the world uninhabitable for water drinking animals and the second, argues that it was due to the arrival of *Homo sapiens*.
3. Two examples of the ancient Australian giants are today's red kangaroos and Tasmanian Devils.

## 36. What the Dinosaurs Left Us

1. Pieces of bone in the coprolite would indicate that it came from a carnivore.
2. Spiral coprolites are indicative of primitive fish, such as sharks and lungfish, which have spiral shaped intestinal valves.
3. The presence of dung beetle burrows indicate that the feces were left at the surface and not buried.

## 37. Fossils Indicate Dinosaurs Nested on Eggs

1. An oviraptor, which is a carnivore.
2. In a circle; about 15 eggs were found.
3. Perhaps a sudden sandstorm overtook the animal as it sat on the eggs.

## 38. Feathers Don't Make the Bird

1. The fossils of *Protarchaeopteryx* were between 150 to 120 million years old.
2. There were 150 shared features listed comparing birds and dinosaurs.
3. The name of the fossil in question is *Caudipteryx*.

## 39. The Dinosaur Detectives

1. There is very little fossil evidence. There is a meager amount of skeletons, and most are usually less than half complete.
2. An ectotherm is a cold-blooded animals such as reptiles, and endotherms are warm-blooded animals such as mammals.
3. The female is larger than the male, and in birds of prey, the dinosaurs' closest living relative, this larger female is a characteristic of monogamy. Also, *T. rex* skeletons have been found in what looks like a family group

## 40. Meteors That Changed the World

1. Only rarely does a meteorite reach the ground.
2. The main impurity of the iron is nickel.
3. A layer of soil above dinosaur fossils contains high amounts of iridium, the presence of osmium and right-handed amino acids, shocked quartz, a large amount of soot, and there is a large crater which dates to the same time as the layer.

## 41. Did Darwin Get It All Right?

1. He originally thought that his results would support the theory of gradual speciation.
2. Species persisted for 2 to 6 million years and speciation events lasted less than 160,000.
3. Natural selection affects a species in many different ways at the same time. The various selection pressures prevent the species from evolving in any one direction. This type of selection pressure is called stabilizing selection, or adaptive gridlock. Speciation events only occur when a small population becomes isolated from the remainder of the species range.

## 42. New Mammal Data Challenge Evolutionary Pulse Theory

1. Elisabeth Vrba studied African antelopes.
2. There was a pulse of extinctions between 2.7 million and 2.5 million years ago.
3. Variations in fossil abundance may have created a false peak in Vrba's data, or the Turkana region in Behrensmeyer's study may not be typical of the whole continent.